Sustainable Agriculture Reviews

Volume 38

Series Editor

Eric Lichtfouse
Aix-Marseille Université, CNRS, IRD, INRA
Coll France, CEREGE
Aix-en-Provence, France

Other Publications by Dr. Eric Lichtfouse

Books

Scientific Writing for Impact Factor Journals
https://www.novapublishers.com/catalog/product_info.php?products_id=42242

Environmental Chemistry
http://www.springer.com/978-3-540-22860-8

Sustainable Agriculture
Volume 1: http://www.springer.com/978-90-481-2665-1
Volume 2: http://www.springer.com/978-94-007-0393-3

Book series

Environmental Chemistry for a Sustainable World
http://www.springer.com/series/11480

Sustainable Agriculture Reviews
http://www.springer.com/series/8380

Journal

Environmental Chemistry Letters
http://www.springer.com/10311

Sustainable agriculture is a rapidly growing field aiming at producing food and energy in a sustainable way for humans and their children. Sustainable agriculture is a discipline that addresses current issues such as climate change, increasing food and fuel prices, poor-nation starvation, rich-nation obesity, water pollution, soil erosion, fertility loss, pest control, and biodiversity depletion.

Novel, environmentally-friendly solutions are proposed based on integrated knowledge from sciences as diverse as agronomy, soil science, molecular biology, chemistry, toxicology, ecology, economy, and social sciences. Indeed, sustainable agriculture decipher mechanisms of processes that occur from the molecular level to the farming system to the global level at time scales ranging from seconds to centuries. For that, scientists use the system approach that involves studying components and interactions of a whole system to address scientific, economic and social issues. In that respect, sustainable agriculture is not a classical, narrow science. Instead of solving problems using the classical painkiller approach that treats only negative impacts, sustainable agriculture treats problem sources.

Because most actual society issues are now intertwined, global, and fast-developing, sustainable agriculture will bring solutions to build a safer world. This book series gathers review articles that analyze current agricultural issues and knowledge, then propose alternative solutions. It will therefore help all scientists, decision-makers, professors, farmers and politicians who wish to build a safe agriculture, energy and food system for future generations.

More information about this series at http://www.springer.com/series/8380

Inamuddin • Abdullah M. Asiri • Eric Lichtfouse
Editors

Sustainable Agriculture Reviews 38

Carbon Sequestration Vol. 2 Materials and Chemical Methods

Springer

Editors
Inamuddin
Chemistry Department, Faculty of Science
King Abdulaziz University
Jeddah, Saudi Arabia

Department of Applied Chemistry, Faculty of Engineering and Technology
Aligarh Muslim University
Aligarh, India

Abdullah M. Asiri
Chemistry Department, Faculty of Science
King Abdulaziz University
Jeddah, Saudi Arabia

Eric Lichtfouse
Aix-Marseille Université, CNRS, IRD, INRA
Coll France, CEREGE
Aix-en-Provence, France

ISSN 2210-4410 ISSN 2210-4429 (electronic)
Sustainable Agriculture Reviews
ISBN 978-3-030-29336-9 ISBN 978-3-030-29337-6 (eBook)
https://doi.org/10.1007/978-3-030-29337-6

This Springer imprint is published by the registered company Springer Nature Switzerland AG.
The registered company address is: Gewerbestrasse 11, 6330 Cham, Switzerland

Preface

Carbon dioxide (CO_2) is a greenhouse gas which is responsible for global warming and consequently environmental changes. Carbon dioxide in the environment is increasing due to fossil fuel combustion. Various strategies are developed for carbon dioxide capture and utilization, including chemical, photochemical, electrochemical, and biological methods. Carbon dioxide capture and utilization exhibit many challenges for the production of value-added products, biofuels, etc.

Moreover, improvement in CO_2 capture and utilization for a sustainable world is essential to produce significant advances in CO_2 conversion to prevent CO_2 increase in the environment. Therefore, the value-added applications of CO_2 capture and utilization had drawn the wise attention of research and development specialists of various disciplines, including environmentalist, engineers, biotechnologists, material scientists, and mechanical engineers. The research in the area of CO_2 capture and utilization has been in progress to use carbon dioxide as an alternative feedstock toward the development of fossil-free technologies. Thus, the CO_2 capture and utilization have an incredible future, but still more research and development studies are needed to commercialize at an enormous scale.

Carbon Sequestration Vol 2: Materials and Chemical Methods discusses cutting-edge research on carbon dioxide capture and utilization. It covers fundamental knowledge on fabrication strategies, properties, and mechanisms of carbon dioxide sequestration. It discusses carbon dioxide capture and utilization by using metal-organic frameworks, ionic liquids, metal oxides, alkali soils, zeolites, and hybrid membranes. The book also supplies knowledge on carbon dioxide capture in adsorbents, nanosponges, chemical solvents, and cryogenics. This book is an archival reference guide for undergraduate and postgraduate students, faculty, R&D professionals, production chemists, food chemists, environmental engineers, and industrial experts. **Based on thematic topics, the book edition contains the following ten chapters:**

Chapter 1 reviews the CO_2 capture by chemisorption on pore-expanded materials.

Chapter 2 summarizes the different types of absorbent, reaction media, and reagents used for carbon dioxide capture and conversion. The one-step reaction or two-step reaction of carbon dioxide capture and subsequent utilization are also discussed in details.

Chapter 3 provides an overview of recent trends in the development of metal oxide-based materials for carbon dioxide capture. The basic principles of adsorption and chemical looping and the different aspects of a carbon dioxide capturing materials based on magnesium oxide, layered double oxides, calcium oxide, and transition metal oxides are also discussed.

Chapter 4 reviews the performance of the mixed matrix membranes fabricated with different types of filler particles with special focus on post- and pre-combustion carbon capture.

Chapter 5 briefly reviews the advantages and disadvantages of using different ionic liquids and their various modifications for sorption of CO_2. Their physical characteristics in their pure state and after absorption of carbon dioxide are also discussed.

Chapter 6 reviews the methods for carbon sequestration in alkaline soils.

Chapter 7 deals with the use of metal organic frameworks as promising agents for carbon dioxide capture. Several MOFs have been presented along with a description of their properties. Metal-organic framework-based derivatives obtained by pyrolysis and their performance compared with the parent MOF are also presented.

Chapter 9 reviews the properties and application of ionic liquids for carbon dioxide capture. The economic aspects of the carbon dioxide capture process in the industrial scale are also discussed.

Chapter 10 summarizes the major technologies and strategies for capturing CO_2. The features of cryogenic routes and the prospect of CO_2 capture are also reviewed.

Jeddah, Saudi Arabia — Inamuddin
Abdullah M. Asiri
Aix-en-Provence, France — Eric Lichtfouse

Contents

Chapter 1
Nanosponges for Carbon Dioxide Sequestration

Enrique Vilarrasa-Garcia, Rafael Morales-Ospino, Rafaelle Gomes Santiago, Juan Antonio Cecilia, Moises Bastos-Neto, and Diana C. S. Azevedo

Abstract The causes of global warming have been subject of controversy, especially in the last 50–100 years. There has been a remarkable effort by the international scientific community to reduce the amounts of CO_2 emitted into the atmosphere. It is well known that the CO_2 capture step accounts for around 60–70% of the whole cost of the carbon capture and storage (CCS) chain. Although some CO_2 capture technologies have been proposed, chemical absorption and adsorption are currently the most suitable ones for post-combustion capture in power plants.

In this review, we focus on CO_2 capture by chemisorption on pore expanded materials, especially silica-based nanosponges reported in recent research. Under post-combustion conditions, i. e., high temperatures and low CO_2 partial pressure, amino functionalized materials show promising performance for CO_2 capture. In recent years, several pore expanded materials from clays, silica, zeolites and carbonaceous matrices have been developed to act as support to host amino-rich polymers. Its high affinity (higher than 40 $kJmol^{-1}$) and selectivity towards CO_2, mild regeneration conditions (lower than 150 °C) and appealing CO_2 uptakes e.g., 5.6 $mmolg^{-1}$ at 0.1 bar reported, justify the effort on materials and process research.

Keywords Adsorption · CO_2 · Amine functionalization · Pore expanded materials · Silicas · Zeolites · Clays · Polymers · Metal organic frameworks · Activated carbons

---

E. Vilarrasa-Garcia · R. Morales-Ospino · R. G. Santiago · M. Bastos-Neto
D. C. S. Azevedo (✉)
Department of Chemical Engineering, LPACO2 –Bl. 731, Federal University of Ceará, Fortaleza, Brazil
e-mail: diana@gpsa.ufc.br

J. A. Cecilia
Department of Inorganic Chemistry, Crystallography and Mineralogy, Universidad de Málaga, Campus de Teatinos, Málaga, Spain

Inamuddin et al. (eds.), *Sustainable Agriculture Reviews 38*, Sustainable Agriculture Reviews 38, https://doi.org/10.1007/978-3-030-29337-6_1

Abbreviations

APTES	3-Aminopropyl triethoxysilane
APTMS	3-Aminopropyl trimethoxysilane
BET	Brunauer-Emmett-Teller
CCS	Carbon Capture and Storage
CNT	Carbon nanotubes
DEN	1,1 dimethylethylenediamine
DT/DETA	Diethylenetriamine
DVB	Divinylbenzene
ED/EDA	Ethylenediamine
EGDMA	Ethylene glycol dimethyl acrylate
ESA	Eletric swing adsorption
FTIR	Fourier-transform infrared
GCMC	Grand Canonical Monte Carlo
HCP-MAAM	Hyper-crosslinked polymeric materials
HMS	Hollow silica microspheres
HPS	Hierarchical porous silica
IPA	Isopropanolamine
IPCC	Intergovernmental Panel on Climate Change
MAAM	Methacrylamide
MCF	Mesoporous cellular foam
MCM-41	Mesoporous silica, Mobil Composition of Matter No. 41
MEA	Monoethanolamine
MEN	1-methylethylenediamine
MENN	N,N′-dimethylethylenediamine
MOF	Metal organic framework
MWCNT	Multi-walled carbon nanotube
NZ	Nanozeolite
OMS	Ordered mesoporous silica
PCH	Porous Clay Heterostructures
PEI	Polyethylenimine
PILC	Pillared Interlayered Clays
PLS	Pillared Layered Structures
PMMA	Polymethyl methacrylate
PSA	Pressure swing adsorption
SBA-15	Mesoporous silica, Santa Barbara Amorphous No. 15
TEM	Transmission electronic microscopy
TEOS	Tetraethyl orthosilicate
TEPA	Tetraethylenepentamine
TETA	Triethylenetetramine
TGA	Thermogravimetric analysis/thermal grvimetric analysis

TIPB	1,3,5- triisopropylbenzene
TMB	1,3,5-trimethylbenzene
TSA	Temperature swing adsorption
UNFCCC	United Nations Framework Convention on Climate Change
WGS	Water-gas shift
XRD	X ray diffraction

1.1 Introduction

1.1.1 General Overview

In the last decades, the world population has become aware that the increasing atmospheric CO_2 levels are directly related to anthropogenic activities. As a consequence of increasing global demand for energy, fossil fuel combustion has been claimed as responsible for the global warming by aggravating the so-called greenhouse effect. Although other molecules, such as SF_6 or CH_4, have greater potential to produce the greenhouse effect, most efforts to control global warming are related to mitigating CO_2 emissions since they account for nearly 70% of the gaseous irradiative force causing the greenhouse effect (Pachauri and Reisinger 2007). Thus, CO_2 emissions have risen from 280 ppmv in preindustrial times to above 400 ppmv at present, which has implied an increase of the earth's surface temperature of 0.7 °C in the last century (Raupach et al. 2007). Considering this trend, the Intergovernmental Panel on Climate Change (IPCC) predicts unpromising prospects since the global temperature may increase between 1.8 and 4.0 °C along this century (Hoegh-Guldberg and Bruno 2010; Bonan and Doney 2018). This continuous increase of the earth's surface temperature triggers harsh consequences on the planet as the acidification of the oceans. Global warming also causes more unstable rain regimes, where long periods of drought are alternated with great floods, leading to depletion of aquifers, diseases and migrations (IPCC 2014).

In order to counteract the consequences of the greenhouse effect, governments have created the United Nations Framework Convention on Climate Change (UNFCCC) to restrict greenhouse gas emissions (UNFCCC 2019). This organism has established a strategy based on reducing the energy demands by means of improving the energy efficiency and productivity, as well as performing a progressive transition to a low-carbon sustainable economy, both aiming at decreasing CO_2 emissions. More sustainable technologies as fuel cells, hydrogen production (steam reforming), the use of gas natural or biomass have emerged to replace the traditional fuels; however, in none of these cases, it is possible to reach the challenge of zero CO_2 emissions in its global process. Bearing in mind zero-emission, CO_2 capture, transport, and long-term storage or sequestration (CCS) is considered to be the most promising strategy to mitigate the CO_2 emissions, mainly in the short- and medium-term (McKee 2002; Leung et al. 2014).

1.1.2 Technologies to Capture CO_2

Among the steps involved in carbon capture, transport and storage (CCS), CO_2 capture is the most expensive step, accounting for 50–90% of the total cost (Pera-Titus 2014). Traditionally, CO_2 capture has been carried out by absorption with alkylamines (Rochelle 2009). In this process, CO_2 is chemically absorbed to form carbamate or bicarbonate species, which are released by vacuum or heating (Pera-Titus 2014; Rochelle 2009). In spite of its high absorption capacity, this methodology has limitations related to amine toxicity and corrosivity as well as the high energy requirements for alkylamine regeneration (Aresta and Dibenedetto 2003).

Several technologies, such as cryogenic distillation, membrane selective permeation, enzyme-based systems or absorption in ionic liquids have been considered as a potential alternatives to replace absorption with alkylamines; however, the high cost of some of these processes or its low efficiency for large scales limits its use for industrial applications (Pera-Titus 2014).

Porous sieves have emerged as promising alternative in pre- and post-combustion processes. These porous structures may act as molecular sieves, capturing CO_2 molecules selectively from gas mixtures containing other molecules, such as CH_4 or N_2, in pre- and post-combustion processes. The use of these porous solids displays several potential advantages in comparison to traditional liquid alkylamines:

1. High thermochemical and mechanical stability, offering longer lifetime and energy savings.
2. Continuous operation upon pressure swings, which suggests the treatment of higher flow rates.
3. High permeability, favoring the design of compact systems and diminishing the costs.

Another important advantage of the porous adsorbents in comparison to liquid alkylamines is that such materials can be easily functionalized, which increases the range of their application in different CO_2 emission scenarios.

CO_2 capture in combustion processes in power plants may follow several strategies (Bredesen et al. 2004):

1. Oxy-combustion – Combustion with pure O_2, so that the only products are CO_2/H_2O.
2. Pre-combustion – Carbon from the fuel can be removed prior to its combustion by converting it to syngas, e.g., autothermal reforming or methane-steam reforming with its subsequent water–gas shift (WGS) reaction. CO_2 may then be separated from H_2, the latter being the energy vector to be burned.
3. Post-combustion – CO_2 is recovered from a flue gas, with a concentration between 5 and 15 vol. % CO_2 concentration.

Among them, CO_2 capture in post-combustion processes poses a technical challenge since the feed of the capture process is a dilute CO_2/N_2 mixture under hot and humid conditions and atmospheric pressure. As most CO_2 emissions in power plants occur under such conditions, CO_2 capture facilities would be readily applied to existing power plants, although a retrofit strategy would be required to adapt prior installations to these new technical considerations.

In porous matrices, CO_2 is primarily adsorbed as a result of confinement effects and physical adsorbate-adsorbent interactions (Modak and Jana 2019). For that reason, a variety of physical adsorbents have been studied in recent years to capture CO_2 in different scenarios (Zhao et al. 2018). The adsorption capacity of physical adsorbents relies heavily on its pore size distribution, surface area, CO_2 partial pressure, temperature and humidity (Kenarsari et al. 2013; Ünveren et al. 2017). Materials like zeolites have earned special attention due to its high CO_2 adsorption capacity and availability at industrial scale. However, its adsorption capacity is drastically impaired in the presence of water (Nie et al. 2018). Metal organic frameworks (MOFs) have become attractive solid sorbents owing to their large surface area and tunability of their structural properties, such as pore shape, size and selectivity (Yaumi et al. 2017). Nevertheless, their synthesis may be complex and expensive. Activated carbons (AC) research aiming at CO_2 capture has been mainly focused on obtaining materials with enhanced adsorptive capacities by adjusting the pore size distribution during their preparation by using different activation methodologies (Chen et al. 2013a, b). These adsorbents show interesting performance at relatively low temperature and are able to withstand moisture. However, for nearly all physisorbents, high temperatures, i.e., above 30 °C, and the presence of other species in the gas mixture lead to drastic drops in CO_2 uptake. In order to improve CO_2 adsorption at moderately high temperatures, i.e., higher than 30 °C, and even in the presence of moisture, amino functionalized adsorbents have been widely reported in the literature in the last few years. Pore expanded structures have been developed recently to improve amino group density, while avoiding pore blocking and improving CO_2 diffusion. Thus, pore expanded silicas, zeolites, activated carbon, polymers, pillared clays among others have become increasingly attractive within the scientific community. The loading of amines by different functionalization methods to these pore expanded materials has become a fast-growing research topic. Hence, in this paper, we review recent advances of several works on this matter for different types of amine-functionalized supports.

1.1.3 *Functionalization*

There are several works that describe functionalization routes onto mesoporous materials (Chang et al. 2009; Li et al. 2013; Yokoi et al. 2012; Rahmat et al. 2010; Zelenak et al. 2008a, b; Chong and Zhao 2003; Hiyoshi et al. 2004; Sanz et al. 2010)

The most commonly reported methods used for the incorporation of amino groups in mesostructured supports are (Alothman 2012):

(a) Direct synthesis or co-condensation. This method was described by two groups in 1996 (Burkett et al. 1996; Macquarrie 1996). In this technique the condensation of the precursor species and the organosilanes compounds occurs simultaneously. The desired functionality into their structure is added. The main advantage of this method is that no post-treatments are needed and the distribution of the organic groups is more homogeneous along the surface of the pores.

(b) Grafting: This method allows to modify chemically the surface of the material by the reaction of organosilane compounds with the silanol (Si-OH) groups that remain in the wall of the siliceous material after its synthesis. Under reflux conditions, covalent bonds of the functional groups are formed on the surface of the material. The original structure of the mesoporous support usually remains after grafting, only a decrease in the pore volume and in the intensity of the X ray diffraction peaks are observed. This technique allows to work with calcined materials and introduce large amounts of organic compound.

(c) Impregnation. This method consists on the physical bonding, by mainly van der Waals forces, between the organic molecules and the siliceous walls of the support, and not by chemical bonds as seen in the previous methods. The addition can be carried out by the incipient wetness method, in which the volume of organic species coincides with the volume of pores; or impregnation with excess dissolution. The molecules are retained, therefore, by weak interactions with the porous surface, so it is preferable to use stable compounds and with sufficiently high boiling temperatures to prevent it from volatilize of the amino source in working conditions.

(d) Double functionalization: This technique combines the impregnation with grafting method of amino molecules. Sanz et al. (2013) reported this methodology for pore expanded supports, since a higher pore size enhances the mobility of impregnated molecules and the diffusion of CO_2. They study the influence of the step sequence in the double functionalization, grafting followed of impregnation or impregnation followed of grafting. With this procedure highly amino loaded pore expanded materials were obtained (Fig. 1.1).

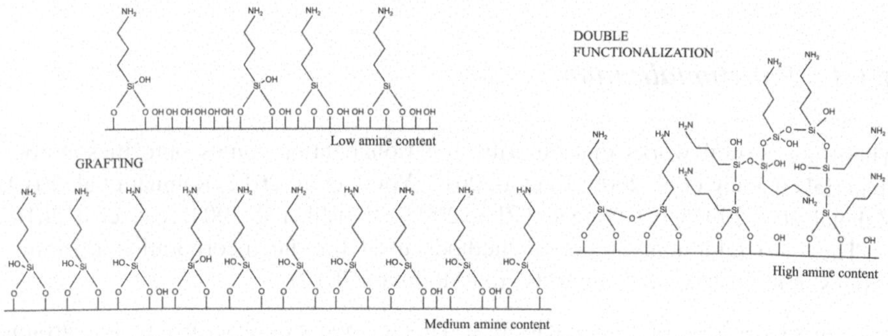

Fig. 1.1 Amine and silanols surface distribution on materials

1.2 Characterization Techniques

1.2.1 N_2 Adsorption/Desorption Isotherms at −196.15 °C

N_2 adsorption/desorption isotherms at −196,15 °C is the technique most widely used to characterize solid sorbents. Specific surface area, average pore size and total pore volume are important properties to have a first track if a sample can exhibit an appropriate performance as CO_2 adsorbent. The pore expanded strategy of e.g. mesoporous silica or zeolites and pillaring strategies of clays can be evidenced through the N_2 adsorption /desorption at −196,15°C analysis. Thus, Lashaki and Sayari (2018) showed the variation of the textural properties when mesoporous silica was synthesized increasing the synthesis temperature from 60 to 150 °C (Table 1.1).

Similarly, Vilarrasa-Garcia et al. (2015) studied the effect of the progressive disorder of the honeycomb arrangement of SBA-15 when it was synthesized with swelling agents. They observe an evolution of the isotherm shape from type IV for SBA-15 to type II (Thommes et al. 2015) for sample synthesized with increasing amounts of ammonium fluoride (Fig. 1.2). They also pointed the addition of a low ammonium fluoride amount produced a shift of the hysteresis loop to higher relative pressure together with a decrease in the surface area and micropore volume.

1.2.2 Transmission Electronic Microscopy

The successful of the synthesis of pore expanded material by using e.g. swelling agents, it can be also observed by transmission electronic microscopy. In the work above cited, Vilarrasa-Garcia et al. (2015) noted the evolution of hexagonal arrangement to mesoporous cellular foam (MCF) by microscopy. After amino functionalization of solid sorbents, to observe changes provoked by adding amino molecules, it is necessary stain the amino groups with a selective and high electronic density reagent. Thus, Sanz et al. (2012) used the selective fixation of RuO_4 over amino groups and consequent reduction to RuO_2 to localize amino groups over silica surface by transmission electronic microscopy.

Table 1.1 Textural properties of mesoporous silica Santa Barbara Amorphous (SBA-15) synthesized at temperatures between 60 and 150 °C (Lashaki and Sayari 2018)

Support name- synthesis temperature	BET surface area ($m^2\ g^{-1}$)	Total pore volume ($cm^3\ g^{-1}$)	Intrawall pore volume ($cm^3\ g^{-1}$)	Pore size (nm)
SBA-15-60	942	1.03	0.235	7.4
SBA-15-80	909	1.16	0.236	8.0
SBA-15-100	696	1.20	0.131	8.6
SBA-15-100-N	527	1.03	0.062	8.6
SBA-15-130	434	1.13	0.045	1.09
SBA-15-150	343	1.10	0.025	1.45

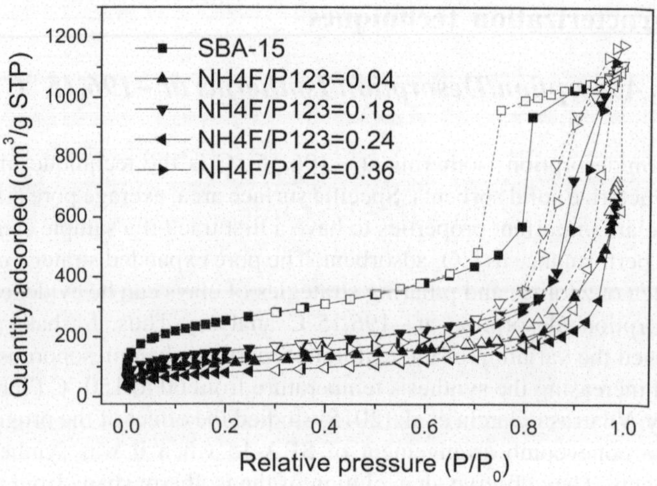

Fig. 1.2 N_2 adsorption/desorption isotherms at −196 °C for mesoporous silica synthesized with different NH_4F/P123 ratio. SBA-15: Mesoporous silica Santa Barbara amorphous, NH_4F: ammonium fluoride and P123: Poly(ethylene glycol)-block-poly(propylene glycol)-bloc

1.2.3 X Ray Diffraction

X ray diffraction is other useful technique to observe variations in the ordering of the long-term range. Thereby Ma et al. (2015) used X ray diffraction to study the influence of 1,3,5-trimethylbenzexne/poly(ethylene glycol)-block-poly(propylene glycol)-block-poly(ethylene glycol), i.e., TMB/P123 ratio on the hexagonal arrangement of SBA-15. They observed that the three characteristic peaks of SBA-15 located below $2\theta = 2°$ remain unaltered up TMB/P123 ratio of 0.15. However for this ratio exhibited a d100 spacing of 11.31 nm compared with 9.19 in the absence of swelling agent. When the ration increases to 0.5 and 2, no peaks it were observed, indicating a disorder pore structure was obtained.

1.2.4 Elemental and Thermogravimetric Analysis

Elemental analysis is usually chosen by many authors to quantify the amount of nitrogen added after samples functionalization. Wang and Yang (2011) used thermogravimetric analysis for this purpose. They compared the performance of the template removing strategy, extraction versus calcination. From thermogravimetric analysis, Wang and Yang (2011) estimated the silanol densities for pure supports and functionalized supports, used the same technique to determine the amount of nitrogen, in mmolg^{-1}, added on the supports. Similarly Mei et al. (2018) used thermogravimetric analysis to determine the nitrogen content on mesoporous silica with cubic symmetry KIT-6 support and related how the characteristic peaks of this support, commonly observed by X ray diffraction, tends to miss as the amino loading increases.

1.2.5 *Nuclear Magnetic Resonance*

An interesting research was developed by Hung et al. (2017) using nuclear magnetic resonance to study the influences of type and loading of polyamines, surfactant template, textural and surface properties of the support on CO_2 adsorption capacity. They concluded the inferior effects of template P123 and surface silanol on CO_2 adsorption on materials synthesized and functionalized in a one pot process. They carried out studies using isotoped –enriched $^{13}CO_2$.

1.2.6 *Infrared Spectroscopy*

The infrared spectroscopy was also used to understand the CO_2- amino groups interactions. Hedin and Bacsik (2019) observed experimentally that ammonium carbamates ion pairs were preferred formed in high amine density materials, while silylpropylcarbamates or carbamic acids tend to form at lower amine densities.

1.2.7 *Calorimetry*

By calorimetry technique, Sanchez-Zambrano et al. (2018, 2019) studied amino grafted and double functionalized porous expanded silica and they related that, under dry conditions, materials with higher amine density showed a proportion of propyl carbamate/silyl carbamate formed higher than for materials with low or medium amine density. Also, Saha (2018) developed a calorimetric study on some amino modified adsorbents. Saha (2018) used mono-, di- and triamine functionalized sorbents and measured the adsorption heats under dry and moist conditions. Saha (2018) observed a decrease on the CO_2 adsorption heat when polyethylenimine (PEI) functionalized polymethyl methacrylate (PMMA) worked on moist conditions and related it with the hydrophobic surface of PMMA, opposite behavior it was observed on PEI functionalized SBA-15 (Fig. 1.3).

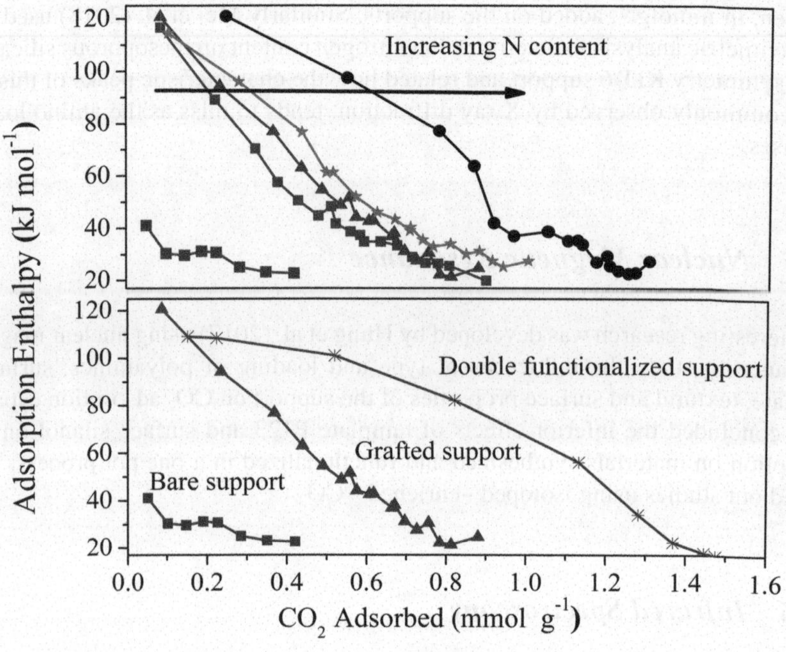

Fig. 1.3 Effect of the nitrogen loading in the shape of calorimetric curves. (Modified after Sanchez-Zambrano 2018)

1.3 Amine-Functionalized Adsorbents

In the following section of this review, recent advances of several works are reported for different types of amine-functionalized supports. The main amine molecules referenced in this review are summarized in Table 1.2.

In the Table 1.3, the main characteristics as support, type of amine, CO_2 uptake capacity and experiment conditions of the chemically modified materials of the cited references are presented.

1.3.1 Amine Functionalized Zeolites

Chatti et al. (2009) worked on the immobilization of various amines (monoethanolamine (MEA), ethylenediamine (ED) and isopropanol amine (IPA)) on synthetic zeolite 13X and assessed their respective CO_2 adsorption capacities. An enhancement in the adsorption capacity was observed as compared to the pristine zeolite matrix, which they attributed to a hybrid absorption–adsorption mechanism. Su et al. (2010) employed a commercial Y-type zeolite with a Si/Al molar ratio of 60 (Y60) and functionalized it with tetraethylenepentamine (TEPA) to study the

Table 1.2 Denomination, molar mass and chemical structure of the main amines used in materials functionalization for CO_2 capture

Molecule (Abreviation)	Molar mass	Chemical structure
Monoethanolamine (MEA)	61.08	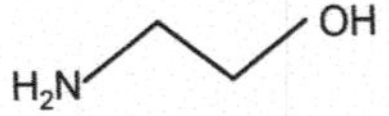
Isopropanolamine (IPA)	75.11	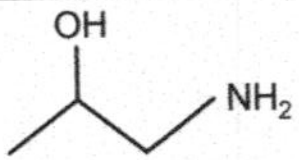
Ethylenediamine (EDA)	60.1	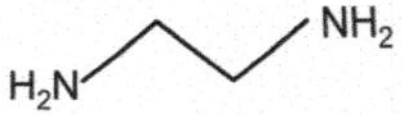
(3-Aminopropyl) trimethoxysilane (APTMS)	179.29	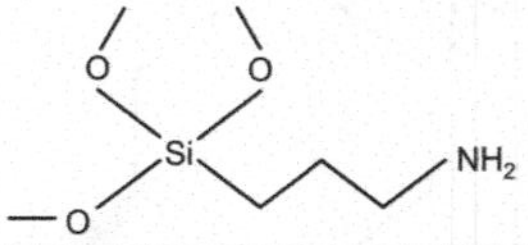
(3-Aminopropyl) triethoxysilane (APTES)	221.37	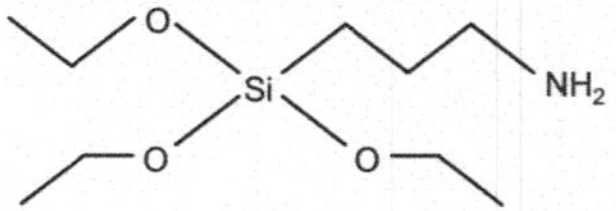
Polyethylenimine (PEI)	–	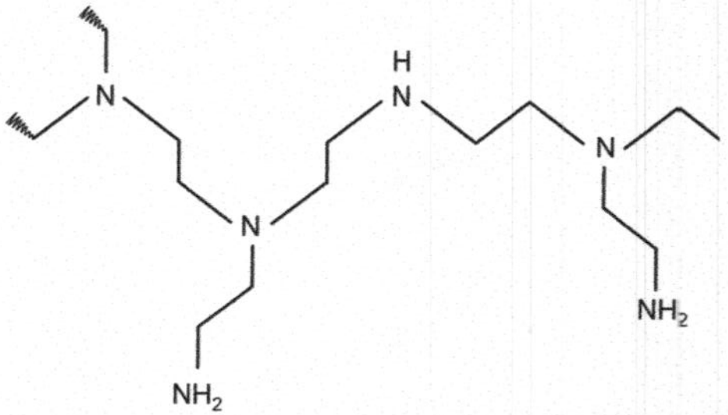

(continued)

Table 1.2 (continued)

Molecule (Abreviation)	Molar mass	Chemical structure
Tetraethylenepentamine (TEPA)	189.30	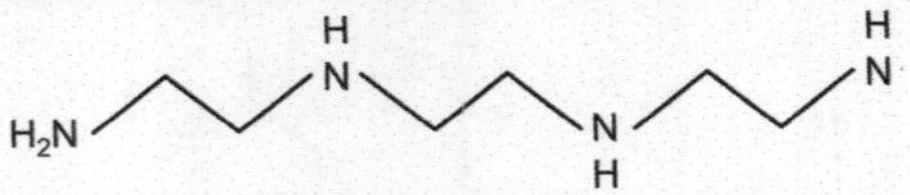
1-methylethylenediamine (MEN)	74.12	
N,N′-dimethylethylenediamine (MENN)	88.15	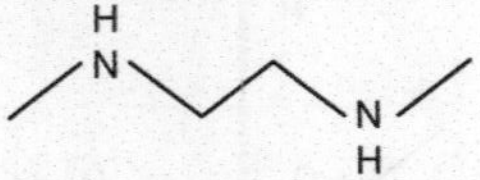
Trans-2,5-Dimethylpiperazine	114.19	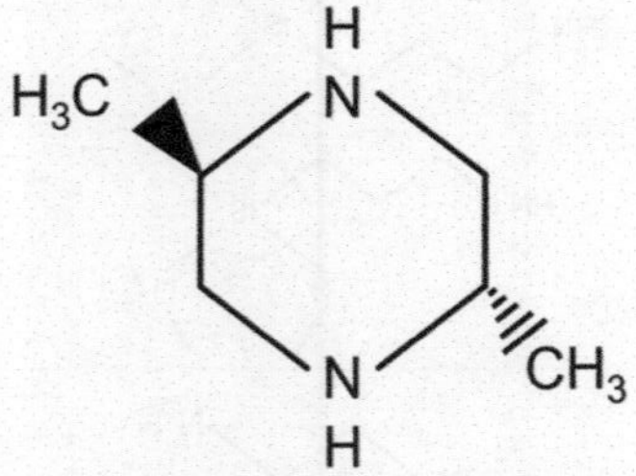
Diethylenetriamine (DT/DETA)	103.17	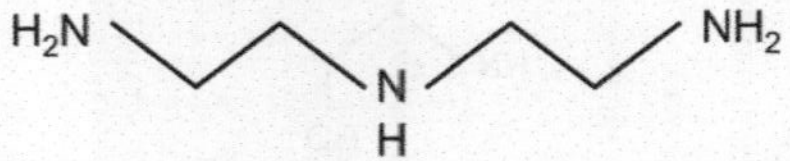

Table 1.3 CO_2 adsorption capacities, amine content and operating conditions for supports

Support	Amine	CO_2 adsorption capacity ($mmolg^{-1}$)	Equilibrium method	CO_2 partial pressure (bar)	T (°C)	References
Zeolite 13X	50 wt.% Monoethanolamine	0.45	BT	0.15	75	Chatti et al. (2009)
Zeolite 13X	50 wt.% Isopropanolamine	0.51	BT	0.15	75	Chatti et al. (2009)
Zeolite 13X	50 wt.% Monoethanolamine	1.11	Vol.	1	75	Chatti et al. 2009)
Zeolite 13X	Monoethanolamine MEA (0.2 vol.% MEA in methanol)	3.18/1.93	Grav.	1	25/75	Bezerra et al. (2014)
Zeolite 13X	Monoethanolamine MEA (10 vol.% MEA in methanol)	0.52/1.66	Grav.	1	25/75	Bezerra et al. (2014)
Meso 13X	Polyethylenimine	1.81	TGA	1	100	Chen et al. (2015)
Zeolite 13X	Polyethylenimine	1.09	TGA	1	100	Chen et al. (2015)
Nano zeolite	Ethylenediamine EDA (10 vol.% EDA in methanol)	7.48	BT	1	70	Pham et al. 2016)
ZSM-5 type zeolite	Tetraethylenepentamine	1.80	TGA	1	100	Wang et al. (2017)
ZSM-5 type zeolite	Polyethylenimine	1.96	TGA	1	120	Wang et al. (2018)
Y type zeolite(Si/Al ratio of 60)	Tetraethylenepentamine	2.50	Vol.	0.15	60	Su et al. (2010)
LTA zeolite	3-Aminopropyl trimethoxysilane	2.30	Vol.	0.15	60	Nguyen et al. (2016)
Activated carbon	Polyethylenimine	2.13	TGA	1	75	Zhang et al. (2004)

(continued)

Table 1.3 (continued)

Support	Amine	CO_2 adsorption capacity ($mmolg^{-1}$)	Equilibrium method	CO_2 partial pressure (bar)	T (°C)	References
Activated carbon	Polyethylenimine	4.82	BT	0.15	75	Wang et al. (2013a)
Mesoporous carbon	Polyethylenimine	4.67/2.80	Vol.	1	30/0	Wang et al. (2013b)
Commercial carbon black	70 wt.% Polyethylenimine	2.22	BT	0.15	28	Pino et al. (2016)
Commercial carbon black	70 wt.% Tetraethylenepentamine	4.96/5.74	BT	0.3/0.8	28	(Pino et al. 2016)
Modified carbon nanotubes	75 wt.% Tetraethylenepentamine	5.00	BT	0.10	60	Irani et al. (2017)
Carbon nanotubes	20 wt.% Polyethylenimine	2.11	Vol.	0.15	25	Keller et al. (2018)
ZIF-8	$-NH_2$	1.90	GMCM simulations	2.5	25	Liu et al. (2012)
ZIF-8	$-(NH_2)_2$	2.50	GMCM simulations	2.5	25	Liu et al. (2012)
Mg_2(dobpdc)	Ethylenediamine	4.61/3.53	Grav.	1/0.15	25/40	Jo et al. (2017)
Mg_2(dobpdc)	1-methylethylenediamine (MEN)	4.50/3.60	Grav.	1/0.15	25/40	Jo et al. (2017)
Mg_2(dobpdc)	1,1-dimethylethylenediamine (DEN)	3.15/2.15	Grav.	1/0.15	25/40	Jo et al. (2017)
Mg_2(dobpdc)	2,2-dimethyl-1,3-diaminopropane (DMPN)	2.90	TGA	0.15	40	Milner et al. (2017)
UiO-66	Polyethylenimine (10 wt.%)/ (30 wt.%)	3.13/1.70	Vol.	1	25/50	Fu et al. (2017)
MIL-101 (Cr, Mg)	Polyethylenimine (10 wt.%)	3.00	TGA	1	25	Gaikwad et al. (2019)
Mn_2(dobpdc)	N,N′-dimethylethylenediamine(MMEN)	3.70	Vol.	1	25	McDonald et al. (2015)

Fe_2(dobpdc)	N,N′-dimethylethylenediamine(MMEN)	3.90	Vol.	1	25	McDonald et al. (2015)
Co_2(dobpdc)	N,N′-dimethylethylenediamine(MMEN)	3.90	Vol.	1	25	McDonald et al. (2015)
Zn2(dobpdc)	N,N′-dimethylethylenediamine(MMEN)	3.00	Vol.	1	25	McDonald et al. (2015)
Carbonyl-incorporated aromatic polymer (CBAP-1)	Ethylenediamine EDA (5 vol.% EDA in methanol)	1.97	Vol.	1	30	Puthiaraj et al. (2017)
Carbonyl-incorporated aromatic polymer (CBAP-2)	Ethylenediamine EDA (5 vol.% EDA in methanol)	1.80	Vol.	1	30	Puthiaraj et al. (2017)
Divinylbenzene and ethylene glycol dimethyl acrylate copolymer (DEA)	30 wt.% Polyethylenimine	3.28	BT	0.1	25	Liu et al. (2017)
Poly[methacrylamide-co-(ethylene glycol dimethacrylate)] (Poly(MAAM-co-EGDMA)).	Inherently amine-functionalized	1.56/0.90	Vol.	1	25/0	Fayemiwo et al. (2018)
Covalent organic polymer (COP-9)	trans-2,6-dimethylpiperazine	79/0.6	Grav./Vol.	200/1	35/25	Ullah et al. (2019)
Covalent organic polymer (COP-10)	2-imidazolidinone	52/0.7	Grav./Vol.	200/1	35/25	Ullah et al. (2019)
Porous polymer (PP2)	Polyethylenimine	1.45/2.60	Vol.	0.15/1	25	Yang et al. (2019)
Porous polymer (PP2)	Ethylenediamine	1.00/2.85	Vol.	0.15/1	25	Yang et al. (2019)
Porous polymer (PP2)	Diethylenetriamine	1.22/2.95	Vol.	0.15/1	25	Yang et al. (2019)

(continued)

Table 1.3 (continued)

Support	Amine	CO_2 adsorption capacity ($mmolg^{-1}$)	Equilibrium method	CO_2 partial pressure (bar)	T (°C)	References
Porous polymer (PP2)	Tetraethylenepentamine	1.25/2.65	Vol.	0.15/1	25	Yang et al. (2019)
Pore expanded-SBA (TiPB)	Diethylenetriamine (7.8 N%)	1.95	Grav.	1	45	Olea et al. (2013)
Pore expanded-SBA (TiPB)	50 wt.% Tetraethylenepentamine (14.1 N%)	3.72	Grav.	1	45	Olea et al. (2013)
Pore expanded-SBA (TiPB)	50 wt.% Polyethylenimine (13.2 N%)	3.13	Grav.	1	45	Olea et al. (2013)
Mesoporous silica	20% 3-Aminopropyl triethoxysilane (4.97 N%)	2.12	Vol.	1	25	Vilarrasa-Garcia et al. (2015)
Mesoporous silica (heptane/F-)	30 wt.% Polyethylenimine (8.43 N%)	1.43	Vol.	1	25	Vilarrasa-Garcia et al. (2015)
Mesoporous silica (heptane/F-)	50 wt.% Polyethylenimine (14.44 N%)	2.27	Vol.	1	25	Vilarrasa-Garcia et al. (2015)
Hollow mesoporous silica	20% 3-Aminopropyl triethoxysilane (5.67 N%)	1.06	Vol.	1	25	Cecilia et al. (2016)
Hollow mesoporous silica (F−)	20% 3-Aminopropyl triethoxysilane (3.27 N%)	1.07	Vol.	1	25	Cecilia et al. (2016)
Hollow mesoporous silica (TiPB)	20% 3-Aminopropyl triethoxysilane (5.18 N%)	1.72	Vol.	1	25	Cecilia et al. (2016)
Hollow mesoporous silica (F−/TiPB)	20% 3-Aminopropyl triethoxysilane (3.92 N%)	1.81	Vol.	1	25	Cecilia et al. (2016)
Hollow mesoporous silica (HMS)	50 wt.% Polyethylenimine (13.9 N%)	1.53	Vol.	1	25	Cecilia et al. (2016)
Hollow mesoporous silica (F−)	50 wt.% Polyethylenimine (16.21 N%)	0.87	Vol.	1	25	Cecilia et al. (2016)

Hollow mesoporous silica (TiPB)	50 wt.% Polyethylenimine (14.70 N%)	1.61	Vol.	1	25	Cecilia et al. (2016)
Hollow mesoporous silica (F–/TiPB)	50 wt.% Polyethylenimine (14.10 N%)	2.06	Vol.	1	25	Cecilia et al. (2016)
Hollow mesoporous silica (HMS)	50 wt.% Tetraethylenepentamine (12.45 N%)	1.28	Vol.	1	25	Cecilia et al. (2016)
Hollow mesoporous silica (F–)	50 wt.% Tetraethylenepentamine (9.22 N%)	0.96	Vol.	1	25	Cecilia et al. (2016)
Hollow mesoporous silica (TiPB)	50 wt.% Tetraethylenepentamine (12.72 N%)	2.17	Vol.	1	25	Cecilia et al. (2016)
Hollow mesoporous silica (F–/TiPB)	50 wt.% Tetraethylenepentamine (14.72 N%)	3.41	Vol.	1	25	Cecilia et al. (2016)
Pore expanded-SBA-15 (TiPB)	30 wt.% Polyethylenimine (9.0 N%)	1.83	Grav.	1	45	Sanz et al. (2013)
Pore expanded -SBA-15 (TiPB)	50 wt.% Polyethylenimine (13.2 N%)	3.13	Grav.	1	45	Sanz et al. (2013)
Pore expanded -SBA-15 (TiPB)	30 wt.% Polyethylenimine/3-Aminopropyl triethoxysilane (10.7 N%)	2.52	Grav.	1	45	Sanz et al. (2013)
Pore expanded -SBA-15 (TiPB)	30 wt.% Tetraethylenepentamine (9.4 N%)	2.16	Grav.	1	45	Sanz et al. (2013)
Pore expanded -SBA-15 (TiPB)	50 wt.% Tetraethylenepentamine (14.1 N%)	3.73	Grav.	1	45	Sanz et al. (2013)
Pore expanded -SBA-15 (TiPB)	50 wt.% Tetraethylenepentamine /3-Aminopropyl triethoxysilane (15.3 N%)	4.88	Grav.	1	45	Sanz et al. (2013)

(continued)

Table 1.3 (continued)

Support	Amine	CO_2 adsorption capacity ($mmolg^{-1}$)	Equilibrium method	CO_2 partial pressure (bar)	T (°C)	References
Ordered mesoporous silica (TMB) 5.6 nm	50 wt.% Tetraethylenepentamine	3.20	BT	0.1	75	Wang et al. (2015)
Ordered mesoporous silica (TMB) 7.6 nm	50 wt.% Tetraethylenepentamine	3.45	BT	0.1	75	Wang et al. (2015)
Ordered mesoporous silica (TMB) 7.6 nm	50 wt.% Tetraethylenepentamine	4.28	BT(wet)	0.1	75	Wang et al. (2015)
Macroporous silica (MaCS)	50 wt.% Polyethylenimine (14.1 N%)	3.84	BT (wet flue gas)	0.15	40	Min et al. (2017)
MCM-41	50 wt.% Polyethylenimine (14.8 N%)	0.91	BT (wet flue gas)	0.15	40	Min et al. (2017)
SBA-15	50 wt.% Polyethylenimine (15.1 N%)	2.53	BT (wet flue gas)	0.15	40	Min et al. (2017)
Mesocellular foam (TMB)	50 wt.% Polyethylenimine (15.1 N%)	2.96	BT (wet flue gas)	0.15	40	Min et al. (2017)
SBA-15 (synthesized at130°C)	TRI (wet conditions)	1.88	TGA	0.05	75	Lashaki and Sayari (2018)
Platalet silica (F-)	50 wt.% Polyethylenimine	5.65	BT (dry flue gas)	0.1	75	Hou et al. (2018)
Sepiolite (microwave acid treatment)	30 wt.% Polyethylenimine	1.70	Vol.	1	65	Vilarrasa-García et al. (2017a, b, c)
Sepiolite (Acid treatment)	30 wt. % Tetraethylenepentamine	1.93	TGA	1	45	Yuan et al. (2018)
Montmorillonite (acid treatment)	50 wt.% Polyethylenimine	3.22/2.54	TGA	1	25/75	Wang et al. (2014)
Sepiolite (acid treatment)	50 wt.% Diethylenetriamine	1.65	TGA	1	35	Liu et al. (2018)

Porous clay Heterostructure	60 wt.% Tetraethylenepentamine (13.39 N%)	1.64	Vol.	1	25	Vilarrasa-Garcia et al. (2017b)
Porous clay Heterostructure	60 wt.% Polyethylenimine (14.41 N%)	1.45	Vol.	1	25	Vilarrasa-Garcia et al. (2017b)
100 chloropropyl - magadiite	Polyethylenimine	0.49	Grav.	1	75	Vieira et al. (2018)
Ferric sludge (Si pillared)	40 v/v % 3-Aminopropyl triethoxysilane (2.4 N%)	1.10	Vol.	1	25	Vilarrasa-García et al. (2017a)
Ferric sludge (Si pillared)	50 wt.% Polyethylenimine (16 N%)	2.50/3.60	Grav.	1	25/75	Vilarrasa-García et al. (2017a)

LTA: Zeolite Linde Type A; ZSM-5: Zeolite Socony Mobil–5; MCM-41: Mobil Composition of Matter No. 41; SBA-15: Mesoporous silica Santa Barbara Amorphous; MIL-101: metal organic framework from *Matérial Institut Lavoisier*-101; UiO-66: metal organic framework from *Universitetet i Oslo*-66; ZIF-8: Zeolitic imidazolate frameworks

TMB: 1,3,5- Trimethylbenzene; TiPB: 1,3,5- Triisopropylbenzene; F-: ammonium fluoride

Equilibrium method, BT: breakthrough, Vol.: volumetric, Grav.: Gravimetric, TGA: Thermal gravimetric analysis and GCMC: Grand Canonical Mcnte Carlo adsorption/desorption properties of CO_2 from gas streams. The CO_2 adsorption capacity of Y60 (TEPA) was studied between 30 and 60 °C and reached the highest value of 2.50 mmolg^{-1} at 60 °C and 0.15 bar. The adsorption capacities and the physicochemical properties of amino rich functionalized Y60 remained nearly the same after 20 successive cycles of adsorption and regeneration, indicating that the amino rich functionalized Y60 might exhibit a promising potential if used in cyclic operation for CO_2 capture from flue gas.

Bezerra et al. (2014) funtionalized zeolite 13X by grafting it with different loads of monoethanolamine (MEA). They found that increasing amine loads from 0.2 to 10 vol. % in methanol the adsorbent tends to reduce the micropore volume by pore blocking with the amine. As a result, amine loaded zeolites adsorbed less CO_2 than the pristine material at a given temperature. In order to overcome the inconvenience of pore blocking by amines in zeolitic materials, some studies proposed expanding the adsorbent pore size before amine functionalization. Chen et al. (2015) synthesized a faujasite (FAU) zeolite with mesoporosity (Meso-13X), by using a mesopore-generating agent, and then modified it with polyethylenimine (PEI) by physical impregnation to yield a hybrid material (Meso-13X-PEI). They also functionalized a conventional microporous zeolite 13X with polyethylenimine to compare its performance as CO_2 adsorbent against Meso-13X-PEI. The latter exhibited higher CO_2 uptake than PEI-modified zeolite 13X due to its larger pore volume, which may host more amine molecules. Moreover, Meso-13X-PEI showed both higher CO_2/ N_2 selectivity and CO_2 uptake at a relatively high temperature, e.g. 100 °C, as compared to the amine-loaded microporous zeolite. Likewise, Nguyen et al. (2016) synthesized a series of hierarchical Linde type A (LTA) zeolites with varying mesopore volume by controlling the amount of surfactant in the synthesis gel. The highest CO_2 uptake was found when 3-aminopropyl trimethoxysilane (APTMS) was grafted onto the hierarchical LTA zeolite having the largest mesopore size. Moreover, the adsorbent showed a negligible CO_2 adsorption capacity loss under dynamic conditions for over 10 adsorption-desorption cycles.

Pham et al. (2016) modified a nanozeolite (NZ) with ethylenediamine (EDA) to enhance CO_2 adsorption–desorption properties. They observed that the CO_2 uptake in such adsorbent increased in a temperature range between 20 °C and 70 °C and then decreased as the temperature increased from 70 °C to 100 °C. The CO_2 uptake measured for NZ-EDA (7.48 mmol g^{-1}) was 2.6 times higher than that of the unmodified nanozeolite. They also concluded that the CO_2 binding mechanism onto NZ is entirely ruled by physical interaction whereas, in the case of the NZ-EDA, chemical interaction between CO_2 and the amine loaded to the adsorbent becomes the main adsorption mechanism. Additionally, adsorbent stability over adsorption–desorption cycles was examined, suggesting that NZ-EDA can be a cost-effective material for CO_2 capture. Wang et al. (2017) reported the synthesis of a mesoporous ZSM-5 type zeolite to be then functionalized with tetraethylenepentamine by wet impregnation so as to obtain a series of adsorbents with increasing amounts of loaded tetraethylenepentamine. The CO_2 adsorption capacities were measured between 40 and 100 °C and the highest uptake was 1.80 mmol g^{-1} at 100 °C and 1 bar for the adsorbent modified by 0.7 g of tetraethylenepentamine per gram of support. This adsorbent also showed good stability after 5 adsorption-desorption cycles, suggesting a promising performance to capture CO_2 from industrial flue gas continuously. In another work, Wang et al. (2018) modified the same mesoporous Zeolite Socony Mobil–5, ZSM-5, with another polyamine, such as polyethylenimine, by via physical impregnation and obtained a series of materials (ZSM-5-PEI-X) with different polyethylenimine loadings. The amine-functionalized adsorbent with 30 wt. % PEI (ZSM-5-PEI-30) adsorbed nearly 5 times more CO_2 than the pristine ZSM-5, reaching 1.96 mmolg^{-1} at 120 °C and 1 bar. The adsorbent also displayed a high CO_2/N_2 adsorption selectivity and good stability, i.e., after adsorption-desorption 10 cycles, CO_2 uptake decreased only 8%.

1.3.2 *Amine Functionalized Activated Carbons*

Zhang et al. (2004) prepared activated carbons from unburned carbon in fly ash to be then modified via polyethylenimine (PEI) impregnation. They observed that amino rich polymer impregnation could improve significantly the CO_2 adsorption, reaching as much as 2.13 mmol of CO_2 g^{-1} at 75 °C against 0.22 mmol of CO_2 g^{-1} of its non-modified counterpart under the same conditions. Similarly, Wang et al. (2013a) used PEI as impregnation agent to develop a high efficiency sorbent for CO_2 capture using mesoporous carbons as support. The pristine adsorbent had a well-developed mesoporosity and large pore volume. They found that a polyethylenimine loading of 65 wt. % was optimal for CO_2 capture in this material, having measured 4.82 mmolg^{-1} in CO_2/N_2 (15/85 v/v) at 75 °C and atmospheric pressure. Additionally, the amino rich impregnated adsorbent required a low regeneration temperature of 100 °C and exhibited a good regeneration performance after 10 cycles with only 4% loss in adsorption capacity. In another work by Wang et al. (2013b), polyethylenimine impregnated mesoporous carbons were prepared aiming at CO_2 capture under low temperature conditions. CO_2 uptakes reached 4.67 mmol of CO_2 g^{-1} at 30 °C and 2.80 mmol of CO_2 g^{-1} at 0 °C. The polyethylenimine modified mesoporous carbon also displayed fast kinetics, a good selectivity for CO_2/N_2 separation, and very reversible and stable CO_2 capture performance at low temperature.

Pino et al. (2016) investigated commercial activated carbons from Akzo Nobel as a support for polyethylenimine and tetraethylenepentamine impregnation. The optimum amine loading for both polyethylenimine and tetraethylenepentamine was found to be 70 wt. %. Polyethylenimine impregnated carbon reached a CO_2 uptake of 2.22 mmol g^{-1} at 28 °C at 0.15 bar. The tetraethylenepentamine impregnated sample reached a breakthrough capacity of 4.96 mmol g^{-1} under a CO_2 concentration of 30 vol. % in the inlet gas, which improved to 5.74 mmol g^{-1} by increasing CO_2 concentration to 80 vol. %. As equally reported by previous works in the literature, such as Sanz-Pérez et al. 2013 and Yue et al. 2006, their research also confirms that the superior performance of tetraethylenepentamine as compared to polyethylenimine -based adsorbents is to be attributed to the higher density of amino-groups and less viscous nature of the former polyamine. Hence, tetraethylenepentamine provides more reactive sites for CO_2 adsorption increasing the capture performance. Irani et al. (2017) also used tetraethylenepentamine to prepare CO_2 capture adsorbents by loading the amine into modified and unmodified carbon nanotubes (CNTs). The modified CNTs were obtained by potassium hydroxide treatment of the pristine CNTs to increase the surface area and pore volume. With unmodified carbon nanotubes as support, a maximum CO_2 adsorption capacity of 2.5 mmol g^{-1} was measured with an optimal tetraethylenepentamine loading of 50 wt. % for 10 vol. % CO_2 in N_2 and 1 vol. % of H_2O at 60 °C. In the case of the modified CNTs, CO_2 adsorption capacity reached 5 mmol g^{-1} for an optimal TEPA loading of 75 wt. %, under the same conditions.

In a recent work conducted by Keller et al. (2018), highly porous multi-walled carbon nanotubes (MWCNT) were produced and functionalized with polyethylenimine, PEI, via wet impregnation. The polyethylenimine modified material attained a maximum uptake of 2.11 mmol g^{-1} at 0.15 bar and 25 °C. The optimal polyethylenimine loading was found to be 20 wt. % They concluded that increasing PEI load leads to an increase in available amine groups and at the same time to a decrease in internal surface area and pore size, so that a trade-off between the number of available amine groups and available internal surface should be met.

1.3.3 Amine Functionalized Metal Organic Frameworks

Liu et al. (2012) worked on Grand Canonical Monte Carlo (GCMC) simulations to investigate the adsorption isotherms and adsorption sites of CO_2 in zeolitic imidazolate framework (ZIF-8) and amine-modified ZIF-8 (ZIF-8-NH_2, and ZIF-8-$(NH_2)_{2}$.). They have shown that simulated CO_2 adsorption isotherms of ZIF-8 were consistent with experimental data. Grand Canonical Monte Carlo simulations demonstrated that CO_2 adsorption uptake in ZIF-8 can be improved by the addition of amino functional groups in the organic linker, following the sequence: ZIF-8-$(NH_2)_2$ > ZIF-8-NH_2 > ZIF-8 at low pressures and ZIF-8-NH_2 > ZIF-8-$(NH_2)_2$ > ZIF-8 at high pressure.

McDonald et al. (2015) synthesized a series of isostructural metal organic frameworks, MOFs, functionalized with N,N-dimethylethylenediamine (mmen) labelled as mmen-M_2(dobpdc) (M = Mg, Mn, Fe, Co, Ni, Zn and dobpdc^{4-} = 4,4-dioxidobiphenyl-3,3′-dicarboxylate) and measured their CO_2 adsorption isotherms in the temperature range between 25 and 75 °C. All of the materials, except that containing Ni, showed sharp isotherm steps that shifted to higher pressure with increasing temperature. In an attempt to understand the mechanism that leads to step-shaped isotherms in the amine modified adsorbent mmen-Mg_2(dobpdc) under flue gas conditions, it has been demonstrated that replacing Mg^{2+} of the functionalized metal organic framework mmen-Mg_2(dobpdc) with other divalent metal ions, such as Mn^{2+}, Fe^{2+}, Co^{2+}, Zn^{2+}, allowed the position of the CO_2 adsorption step to be controlled by the metal-amine bond strength. The resulting amine modified

MOFs were named as 'phase-change' adsorbents and exhibited interesting characteristics that rendered them superior performance for CO_2 capture as compared to other solid or liquid sorbents. Jo et al. (2017) also employed Mg_2(dobpdc) as support for grafting diamines such as ethylenediamine (EDA), 1-methylethylenediamine (MEN), and 1,1 dimethylethylenediamine (DEN)). In an attempt to maximize the amine density on the open metal sites of the metal organic framework, they included a combination of sonication and microwave- assisted treatment during the preparation of the materials. The DEN-grafted MOF showed superb recyclability and stability after exposure to O_2, moisture, and SO_2. CO_2 adsorption capacities of 3.60, 3.53 and 2.15 mmol g^{-1} were attained for MEN, EDA and DEN-grafted MOFs, respectively, at 15 vol. % CO_2 at 1 bar and 40 °C.

Milner et al. (2017) developed a diamine-appended metal organic framework (DMPN–Mg_2(dobpdc)) that exhibited a relatively high enthalpy of adsorption (−74 kJmol^{-1}). The results of their investigation indicated a CO_2 working capacity of 2.42 mmol g^{-1} for the functionalized metal organic framework under temperature swings between 40 °C for adsorption and 100 °C for desorption. Furthermore, according to their thermogravimetric analysis and breakthrough experiments, the DMPN–Mg_2(dobpdc) was able to capture CO_2 effectively under humid conditions and could withstand 1000 adsorption/desorption cycles with negligible degradation. Fu et al. (2017) synthesized UiO-66 hydrothermally and further modified it by polyethylenimine, PEI, impregnation to study the CO_2 adsorption of the chemically modified synthesized material. The results indicated that the UiO-66/PEI is thermally stable below 350 °C, which is a higher than most reported PEI-modified materials. On the other hand, CO_2 uptakes at 25 °C were up to 3.13 mmol g^{-1} at an optimal PEI loading of 10 wt. %, which is 1.5 times greater than that of the unmodified UiO-66. Gaikwad et al. (2019) worked initially on the synthesis of a bimetallic MIL-101 (Cr, Mg) metal organic framework by doping Mg on pristine MIL-101(Cr). They observed that CO_2 adsorption capacity increased from 1.6 to 2.0 mmol g^{-1} after Mg doping due to the formation of new CO_2 adsorption sites on the surface of MIL-101(Cr, Mg). Subsequently, the synthesized bimetallic material was impregnated with polyethylenimine under increasing loads seeking for an improved CO_2 uptake. The optimal PEI loading was found to be 20 wt.%, in which case CO_2 uptake was 3.0 mmol g^{-1} at 25 °C and 1 bar, 50% higher than that of the parent bimetallic metal organic framework. They also analyzed the stability of the PEI-MIL-101(Cr, Mg) adsorbent when exposed to humid air and acid gases, and concluded that the CO_2 adsorption capacity may be reduced down to 35%, but no significant changes in adsorbent crystallinity were observed.

1.3.4 Amine-Functionalized Polymers

Porous polymers own the advantage of displaying enduring porosity with reasonably high surface area and their preparation can be performed by means of innumerous chemical synthetic methods, when compared to inorganic adsorbents like zeolites or silicas. In light of these specific features, their porous structure may be tuned to exhibit enhanced functionalities for CO_2 capture (Nie et al. 2018). Puthiaraj et al. (2017) synthesized two aromatic polymers with carbonyl functionality, namely CBAP-1 and CBAP-2. The porous polymers were produced via a Friedel-Crafts benzoylation reaction and they were then chemically modified by ethylenediamine, EDA, to finally obtain CBAP-1-EDA and CBAP-2-EDA, respectively. CO_2 adsorption isotherms indicated that the functionalized CBAPs had slightly lower CO_2 adsorption capacities but considerably improved CO_2/N_2 selectivities, as compared to CBAP-1 and CBAP-2. In addition, they also led to activated carbons by pyrolization of CBAP-1 and its amine-functionalized counterpart at 800 °C to study their CO_2 capture performance. The carbonized materials, on the contrary, exhibited significantly increased CO_2 uptakes (ca. 57% increase) than that of the functionalized CBAPs but at the cost of a reduced CO_2/N_2 adsorption selectivity. Liu et al. (2017) studied the synthesis of porous polymers and subsequent functionalization using polyethylenimine, PEI, for CO_2 capture. The porous support materials were obtained by the polymerization of divinylbenzene, DVB, and ethylene glycol dimethyl acrylate, EGDMA. The functionalized copolymer with best CO_2 adsorption performance, i.e., 3.28 mmol g^{-1} at 25 °C and 10 vol. % CO_2 at 1 bar, was obtained for 30 wt. % PEI loading. It was concluded that polyethylenimine loadings beyond 30 wt. % lead to adsorbent aggregation and pore blocking. The solid amine adsorbent was outgassed at 75 °C and showed nearly constant CO_2 uptake for eight adsorption-desorption cycles. Fayemiwo et al. (2018) prepared a series of nitrogen-rich hyper-crosslinked polymeric materials, HCP-MAAM, by means of the copolymerization of methacrylamide, MAAM, and ethylene glycol dimethacrylate, EGDMA. The synthesized polymers were intrinsically nitrogen-enriched so as to enhance affinity towards CO_2 at low pressure. They observed that an increase in the density of amide groups within the polymer network led to a higher affinity towards CO_2, but resulted in a superior heat of adsorption and a reduced CO_2 uptake capacity. The maximum CO_2 adsorption capacity was obtained at 0 °C reaching 1.56 mmol g^{-1} and decreased down to 0.90 mmol g^{-1} at 25 °C and 1 bar. The thermogravimetric analysis demonstrated that the polymers were thermally stable up to 260 °C. Moreover, authors also suggested that the cost and performance of a HCP-MAAM-based CO_2 capture system applied to a 580-MW coal-fired power plant was comparable to those of chemical solvent scrubbing.

In a recent publication, Ullah et al. (2019) evaluated the performance of two custom design porous polymers, namely covalent organic polymers COP-9 and COP-10, for CO_2, N_2 and H_2 adsorption by inserting amine and amide functionalities in the final structures. The study included adsorption/desorption isotherms at 35, 50 and 65 °C of the above-mentioned gases in a wide pressure range, up to 200 bar, to cover both pre- and post-combustion capture scenarios. The covalent organic polymers exhibited hysteresis loop, notably at 35 °C, for the CO_2 adsorption/desorption isotherms at high pressure, i.e., between 50 and 125 bar. It was found that COP-10 adsorbs slightly more CO_2 than COP-9 at atmospheric pressure and room temperature owing to its larger pore volume. On the other hand, at high pressure, the CO_2 adsorption capacity becomes larger for COP-9 in comparison to that of COP-10, i.e., 82 mmol g^{-1} vs. 52 mmol g^{-1}, at 200 bar and 35 °C. Unlike low pressure performance, the selectivity of both materials is comparable at high pressure and varies marginally at different temperatures. The researchers concluded that the improved performance of COP-9 at higher pressure might indicate that this adsorbent requires further investigation for large-scale applications both in pre-combustion and post-combustion systems. Yang et al. (2019) developed three kinds of porous organic polymers, denoted as PP-x, via Friedel-Crafts alkylation reaction of dichloro-p-xylene. The synthesized materials labelled as PP-1, PP-2 and PP-3, were produced by employing different polymerization temperatures. They were then covalently grafted with polyamines of strong affinity towards CO_2, i.e., polyethylenimine (PEI), ethylenediamine (EDA), diethylenetriamine (DETA) and tetraethylenepentamine (TEPA). Among the synthesized adsorbents, PP-2-DETA and PP-2-TEPA exhibited high CO_2 adsorption capacities, 2.95 and 2.65 mmol g^{-1} at 1 bar and 25 °C respectively, exceptional CO_2 selectivity over N_2, and suitable isosteric heats of adsorption. Additionally, the adsorbents showed excellent stability at both humid conditions and adsorption/desorption dynamic cycles. Furthermore, the chemically modified polymers displayed a much faster adsorption-desorption kinetics than mmen-Mg_2(dobpdc) and MCM-41-NH_2, which are two benchmark amine-functionalized adsorbents.

1.3.5 *Amine-Functionalized Pore Expanded Silicas and Silica Nanosponges*

Several studies focusing on the synthesis and modification of silica materials with mesoporous texture for efficient CO_2 capture have been reported in the last 15 years (Alkhabbaz et al. 2014; Harlick and Sayari 2007; Sanz et al. 2010; Khatri et al. 2006; Mebane et al. 2013; Zelenak et al. 2008a, b; Li et al. 2008; Knofel et al. 2009; Liu et al. 2010; Yue et al. 2006; Yang et al. 2014; Chen et al. 2009). Son et al. (2008) modified a series of mesoporous silica with 50 wt. % polyethylenimine to assess CO_2 capture performance. They concluded that CO_2 uptakes increased with increasing average pore diameter of the bare support. Thus, they observed that the CO_2 capacities varied following the sequence KIT-6˃SBA-15 ≈ SBA-16˃ MCM-48˃ MCM-41. The same sequence was observed in terms of average pore diameter. Generally speaking, CO_2 adsorption capacity on this kind of materials loaded with amino rich polymers is controlled by the textural properties of the support. Because molecular diffusion in SBA-15 type structure may be hindered by the length of the channels, expanding or swelling agents have been used in the synthesis of mesoporous silica in order to enlarge the size of the micelles formed and obtain pore expanded mesoporous silica hopefully with shorter channel length (see Fig. 1.4).

With this aim, alkanes began to be added to the synthesis gel (Nagarajan et al. 1986; Nagarajan 1999) and mesoporous silicas with larger pore size and interplanar space (1 0 0) were obtained as the chain length of the alkane decreased from C10 to

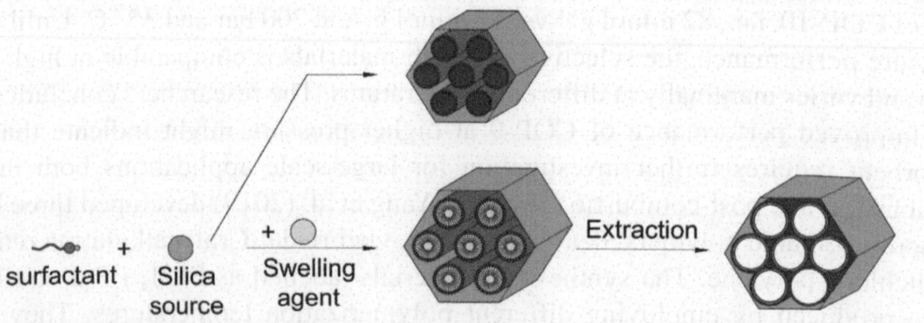

Fig. 1.4 Synthetic scheme of pore expanded sílicas

C5 (Sun et al. 2005). Since short-chain alkanes could not be used in liquid phase under usual synthesis conditions, the possibility of using aromatic hydrocarbons arose, although it was necessary to improve their solubility. The solubility of aromatic compounds decreases as the number and length of benzene ring substituents increase (Nagarajan et al. 1986; Nagarajan 1999). In the case of SBA-15, both 1,3,5-trimethylbenzene, TMB, and 1,3,5-triisopropylbenzene, TIPB, have been used (Zhao et al. 1998a, b) as swelling agents (Fig. 1.4). Kruk research group, at the University of New York, had successfully synthesized large pore size SBA-15 by varying the ratio of surfactant, silica source and swelling agent besides adding a salt to favor its dissolution. Likewise, various hydrolysis temperatures have been used for the mixture. They obtained a series of materials with a pore diameter up to 34 nm (Cao et al. 2009).

Olea et al. (2013) synthesized a pore expanded SBA-15, labelled as PE-SBA-15, and functionalized it by grafting with diethylenetriamine-trimethoxysilane, DT, and by impregnation with polyethylenimine, PEI, and tetraethylenepentamine, TEPA. They used 1,3,5-triisopropylbenzene TIPB (P123:TIPB molar ratio of 2.4:1) as swelling agent and ammonium fluoride as a solubility enhancer. They also compared the influence on CO_2 capture performance of the surfactant removal method, namely calcination versus ethanol extraction, and the hydrolysis temperature. The most outstanding findings in terms of CO_2 adsorption capacity were 1.95 mmol CO_2 g^{-1} for pore expanded SBA-15 functionalized by grafting and 3.72 mmol CO_2 g^{-1} for pore expanded SBA impregnated with tetraethylenepentamine both at 45 °C and 1 bar. They attributed these values to the higher accessibility of organic molecules and the enhanced CO_2 diffusion through filled pores in the pore-expanded materials. At last, under adsorption-desorption cycles, both grafted and impregnated materials showed good stability.

Vilarrasa-Garcia et al. (2015) synthesized mesoporous silica by varying the molar ratio NH_4F/SiO_2 from 0 to 0.065. They used heptane as swelling agent. A progressive disorder of the honeycomb structure is reported as the NH_4F/S_iO_2 increases, leading to materials with higher pore diameter (from 7.95 to 17.3 nm) and pore volume (from 1.70 to 2.33 cm^3 g^{-1}). Polyethylenimine impregnation yield reaches the optimum value for an intermediate NH_4F/S_iO_2 molar ratio. The open porous structure achieved for larger NH_4F/SiO_2 ratios better exposes PEI, which explains the highest CO_2 uptake being observed for a highly disordered mesocellular foam impregnated with 50 wt.%, i.e. 2.3 mmol CO_2 g^{-1} at 25 °C and 1 bar.

A similar study was carried out by Cecilia et al. (2016). They modified hollow silica microspheres, HMS, by adding ammonium fluoride, HMS-F, or 1,3,5- triisoprpylbenzene, HMS-TiPB. They also synthesized HMS by mixing 1,3,5- triisoprpylbenzene and ammonium fluoride, HMS-F-TiPB. Impregnation with amino rich polymers, such as tetraethylenepentamine (TEPA) or polyethylenimine (PEI), was more efficient in adsorbents with high external surface area. The addition of swelling agents improved CO_2 adsorption after amino rich impregnation, in which case 3.4 mmol g^{-1} were adsorbed at 25 °C and 1 bar. Increasing temperature led to an improvement in CO_2 uptake for amino rich polymers possibly due to a decrease in diffusional resistances and the reordering of the amine polymer, as observed by Heydari-Gorji.

Wang et al. (2015) synthesized ordered mesoporous silica, OMS, with average pore diameter from 5.6 to 7.6 nm by adding 1,3,5-trimethylbenzene, TMB, and found that the pore size of silica had a dramatic influence on the CO_2 adsorption performance of the respective amino functionalized sorbents. The maximum total CO_2 capacity obtained by Wang et al. (2015) at 1 bar and 75 °C was 4.28 mmol g^{-1} for the OMS synthesized with TMB and functionalized with 50 wt.% tetraethylenepentamine and they also observed the influence of amino loading on the shape of the breakthrough curves. High amine loading increase the mass transfer resistance and disperse the breakthrough curve whereas low amine content leads to curves with a sharper slope.

Min et al. (2017) prepared a macroporous silica, MacS, by assembling and sintering fractal-like fumed silica particles into a three-dimensional disordered porous network using a spray-drying/calcination method. The resulting materials exhibited an average pore size of 56 nm. Upon impregnation with 50 wt.% polyethylenimine, a CO_2 uptake of 3.84 mmol g^{-1} was obtained at 40 °C and 0.15 bar, which is higher than any other mesoporous silica studied by this group, such as SBA-15, MCF and MCM-41. They remarked the importance of the residual porosity on impregnated mesoporous silica to act as a wide channel for rapid CO_2 diffusion.

Recently, Lashaki and Sayari (2018) studied the combined effect of SBA-15 pore size and intrawall pore volume on the amino loading. A set of SBA-15 was synthesized by varying the aging temperature from 60 to 150 °C. They observed a decrease in the intrawall pore volume as the temperature increases. The opposite effect was observed with the pore size diameter, varying from 7.4 to 14.5 nm when temperature increases from 60 to 150 °C. The supports thus synthesized were functionalized via dry and wet grafting of a triamine (3-[2-(2-Aminoethylamino)ethylamino]propyl trimethoxysilane). In general, wet grafting led to higher yields than dry grafting. The silanol density on the silica surface is enhanced under wet conditions, and these surface groups are crucial to react with aminosilanes (Wang and Yang 2011). With respect to the influence of textural properties, they concluded that the supports with large pore exhibited the higher increase in the surface amine density due to the reduced steric hindrance. This resulted in the highest CO_2 capacities i.e., 1.88 mmol g^{-1} at 75 °C and 0.05 bar for SBA synthesized at 130 °C and functionalized under wet conditions and fastest adsorption kinetics among the grafted samples, providing experimental evidence to confirm the positive effect of the large pore size and high intrawall pore volume on adsorptive properties and adsorption kinetics during CO_2 capture.

Hou et al. (2018) reported a new route to synthesize solid amine sorbents. They prepared mesoporous silicas by emulsion synthesis followed by simple solvent extraction, using ammonium fluoride as etching agent. Mesoporous silica with pore size diameter between 13 and 14 nm were obtained and functionalized with 10, 30 and 50 wt. % polyethylenimine. Outstanding CO_2 adsorption capacities were found at 75 °C and 1 bar: 5.65 mmol g^{-1} for the sample functionalized with 50 wt.% polyethylenimine. This is the highest uptake ever reported with this kind of materials. They provide insights about the optimization of the texture and amino/silica ratio in an ideal CO_2 adsorbent, concluding that a mesoporous structure made of short and wide channels favors amine immobilization and CO_2 diffusion into the pores.

Zhang et al. (2018) also reported the influence of the textural properties of the support on CO_2 adsorption performance. They compared the performance on CO_2 capture by using a fixed bed reactor for SBA-15, MCM-41, hierarchical porous silica, HPS, and a novel three dimensional disordered porous silica (3dd), all of which were functionalized with 40 to 65 wt.% tetraethylenepentamine (TEPA). For supports with a higher TEPA loading, the amounts of CO_2 captured followed the order 3dd > HPS > SBA-15 > MCM-41 at 75 °C. The adsorption capacities observed were 5.1, 4.9, 4.6, and 2.5 mmol g^{-1}, respectively. Thereby, these results further corroborate that a larger pore volume and disordered short channels can promote the dispersion of amine species to expose more active sites for CO_2 capture. Thus, for amino rich functionalized supports, the larger pore size can decrease the CO_2 diffusion resistance and high surface area is no longer the most important factor in determining capture performance.

1.3.6 Amine-Functionalized Modified Clays

Clay minerals are another interesting class of materials for CO_2 adsorption. They are low cost materials due to its wide availability on the Earth surface and promising textural properties, especially after chemical and/or structural modifications. Several clay minerals have been reported as CO_2 adsorbents, such as kaolinite (Chen and Lu 2014, 2015), montmorillonite (Pinto et al. 2011; Chen et al. 2013a, b) and sepiolite (Irani et al. 2015; Vilarrasa-García et al. 2017b, c; Cecilia et al. 2018). Chemical modifications are usually employed to improve the textural properties of naturally occurring clay minerals. Some of them are summarized in Fig. 1.5: chemical modifications by heat and acid treatment; intercalation of organic compounds, synthesis of porous clay heterostructures (PCH) and pillared interlayered clays (PILCs).

The modification by heat and/or acid treatment is well documented in the literature (Balci 1996; Komadel and Madejová 2006; Franco et al. 2014). These treat-

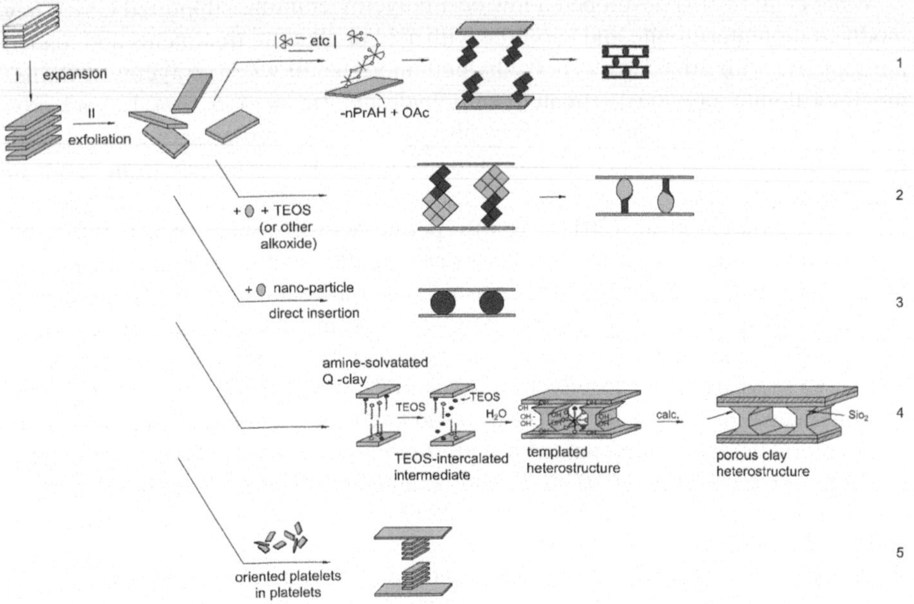

Fig. 1.5 Methods for pillaring clays. 1. Hydrolysis/polymerisation reaction for preparing chromia pillared layered structures PLS (Jones et al. 1995), 2. Intercalated via sol-gel, both homogeneous and heterogeneous (Sterte and Shabtai 1987). 3. Direct intercalation of externally generated known-size nanoparticles (Yamaguchi et al. 2003). 4. Assisting interpillar spacing via amines and surfactants (Tanev and Pinnavaia, 1995). 5. Possible use of platelets themselves as pillars. TEOS: Tetraethyl orthosilicate; PLS: pillared layered structures. Modified after De Stefanis and Tomlinson (2006)

ments are generally intended to develop porosity and increase surface area. Balci (1996) reported that acid pre-treatment of sepiolite caused a change in pore size distribution, increasing the surface area approximately 2.5 fold with respect to the original mineral. Franco et al. (2014) employed microwave-assisted acid treatment to reduce the activation time from 48 h to a few minutes. They treated sepiolite with hydrochloric acid or nitric acid solutions under microwave radiation for different times. Surface area increased approximately 3–4 times respect to the raw sepiolite in just 16 min treatment, accompanied by a significant rise in total pore volume, mainly due to the progressive amorphization and cation depletion of the octahedral sheets. The increase in surface area and pore volume allow for the effective immobilization of amine molecules. Chen and Lu (2015) used kaolinite as starting material and modified it with sulfuric acid under different concentrations. They studied the reaction mechanism of CO_2 with pure clay minerals by using X ray diffraction, Fourier transform infrared spectroscopy and CO_2 adsorption isotherms. CO_2 adsorption mechanism was found to be mainly physical. No interaction of CO_2 with cations of the clay network was observed and acid treatment had a direct impact on CO_2 adsorption performance.

Wang et al. (2014) developed a low cost polyethylenimine-supported clay. They modified montmorillonite and kaolinite with acid or alkaline treatments and loaded the supports with 30–65 wt.% polyethylenimine. The 50 wt. % polyethylenimine-montmorillonite previously treated with hydrochloric acid 6 mol L^{-1} adsorbs 2.54 mmol g^{-1} at 75 °C and 1 bar. They also assessed the influence of the moisture. In the presence of 3 vol. % moisture, CO_2 uptake increases from 2.54 to 3.22 mmol g^{-1}.

Vilarrasa-Garcia et al. (2017a, b, c) reported acid treatment of sepiolite with nitric acid 0.2 mol L^{-1} assisted by microwave radiation and investigated the effects on CO_2 adsorption. They reported that acid treatment improves CO_2 adsorption reaching 1.07 mmol g^{-1} at 1 bar and 25 °C, twice as much as the starting sepiolite. The modified sepiolites were impregnated with 30 wt. % polyethylenimine and a rise in CO_2 adsorption capacity to 1.70 mmol g^{-1} was observed at 1 bar and 65 °C. Likewise, Yuan et al. (2018) also modified industrial-grade sepiolite with hydrochloric acid and impregnated it with triethylenetetramine, TETA. They optimized the TETA loading at 30 wt. % and obtained an adsorbent capable to adsorb 1.93 mmol g^{-1} at 50 °C and 1 bar. This TETA-sep also showed fast kinetics and good regenerability.

Liu et al. (2018) prepared acid-activated sepiolite with 20% hydrochloric acid and functionalized it with different loads of diethylenetriamine, DETA. Thermal gravimetric analysis was the technique employed to assess the CO_2 capacity of the adsorbents. They reported that 45 wt. % DETA-sep at 35 °C and 1 bar adsorbs 1.65 mmol g^{-1}. Stability results also demonstrated that 45 wt. % DETA-sep maintained around 95% of its adsorption capacity after 4 adsorption-desorption cycles and CO_2 could be almost completely desorbed at 75 °C under N_2 flow. The cost of the several adsorbents estimated by the authors of this work evidenced that functionalized clays are cheaper than other commonly used amine-based mesoporous silicas, despite their smaller adsorption capacity.

Porous clay heterostructures, PCH, can be synthesized from clay minerals. The first synthesis was reported in 1995 by Galarneau et al. They used surfactants commonly employed in the synthesis of mesoporous silicas together with aqueous silicate species. Surfactants are intercalated into layered inorganic materials and silicate species forming a silica framework between the layers. After removing the surfactant by calcination, an expanded structure can be obtained. Thus the synthesis of PCHs involves several parameters as the starting clay mineral, choice of surfactant and co-surfactant as well as the precursor employed to form the silica pillars and the method to remove both surfactant and co surfactant. On basis to these choices, several materials can be synthesized with suitable textural properties and they may be further functionalized with amino groups.

Pinto et al. (2011) employed porous clay heterostructures functionalized with 3-aminopropyl triethoxysilane and assessed CO_2 adsorption. They also used microcalorimetry and nuclear magnetic resonance to understand the CO_2 binding mechanism on amine-loaded adsorbents. They observed a high adsorption heat at low coverage, i.e. 125 kJ mol^{-1}, although adsorption was reversible. They claimed the formation of carbamate and carbamic acid from nuclear magnetic resonance data, this being the first time such technique was used to infer about chemisorption products of CO_2 interacting with amines attached to the surface of a sorbent.

Previously, Pinto et al. (2008) had compared the performance of various porous clay heterostructures, PCH, synthesized from a natural clay, two synthesized with tetraethoxysilane or phenyltriethoxysilane and other two with 3-aminopropyl triethoxysilane and tetraethoxysilane. Also, they assessed a pillared clay with aluminum oxide pillars and mesoporous MCM-41. The high pressure adsorption isotherms revealed that some of these PCH were more selective for CO_2 than mesoporous MCM-41 or pillared clays. In the case of PCH grafted with 3-aminopropyl triethoxysilane, CO_2 uptake reached 1.17 mmol g^{-1} at 25 °C and 1 bar. The calculated selectivities ranged from 60 to 145 for the CO_2/CH_4 mixture, and from 50 and 28 for CO_2/C_2H_6 mixture, up to 10 bar.

Vilarrasa-Garcia et al. (2017a, b, c) synthesized porous clay heterostructures, PCH, from raw bentonite, obtaining nanoporous materials with higher surface area and pore volume than the starting clay. They functionalized PCH with 3-aminopropyl triethoxysilane (APTES), tetraethylenepentamine (TEPA) and polyethylenimine (PEI) and observed the enhanced performance of these sorbents for CO_2 capture. First of all, CO_2 uptake for the raw bentonite and the pure PCH rose from 0.11 to 0.64 mmol g^{-1} at 25 °C and 1 bar. CO_2 adsorption on APTES –PCH (20 vol. %) suggested the coexistence of physical and chemical adsorption, reaching 1.02 mmol g^{-1} at the same conditions. On the other hand, CO_2 adsorption on PCH impregnated with PEI and TEPA was governed mainly by chemical interactions reaching 1.46 and 1.64 mmol g^{-1}, respectively, both with 60 wt.% of amino-rich polymers.

Vieira et al. (2018) synthesized an inorganic-organic hybrid adsorbent composed of magadiite grafted with polyethylenimine using different alkoysilanes as pending groups. They used different loadings of polyethylenimine and tested the as-synthesized sorbents for CO_2/N_2 and CO_2/CH_4 separation. They observed that the increase in N content upon polyethylenimine impregnation was proportional to the loading of pending groups. The selectivity estimated from the pure and binary isotherms at 1 bar were in the range of 111–279 for CO_2/N_2 and 10–83 for CO_2/CH_4, values comparable to those found in zeolites and metal organic frameworks.

Solid wastes with a high content of clay mineral may also be used as starting material to synthesize CO_2 adsorbents upon thermal treatment and a pillaring strategy. Vilarrasa-Garcia et al. (2017a, b, c) reported the use of wastes from the flocculation step of a drinking water plant, which showed a high content in Fe clay minerals. The thermal and pillaring strategy successfully increased the specific surface area from 60 to 380 m^2 g^{-1}. The pillared clay-rich waste, denoted as SiFe, was functionalized by 3-aminopropyl triethoxysilane grafting and by polyethylenimine impregnation. The amount of loaded polyethylenimine varied from 10 to 50 wt. %, showing an increase in N content from 3.2 to 16%. This has a direct impact in the CO_2 adsorption capacity of these functionalized materials, which rose from 0.6 to 1.8 mmol g^{-1} at 25 °C and 1 bar. SiFe-50P, which was 50 wt.% impregnated with polyethylenimine, reached 3.6 mmol CO_2 g^{-1} at 75 °C and 1 bar and exhibited an extremely high CO_2/N_2 selectivity, 155 mol CO_2/mol N_2 for 15 vol. % CO_2.

For these high amino rich adsorbents, selectivity is higher at low pressures and nearly independent on CO_2 concentration in the gas phase. Nonetheless, polyethylenimine loaded adsorbents still exhibit a lower amine efficiency, in terms of mol of CO_2 adsorbed per mol of N, as compared to 3-aminopropyl triethoxysilane-grafted counterparts.

1.4 Final Remarks

In light of the increasing concentration of CO_2 in the atmosphere, different technologies are currently under development to capture, store and use the CO_2 released by anthropogenic activities especially those related to power generation and industrial processes. Chemical absorption by liquid amines continues to be nowadays the most technically mature CO_2 capture technology applied at industrial scale, despite its costly solvent regeneration and potential environmental hazards. CO_2 capture by solid adsorbents (a dry technology) shows promise to be a more environmentally friendly technology, though it is yet to be consolidated, particularly regarding materials research. Most of the research on the CO_2 adsorption domain is focused primarily on three main areas: CO_2 capture adsorbent synthesis and shaping, development of suitable adsorption/desorption continuous processes (PSA, TSA, ESA, and hybrids) and integrated materials/process optimization.

Regarding adsorbent synthesis, there are challenges/issues that still require attention and further improvement:

- Reports on novel adsorbents are predominantly devoted to measuring pure CO_2 adsorption capacity and often lack information on adsorption enthalpy and selectivity with respect to competing species.
- Studies on the stability of the adsorbent are often limited to a few tenths of adsorption-desorption cycles with very small sample amounts where hydrodynamics and heat transfer issues do not play a role, unlike large-scale operation.
- The impact of impurities present in the flue gas on the performance of the adsorbent has been scarcely addressed. For instance, he effect of moisture on the materials performance is not properly assessed in current commercially available experimental devices.
- Energy requirements for adsorbent regeneration are scarcely reported.
- The cost of novel materials is not usually reported or even roughly estimated, taking into account the cost of chemicals and yields of the various procedures, including the synthesis per se and post-synthesis functionalization.
- Integrated technical-economic analysis of a given combination of adsorbent and process (VPSA, TSA, ESA) is required, taking into account the different scenarios for CO_2 emissions, since this a key information to compare the viability against current commercial CO_2 capture technologies.

- Molecular simulation tools are highly useful and should be more intensively used to guide synthesis efforts of new adsorbents and provide new insight into the nature of CO_2 binding mechanism

With regards to the amine-functionalized adsorbents, the literature agrees that there is an optimal loading of amine groups onto the surface of a given porous material, which maximizes CO_2 uptake as compared to the pristine material (support). Some of these supports benefit from pore expansion strategies that prevent pore blockage as the amine fills the pores but must also allow for CO_2 diffusion. Based on this reasoning, numerous synthesis methodologies and several chemical surface modification routes with amines have been reported on different adsorbents to find a suitable material able to efficiently capture CO_2 under different emission scenarios. On the other hand, a more integrated approach between adsorbent and processes development is vital to translate results from the bench to industrial scale. Simulations (both at a molecular and macroscopic scales) play also a key role in the adsorbent screening and to assess CO_2 capture performance in conditions close to reality.

Acknowledgements We gratefully acknowledge *Conselho Nacional de Desenvolvimento Científico e Tecnológico* (CNPq, Ministry of Science and Technology, Brazil) and *Coordenação de Aperfeiçoamento de Pessoal de Nível Superior* (CAPES, Ministry of Education, Brazil) for financial support.

References

Alkhabbaz MA, Khunsupat R, Jones CW (2014) Guanidinylated poly(allylamine) supported on mesoporous silica for CO2 capture from flue gas. Fuel 121:79–85. https://doi.org/10.1016/j.fuel.2013.12.018

Alothman ZA (2012) A review: fundamental aspects of silicate mesoporous materials. Materials 5(12):2874–2902. https://doi.org/10.3390/ma5122874

Aresta M, Dibenedetto A (2003) Chapter 9: Carbon dioxide fixation into organic compounds. In: Aresta M (ed) Carbon dioxide recovery and utilization. Kluwer Academic Publishers, Dordrecht, pp 211–260

Balci S (1996) Thermal decomposition of sepiolite and variations in pore structure with and without acid pre-treatment. J Chem Technol Biotechnol 66:72–78. https://doi.org/10.1002/(SICI)1097-4660(199605)66:1<72::AID-JCTB442>3.0.CO;2-T

Bezerra DP, Silva FWM, Moura PAS, Sousa AGS, Vieira RS, Rodriguez-Castellon E, Azevedo DCS (2014) CO2 adsorption in amine-grafted zeolite 13X. Appl Surf Sci 314:314–332. https://doi.org/10.1016/j.apsusc.2014.06.164

Bonan GB, Doney SC (2018) Climate, ecosystems, and planetary futures: the challenge to predict life in Earth system models. Science 359(6375):533–544. https://doi.org/10.1126/science.aam8328

Bredesen R, Jordal K, Bolland O (2004) High-temperature membranes in power generation with CO2 capture. Chem Eng Process 43(9):1129–1158. https://doi.org/10.1016/j.cep.2003.11.011

Burkett SL, Sims SD, Mann S (1996) Synthesis of hybrid inorganic–organic mesoporous silica by co-condensation of siloxane and organosiloxane precursors. Chem Commun:1367–1368. https://doi.org/10.1039/CC996000136

Cao L, Man T, Kruk M (2009) Synthesis of ultra-large-pore SBA-15 silica with two-dimensional hexagonal structure using Triisopropylbenzene as micelle expander. Chem Mater 21(6):1144–1153. https://doi.org/10.1021/cm8012733

Cecilia JA, Vilarrasa-Garcia E, Garcia-Sancho C, Saboya RMA, Azevedo DCS, Cavalcante CL Jr, Rodríguez-Castellón E (2016) Functionalization of hollow silica microspheres by impregnation or grafted of amine groups for the CO2 capture. Int J Greenhouse Gas Control 52:344–356. https://doi.org/10.1016/j.ijggc.2016.07.018

Cecilia JA, Vilarrasa-García E, Cavalcante CL Jr, Azevedo DCS, Franco F, Rodríguez-Castellón E (2018) Evaluation of two fibrous clay minerals (sepiolite and palygorskite) for CO2 capture. J Environ Chem Eng 6:4573–4587. https://doi.org/10.1016/j.jece.2018.07.001

Chang FY, Chao KJ, Cheng HH, Tan CS (2009) Adsorption of CO2 onto amine-grafted mesopo rous silicas. Sep Purif Technol 70:87–95. https://doi.org/10.1016/j.seppur.2009.08.016

Chatti R, Bansiwal AK, Thote JA, Kumar V, Jadhav P, Lokhande SK, Rayalu SS (2009) Amine loaded zeolites for carbon dioxide capture: amine loading and adsorption studies. Microporous Mesoporous Mater 121(1–3):84–89. https://doi.org/10.1016/j.micromeso.2009.01.007

Chen YH, Lu DL (2014) Amine modification on kaolinites to enhance CO2 adsorption J. Colloid Interface Sci 436:47–51. https://doi.org/10.1016/j.jcis.2014.08.050

Chen YH, Lu DL (2015) CO2 capture by kaolinite and its adsorption mechanism. Appl Clay Sci 104:221–228. https://doi.org/10.1016/j.clay.2014.11.036

Chen C, Yang ST, Ahn WS, Ryoo R (2009) Amine-impregnated silica monolith with a hierarchical pore structure: enhancement of CO2 capture capacity. Chem Commun 24:3627–3629. https://doi.org/10.1039/b905589d

Chen Z, Deng S, Wei H, Wang B, Huang J, Yu G (2013a) Activated carbons and amine-modified materials for carbon dioxide capture — a review. Front Environ Sci Eng 7(3):326–340. https://doi.org/10.1007/s11783-013-0510-7

Chen C, Park DW, Ahn WS (2013b) Surface modification of a low cost bentonite for post-combustion CO2 capture. Appl Surf Sci 283:699–704. https://doi.org/10.1016/j.apsusc.2013.07.005

Chen C, Kim S-S, Cho W-S, Ahn W-S (2015) Polyethylenimine-incorporated zeolite 13X with mesoporosity for post-combustion CO2 capture. Appl Surf Sci 332:167–171. https://doi.org/10.1016/j.apsusc.2015.01.106

Chong ASM, Zhao XS (2003) Functionalization of SBA-15 with APTES and characterization of functionalized materials. J Phys Chem B 107(46):12650–12657. https://doi.org/10.1021/jp035877+

De Stefanis A, Tomlinson AAG (2006) Towards designing pillared clays for catalysis. Catal Today 104:126–141. https://doi.org/10.1016/j.cattod.2006.01.019

Fayemiwo KA, Vladisavljević GT, Nabavi SA, Benyahia B, Hanak DP, Loponov KN, Manović V (2018) Nitrogen-rich hyper-crosslinked polymers for low-pressure CO2 capture. Chem Eng J 334:2004–2013. https://doi.org/10.1016/j.cej.2017.11.106

Franco F, Pozo M, Cecilia JA, Benítez-Guerrero M, Pozo E, Rubí M (2014) Microwave assisted acid treatment of sepiolite: the role of composition and "crystallinity". App Clay Sci 102:15–27. https://doi.org/10.1016/j.clay.2014.10.013

Fu Q, Ding J, Wang W, Lu J, Huang Q (2017) Carbon dioxide adsorption over amine-functionalized MOFs. Energy Procedia 142:2152–2157. https://doi.org/10.1016/j.egypro.2017.12.620

Gaikwad S, Kim S-J, Han S (2019) CO2 capture using amine-functionalized bimetallic MIL-101 MOFs and their stability on exposure to humid air and acid gases. Microporous Mesoporous Mater 277:253–260. https://doi.org/10.1016/j.micromeso.2018.11.001

Galarneau A, Barodawall A, Pinnavaia TJ (1995) Porous clay heterostructures formed by gallery-templated synthesis. Nature 374:529–531. https://doi.org/10.1038/374529a0

Harlick PJE, Sayari A (2007) Applications of pore-expanded Mesoporous silica. 5. Triamine grafted material with exceptional CO2 dynamic and equilibrium adsorption performance. Ind Eng Chem Res 46(1):446–458. https://doi.org/10.1021/ie060774

Hedin N, Bacsik Z (2019) Perspectives on the adsorption of CO2 on amine-modified silica studied by infrared spectroscopy. Curr Opin Green Sustain Chem 16:13–19. https://doi.org/10.1016/j.cogsc.2018.11.010

Hiyoshi N, Yogo K, Yashima T (2004) Adsorption of carbon dioxide on modified mesoporous materials in the presence of water vapor. Stud Surf Sci Catal 154:2995–3002. https://doi.org/10.1016/S0167-2991(04)80583-4

Hoegh-Guldberg O, Bruno JF (2010) The impact of climate change on the world's marine ecosystems. Science 328(5985):1523–1528. https://doi.org/10.1126/science.1189930

Hou X, Zhuang L, Ma B, Chen S, He H, Yin F (2018) Silanol-rich platelet silica modified with branched amine for efficient CO2 capture. Chem Eng Sci 181:315–325. https://doi.org/10.1016/j.ces.2018.02.015

Hung CT, Yang CF, Lin JS, Huang SJ, Chang YC, Liu SB (2017) Capture of carbon dioxide by polyamine-immobilized mesostructured silica: a solid-state NMR study. Microporous Mesoporous Mater 238:2–13. https://doi.org/10.1016/j.micromeso.2016.03.006

Intergovernmental Panel on Climate Change (2014) Climate change: impacts, adaptation, and vulnerability. Cambridge University Press, Cambridge/New York

Irani M, Fan M, Ismail H, Tuwati A, Dutcher B, Russell AG (2015) Modified nanosepiolite as an inexpensive support of tetraethylenepentamine for CO2 sorption. Nano Energy 11:235–246. https://doi.org/10.1016/j.nanoen.2014.11.005

Irani M, Jacobson AT, Gasem KAM, Fan M (2017) Modified carbon nanotubes/tetraethylenepentamine for CO2 capture. Fuel 206:10–18. https://doi.org/10.1016/j.fuel.2017.05.087

Jo H, Lee WR, Kim NW, Jung H, Lim KS, Kim JE, Hong CS (2017) Fine-tuning of the carbon dioxide capture capability of Diamine-grafted metal-organic framework adsorbents through amine functionalization. ChemSusChem 10(3):541–550. https://doi.org/10.1002/cssc.201601203

Jones D, Roziere J, Maireles-Torres P, Jimenez-Lopez A, Olivera-Pastor P, Rodriguez-Castellon E, Tomlinson AAG (1995) Chromia Pillaring in .Alpha.-zirconium phosphate: a structural investigation using X-ray absorption spectroscopy. Inorg Chem 34(18):4611–4617. https://doi.org/10.1021/ic00122a018

Keller L, Ohs B, Lenhart J, Abduly L, Blanke P, Wessling M (2018) High capacity polyethylenimine impregnated microtubes made of carbon nanotubes for CO2 capture. Carbon 126:338–345. https://doi.org/10.1016/j.carbon.2017.10.023

Kenarsari SD, Yang D, Jiang G, Zhang S, Wang J, Russell AG, Fan M (2013) Review of recent advances in carbon dioxide separation and capture. RSC Adv 3(45):22739–22773. https://doi.org/10.1039/c3ra43965h

Khatri RA, Chuang SSC, Soong Y, Gray M (2006) Thermal and chemical stability of regenerable solid amine sorbent for CO2 capture. Energy Fuel 20(4):1514–1520. https://doi.org/10.1021/ef050402y

Knofel C, Martin C, Hornebecq V, Llewellyn PL (2009) Study of carbon dioxide adsorption on Mesoporous Aminopropylsilane-functionalized silica and Titania combining Microcalorimetry and in situ infrared spectroscopy. J Phys Chem C 113(52):21726–21734. https://doi.org/10.1021/jp907054h

Komadel P, Madejová J (2006) Chapter 7.1: Acid activation of clay minerals. In: Bergaya F, Theng BKG, Lagaly G (eds) Handbook of clay science, vol 1, 1st edn. Elsevier Science, Amsterdam/Oxford, pp 263–287. Hardcover ISBN: 9780080441832, eBook ISBN: 9780080457635

Lashaki MJ, Sayari A (2018) CO2 capture using triamine-grafted SBA-15: the impact of the support pore structure. Chem Eng J 334:1260–1269. https://doi.org/10.1016/j.cej.2017.10.103

Leung DYC, Caramanna G, Maroto-Valer MM (2014) An overview of current status of carbon dioxide capture and storage technologies. Renew Sust Energ Rev 39:426–443. https://doi.org/10.1016/j.rser.2014.07.093

Li P, Ge B, Zhang S, Chen S, Zhang Q, Zhao Y (2008) CO2 capture by Polyethylenimine-modified fibrous adsorbent. Langmuir 24(13):6567–6574. https://doi.org/10.1021/la800791s

Li Y, Sun N, Li L, Zhao N, Xiao F, Wei W, Sun Y, Huang W (2013) Grafting of amines on ethanol-extracted SBA-15 for CO2 adsorption. Materials 6:981–999. https://doi.org/10.3390/ma6030981

Liu Y, Shi J, Chen J, Ye Q, Pan H, Shao Z, Shi Y (2010) Dynamic performance of CO2 adsorption with tetraethylenepentamine-loaded KIT-6. Microporous Mesoporous Mater 134(1–3):16–21. https://doi.org/10.1016/j.micromeso.2010.05.002

Liu D, Wu Y, Xia Q, Li Z, Xi H (2012) Experimental and molecular simulation studies of CO2 adsorption on zeolitic imidazolate frameworks: ZIF-8 and amine-modified ZIF-8. Adsorption 19(1):25–37. https://doi.org/10.1007/s10450-012-9407-1

Liu F, Chen S, Gao Y (2017) Synthesis of porous polymer based solid amine adsorbent: effect of pore size and amine loading on CO2 adsorption. J Colloid Interface Sci 506:236–244. https://doi.org/10.1016/j.jcis.2017.07.049

Liu L, Chen H, Shiko E, Fan X, Zhou Y, Zhang G, Luo X, Hu X (2018) Low-cost DETA impregnation of acid-activated sepiolite for CO2 capture. Chem Eng J 353:940–948. https://doi.org/10.1016/j.cej.2018.07.086

Ma J, Liu Q, Chen D, Wen S, Wang T (2015) Synthesis and characterisation of pore-expanded mesoporous silica materials. Micro Nano Lett 10(2):140–144. https://doi.org/10.1049/mnl.2014.0413

Macquarrie DJ (1996) Direct preparation of organically modified MCM-type materials. Preparation and characterisation of aminopropyl–MCM and 2-cyanoethyl–MCM. Chem Commun:1961–1962. https://doi.org/10.1039/CC9960001961

McDonald TM, Mason JA, Kong X, Bloch ED, Gygi D, Dani A, Long JR (2015) Cooperative insertion of CO2 in diamine-appended metal-organic frameworks. Nature 519(7543):303–308. https://doi.org/10.1038/nature14327

McKee B (2002) Solutions for the 21st century: zero emissions technology for fossil fuels IEA, Com. Energy Research and Technology. OECD/IEA, Paris

Mebane DS, Kress JD, Storlie CB, Fauth DJ, Gray ML, Li K (2013). https://doi.org/) Transport, zwitterions, and the role of water for CO2 adsorption in Mesoporous silica-supported amine sorbents. J Phys Chem C 117(50):26617–26627. https://doi.org/10.1021/jp4076417

Mei D, Geng L, Lin Z, Chen S, Wei J, Liao L (2018) CO2 adsorption properties of mixed-amine functionalized mesoporous molecular sieve KIT-6. Mater Res Express 5(6). https://doi.org/10.1088/2053-1591/aacc7c

Milner PJ, Siegelman RL, Forse AC, Gonzalez MI, Runcevski T, Martell JD, Long JR (2017) A Diaminopropane-appended metal-organic framework enabling efficient CO2 capture from coal flue gas via a mixed adsorption mechanism. J Am Chem Soc 139(38):13541–13553. https://doi.org/10.1021/jacs.7b07612

Min K, Choi W, Choi M (2017) Macroporous silica with thick framework for steam-stable and high-performance poly(ethyleneimine)/silica CO2 adsorbent. ChemSusChem 10(11):2518–2526. https://doi.org/10.1002/cssc.201700398

Modak A, Jana S (2019) Advancement in porous adsorbents for post-combustion CO2 capture. Microporous Mesoporous Mater 276:107–132. https://doi.org/10.1016/j.micromeso.2018.09.018

Nagarajan R (1999) Solubilization of hydrocarbons and resulting aggregate shape transitions in aqueous solutions of Pluronic® (PEO–PPO–PEO) block copolymers. Colloid Surf B Biointerfaces 16(1–4):55–72. https://doi.org/10.1016/S0927-7765(99)00061-2

Nagarajan R, Barry M, Ruckenstein E (1986) Unusual selectivity in solubilization by block copolymer micelles. Langmuir 2(2):210–215. https://doi.org/10.1021/la00068a017

Nguyen TH, Kim S, Yoon M, Bae TH (2016) Hierarchical zeolites with amine-functionalized Mesoporous domains for carbon dioxide capture. ChemSusChem 9(5):455–461. https://doi.org/10.1002/cssc.201600004

Nie L, Mu Y, Jin J, Chen J, Mi J (2018) Recent developments and consideration issues in solid adsorbents for CO2 capture from flue gas. Chin J Chem Eng 26(11):2303–2317. https://doi.org/10.1016/j.cjche.2018.07.012

Olea A, Sanz-Pérez ES, Arencibia A, Sanz R, Calleja G (2013) Amino-functionalized pore-expanded SBA-15 for CO2 adsorption. Adsorption 19(2–4):589–600. https://doi.org/10.1007/s10450-013-9482-y

Pachauri RK, Reisinger A (eds) (2007) Fourth assessment report on climate change. International Panel on Climate Change, Geneva

Pera-Titus M (2014) Porous inorganic membranes for CO_2 capture: present and prospects. Chem Rev 114(2):1413–1492. https://doi.org/10.1021/cr400237k

Pham T-H, Lee B-K, Kim J (2016) Novel improvement of CO 2 adsorption capacity and selectivity by ethylenediamine-modified nano zeolite. J Taiwan Inst Chem Eng 66:239–248. https://doi.org/10.1016/j.jtice.2016.06.030

Pino L, Italiano C, Vita A, Fabiano C, Recupero V (2016) Sorbents with high efficiency for CO2 capture based on amines-supported carbon for biogas upgrading. J Environ Sci (China) 48:138–150. https://doi.org/10.1016/j.jes.2016.01.029

Pinto ML, Pires J, Rocha J (2008) Porous materials prepared from clays for the upgrade of landfill gas. J Phys Chem C 112:14394–14402. https://doi.org/10.1021/jp803015d

Pinto ML, Mafra L, Guil JM, Pires J, Rocha J (2011) Adsorption and activation of CO2 by amine-modified Nanoporous materials studied by solid-state NMR and 13CO2 adsorption. Chem Mater 23(6):1387–1395. https://doi.org/10.1021/cm1029563

Puthiaraj P, Lee Y-R, Ahn W-S (2017) Microporous amine-functionalized aromatic polymers and their carbonized products for CO2 adsorption. Chem Eng J 319:65–74. https://doi.org/10.1016/j.cej.2017.03.001

Rahmat N, Abdullah AZ, Mohamed AR (2010) A review: Mesoporous Santa Barbara Amorphous-15, types, synthesis and its applications towards biorefinery production. Am J Appl Sci 7(12):1579–1586. https://doi.org/10.3844/ajassp.2010.1579.1586

Raupach MR, Marland G, Ciais P, Le Quéré C, Canadell JG, Klepper G, Field CB (2007) Proc Natl Acad Sci U S A 104(24):10288–10293. https://doi.org/10.1073/pnas.0700609104

Rochelle GT (2009) Amine scrubbing for CO2 capture. Science 325(5948):1652–1654. https://doi.org/10.1126/science.1176731

Saha A (2018) Structure-function, recyclability and calorimetry studies of CO2 adsorption on some amine modified type I & type II sorbents. Int J Greenhouse Gas Control 78:198–209. https://doi.org/10.1016/j.ijggc.2018.07.028

Sanchez-Zambrano KS, Duarte LL, Maia DAS, Vilarrasa-Garcia E, Bastos-Neto M, Rodríguez-Castellón E, Azevedo DCS (2018) CO2 capture with Mesoporous Silicas modified with amines by double functionalization: assessment of adsorption/desorption cycles. Materials 11(6):887–906. https://doi.org/10.3390/ma11060887

Sanchez-Zambrano KS, Vilarrasa-Garcia E, Maia DAS, Bastos-Neto M, Rodríguez-Castellón E, Azevedo DCS (2019) Adsorption microcalorimetry as a tool in the characterization of amine-grafted mesoporous silicas for CO2 capture. Adsorption (Accepted Manuscript). https://doi.org/10.1007/s10450-019-00064-y

Sanz R, Calleja G, Arencibia A, Sanz-Pérez ES (2010) CO2 adsorption on branched polyethyleneimine-impregnated mesoporous silica SBA-15. Appl Surf Sci 256(17):5323–5328. https://doi.org/10.1016/j.apsusc.2009.12.070

Sanz R, Calleja G, Arencibia A, Sanz-Perez ES (2012) Amino functionalized mesostructured SBA-15 silica for CO2 capture: exploring the relation between the adsorption capacity and the distribution of amino groups by TEM. Microporous Mesoporous Mater 158:309–317. https://doi.org/10.1016/j.micromeso.2012.03.053

Sanz R, Calleja G, Arencibia A, Sanz-Perez ES (2013) Development of high efficiency adsorbents for CO2 capture based on a double-functionalization method of grafting and impregnation. J Mater Chem A 1:1956–1962. https://doi.org/10.1016/10.1039/c2ta01343f

Son WJ, Choi JS, Ahn WS (2008) Adsorptive removal of carbon dioxide using Polyethyleneimine-loaded Mesoporous silica materials. Microporous Mesoporous Mater 113(1–3):31–40. https://doi.org/10.1016/j.micromeso.2007.10.049

Sterte JP, Shabtai J (1987) Cross-linked Smectites. V. Synthesis and properties of Hydroxy-silicoaluminum montmorillonites and fluorhectorites. Clays Clay Miner 35(6):429–439. https://doi.org/10.1346/CCMN.1987.0350603

Su F, Lu C, Kuo S-C, Zeng W (2010) Adsorption of CO2on amine-functionalized Y-type zeolites. Energy Fuel 24(2):1441–1448. https://doi.org/10.1021/ef901077k

Sun J, Zhang H, Ma D, Chen Y, Bao X, Klein-Hoffmann A, Pfänder N, Su DS (2005) Alkanes-assisted low temperature formation of highly ordered SBA-15 with large cylindrical mesopores. Chem Commun 42:5343–5345. https://doi.org/10.1039/b509713d

Tanev PT, Pinnavaia TJ (1995) A neutral templating route to mesoporous molecular sieves. Science 267(5199):865–867. https://doi.org/10.1126/science.267.5199.865

Thommes M, Kaneko K, Neimark AV, Olivier JP, Rodríguez-Reinoso F, Rouquerol J, Sing KSW (2015) Physisorption of gases, with special reference to the evaluation of surface area and pore size distribution (IUPAC Technical Report). Pure Appl Chem 87(9–10):1051–1069. https://doi.org/10.1515/pac-2014-1117

Ullah R, Patel H, Aparicio S, Yavuz CT, Atilhan M (2019) A combined experimental and theoretical study on gas adsorption performance of amine and amide porous polymers. Microporous Mesoporous Mater 279:61–72. https://doi.org/10.1016/j.micromeso.2018.12.011

UNFCCC Kyoto Protocol, revised 2019. https://unfccc.int/process/the-kyoto-protocol

Ünveren EE, Monkul BÖ, Sarıoğlan Ş, Karademir N, Alper E (2017) Solid amine sorbents for CO 2 capture by chemical adsorption: a review. Petroleum 3(1):37–50. https://doi.org/10.1016/j.petlm.2016.11.001

Vieira RB, Moura PAS, Vilarrasa-García E, Azevedo DCS, Pastore HO (2018) Polyamine-grafted Magadiite: high CO2 selectivity at capture from CO2/N2 and CO2/CH4 mixtures. J Co2 Util 23:29–41. https://doi.org/10.1016/j.jcou.2017.11.004

Vilarrasa-Garcia E, Ortigosa-Moya EM, Cecilia JA, Cavalcante CL Jr, Jiménez-Jiménez JJ, Azevedo DCS, Rodríguez-Castellón E (2015) CO2 adsorption on amine modified mesoporous silicas: effect of the progressive disorder of the honeycomb arrangement. Microporous Mesoporous Mater 209:172–183. https://doi.org/10.1016/j.micromeso.2014.08.032

Vilarrasa-García E, Cecilia JA, Bastos-Neto M, Cavalcante CL Jr, Azevedo DCS, Rodríguez-Castellón E (2017a) Microwave-assisted nitric acid treatment of sepiolite and functionalization with polyethylenimine applied to CO2 capture and CO2/N2 separation. Appl Surf Sci 410:315–325. https://doi.org/10.1016/j.apsusc.2017.03.054

Vilarrasa-García E, Cecilia JA, Azevedo DCS, Cavalcante CL Jr, Rodríguez-Castellón E (2017b) Evaluation of porous clay heterostructures modified with amine species as adsorbent for the CO2 capture. Microporous Mesoporous Mater 249:25–33. https://doi.org/10.1016/j.micromeso.2017.04.049

Vilarrasa-García E, Cecilia JA, Rodriguez Aguado E, Jimenez-Jimenez J, Cavalcante CL Jr, Azevedo DCS, Rodriguez-Castellon E (2017c) Amino-modified pillared adsorbent from water-treatment solid wastes applied to CO2/N2 separation. Adsorption 23(2–3):405–421. https://doi.org/10.1007/s10450-017-9871-8

Wang L, Yang RT (2011) Increasing selective CO2 adsorption on amine-grafted SBA-15 by increasing silanol density. J Phys Chem C 115:21264–21272. https://doi.org/10.1021/jp206976d

Wang J, Chen H, Zhou H, Liu X, Qiao W, Long D, Ling L (2013a) Carbon dioxide capture using polyethylenimine-loaded mesoporous carbons. J Environ Sci (China) 25(1):124–132. https://doi.org/10.1016/s1001-0742(12)60011-4

Wang J, Wang M, Zhao B, Qiao W, Long D, Ling L (2013b) Mesoporous carbon-supported solid amine sorbents for low-temperature carbon dioxide capture. Ind Eng Chem Res 52(15):5437–5444. https://doi.org/10.1021/ie303388h

Wang W, Xiao J, Wei X, Ding J, Wang X, Song C (2014) Development of a new clay supported polyethylenimine composite for CO2 capture. Appl Energy 113:334–341. https://doi.org/10.1016/j.apenergy.2013.03.090

Wang L, Yao M, Hu X, Hu G, Lu J, Luo M, Fan M (2015) Amine-modified ordered mesoporous silica: the effect of pore size on CO2 capture performance. Appl Surf Sci 324:286–292. https://doi.org/10.1016/j.apsusc.2014.10.135

Wang Y, Du T, Song Y, Che S, Fang X, Zhou L (2017) Amine-functionalized mesoporous ZSM-5 zeolite adsorbents for carbon dioxide capture. Solid State Sci 73:27–35. https://doi.org/10.1016/j.solidstatesciences.2017.09.004

Wang Y, Du T, Qiu Z, Song Y, Che S, Fang X (2018) CO2 adsorption on polyethyleniminemodified ZSM-5 zeolite synthesized from rice husk ash. Mater Chem Phys 207:105–113. https://doi.org/10.1016/j.matchemphys.2017.12.040

Yamaguchi EGT, Kitajima K, Sakai E, Daimon M (2003) Properties of ZrO2-pillared fluorine micas synthesized using poly(vinyl alcohol) as a template agent. J Ceramic Soc Jpn 111(1296):567–571. https://doi.org/10.2109/jcersj.111.567

Yang M, Song Y, Jiang L, Wang X, Liu W, Zhao Y, Liu Y, Wang S (2014) Dynamic measurements of hydrate based gas separation in cooled silica gel. J Ind Eng Chem 20:322–330. https://doi.org/10.1016/j.jiec.2013.03.031

Yang Y, Chuah CY, Bae T-H (2019) Polyamine-appended porous organic polymers for efficient post-combustion CO2 capture. Chem Eng J 358:1227–1234. https://doi.org/10.1016/j.cej.2018.10.122

Yaumi AL, Bakar MZA, Hameed BH (2017) Recent advances in functionalized composite solid materials for carbon dioxide capture. Energy 124:461–480. https://doi.org/10.1016/j.energy.2017.02.053

Yokoi T, Kubota Y, Tatsumi T (2012) Amino-functionalized mesoporous silica as base catalyst and adsorbent. Appl Catal A Gen 421–422:14–37. https://doi.org/10.1016/j.apcata.2012.02.004

Yuan M, Gao G, Hu X, Luo X, Huang Y, Jin B, Liang Z (2018) Premodified Sepiolite functionalized with Triethylenetetramine as an effective and inexpensive adsorbent for CO2 capture. Ind Eng Chem Res 57:6189–6200 . (DRX despues de adsorcion de CO2…). https://doi.org/10.1021/acs.iecr.8b00348

Yue MB, Chun Y, Cao Y, Dong X, Zhu JH (2006) CO2 capture by as-prepared SBA-15 with an occluded organic template. Adv Funct Mater 16:1717–1722. https://doi.org/10.1002/adfm.200600427

Zelenak V, Badanicova M, Halamova D, Cejka J, Zukal A, Murafa N, Goerigk G (2008a) Amine-modified ordered Mesoporous silica: effect of pore size on carbon dioxide capture. Chem Eng J 144:336–342. https://doi.org/10.1016/j.cej.2008.07.025

Zelenak V, Halamova D, Gaberova L, Bloch E, Llewellyn P (2008b) Amine-modified SBA-12 mesoporous silica for carbon dioxide capture: effect of amine basicity on sorption properties. Microporous Mesoporous Mater 116(1–3):358–364. https://doi.org/10.1016/j.micromeso.2008.04.023

Zhang Y, Maroto-Valer MM, Tang Z (2004) Microporous activated carbons produced from unburned carbon in fly ash and their application for CO2 capture. Fuel Chem Div 49(1):304–305

Zhang G, Zhao P, Xu Y, Yang Z, Cheng H, Zhang Y (2018) Structure property-CO_2 capture performance relations of amine-functionalized porous silica composite adsorbents. Appl Mater Interfaces 10:34340–34354. https://doi.org/10.1021/acsami.8b13069

Zhao D, Feng J, Huo Q, Melosh N, Fredrikson GH, Chmelka BF, Stucky GD (1998a) Triblock copolymer syntheses of Mesoporous silica with periodic 50 to 300 angstrom pores. Science 279:548–552. https://doi.org/10.1126/science.279.5350.548

Zhao D, Huo Q, Feng J, Chmelka BF, Stucky GD (1998b) Nonionic Triblock and star Diblock copolymer and Oligomeric surfactant syntheses of highly ordered, hydrothermally stable, Mesoporous silica structures. J Am Chem Soc 120:6024–6036. https://doi.org/10.1021/ja974025i

Zhao X, Cui Q, Wang B, Yan X, Singh S, Zhang F, Li Y (2018) Recent progress of amine modified sorbents for capturing CO2 from flue gas. Chin J Chem Eng 26(11):2292–2302. https://doi.org/10.1016/j.cjche.2018.04.009

Chapter 2
Absorbents, Media, and Reagents for Carbon Dioxide Capture and Utilization

Hanan Mohamed Mohsin, Khairiraihanna Johari, and Azmi Mohd Shariff

Abstract In the last few decades, climate change phenomena namely floods, droughts, and cyclones commonly appeared in the news headlines. These phenomena were mainly driven by global warming. Anthropogenic emissions of carbon dioxide resulted in its accumulation in the atmosphere which enhanced the greenhouse effect. Carbon dioxide capture and utilization was introduced as a method to mitigate the effects of global warming. Absorption is currently the most well-established type of technology available in the market. Different types of solvent reported as absorbents for carbon dioxide capture along with media and reagents for utilization process were reviewed. Among all the absorbents, amine solution was the most widely reported type of absorbent. Despite its effectiveness in capturing carbon dioxide, amine solution has some disadvantages such as high volatility, low thermal stability, and high oxidation degradation. Emerging solvents such as ionic liquid, amino acid salt, and weak base were suggested as alternative solvents for carbon dioxide capture. For carbon dioxide conversion process, metal ions and organic reagents were added into carbon dioxide-saturated absorbents to convert carbon dioxide into value-added products at relatively lower temperature and

H. Mohamed Mohsin
Chemical Engineering Department, Universiti Teknologi PETRONAS,
Bandar Seri Iskandar, Perak, Malaysia

K. Johari (✉)
Chemical Engineering Department, Universiti Teknologi PETRONAS,
Bandar Seri Iskandar, Perak, Malaysia

Centre of Contaminant Control & Utilization (CenCoU),
Institute of Contaminant Management (ICM), Universiti Teknologi PETRONAS,
Bandar Seri Iskandar, Perak, Malaysia
e-mail: khairiraihana.j@utp.edu.my

A. Mohd Shariff
Chemical Engineering Department, Universiti Teknologi PETRONAS,
Bandar Seri Iskandar, Perak, Malaysia

Carbon Dioxide Research Centre (CO2RES), Institute of Contaminant Management (ICM),
Universiti Teknologi PETRONAS, Bandar Seri Iskandar, Perak, Malaysia

Inamuddin et al. (eds.), *Sustainable Agriculture Reviews 38*, Sustainable Agriculture Reviews 38, https://doi.org/10.1007/978-3-030-29337-6_2

pressure compared to conventional utilization process. This reduces the energy requirement during carbon dioxide capture and utilization process.

Keywords Absorption · Carbamates · Carbonates · Chemical · Conversion · Solvent

2.1 Introduction

The adverse effects of global warming on the environment are well documented throughout the world such as increasing of the Earth's surface temperature, melting of ice glaciers, droughts, and floods. These phenomena are mainly driven by excessive anthropogenic emissions of carbon dioxide. Carbon capture and storage is proposed as one of the methods for carbon dioxide mitigation to reduce the environmental impacts of global warming. Carbon capture and storage technology originated from the process of separating carbon dioxide from natural gas and injecting the captured gas into oil wells for enhanced oil recovery (Metz et al. 2005). By using a similar method, carbon dioxide can be captured from large point sources, such as power plants and industrial facilities, to reduce carbon emissions. Several methods were reported for carbon dioxide capture process, such as absorption process (Rochelle 2009), membrane technology (Basile et al. 2010), adsorption process (Gibson et al. 2016), and cryogenic distillation (Xu et al. 2014).

Absorption process is the most well understood technology and is extensively used in the natural gas purification industry (Rochelle 2009). The absorption process relies on physical or chemical interactions between the absorbent and carbon dioxide molecules (Ravanchi and Sahebdelfar 2014). The acidic nature of carbon dioxide allows the molecules to react chemically with an alkaline solution, and hence, separate carbon dioxide molecules from the feed gas. On the other hand, organic solvent can also be used to capture carbon dioxide through physical binding (Ravanchi and Sahebdelfar 2014). Commercial solvents, which are used for chemical absorption process, include aqueous amine and aqueous ammonia solutions (Budzianowski 2015). Meanwhile, physical absorption process utilizes glycols or methanol as absorbents (Budzianowski 2015). Although absorption process is proven to be an efficient method for carbon dioxide separation process, the energy intensive process during carbon dioxide desorption process is a major drawback of this system. During regeneration process, carbon dioxide is separated from the absorbent to allow the solvent to be recycled back into the system. Due to the strong chemical bonding between carbon dioxide and the absorbent, a higher amount of energy is required for chemical absorption as compared to physical absorption process (Budzianowski 2015).

Based on the carbon capture and storage technology, high purity carbon dioxide is produced in large quantities and needs to be stored underground to prevent its accumulation in the atmosphere. In the case of natural gas sweetening process, carbon capture and storage technology is seen as an opportunity such that the carbon dioxide waste produced during natural gas purification process is utilized to enhance the production of crude oil, particularly in mature oil fields where output decreases over time. Therefore, the high cost of transportation and maintenance for storage can be offset by the increase in oil revenue. However, the scenario is different when carbon dioxide is captured from large point sources at low pressure. Moreover, the carbon dioxide capture sites are usually located further away from oil fields, which increase the cost for storage of carbon dioxide in underground geological sites (De Falco et al. 2013).

Carbon capture and utilization process is an emerging technology which is aimed to complement carbon capture and storage technology in combating climate change. Apart from enhanced oil recovery, carbon dioxide is being utilized as building blocks for carbon-based chemicals, which include urea, methanol, salicylic acid, and carbonates (Alper and Yuksel Orhan 2017). The downside of conventional carbon capture and utilization technology is the high energy requirement during conversion process due to high thermodynamic stability of carbon dioxide molecules. The main product of carbon dioxide conversion produced in large quantity is urea, which is synthesized at temperature of 185 °C to 190 °C and pressure of 180 atmosphere to 200 atmosphere (Alper and Yuksel Orhan 2017). One of the strategies identified to reduce the energy requirement during conversion process is by reacting carbon dioxide in its aqueous form (Yang et al. 2012). This can be achieved by capturing carbon dioxide from the gas source by using absorption technology, which in turn will reduce or possibly eliminate the high energy requirement during the solvent regeneration process.

The conversion of metal ions into carbonated products originated from natural mineralization process. Natural silicates such as olivine, serpentine, and wollastonite contain metal ions which will react with carbon dioxide over a period of time (Mazzotti et al. 2005). The injection of carbon dioxide into geological reservoirs resulted in an in-situ carbonation process, which occurs between hundreds to millions of years (Teir 2008). For industrial application, ex-situ carbonation takes place in a chemical plant, where metal ions are directly reacted with carbon dioxide gas. The ex-situ carbonation process can be speed up by increasing the temperature, grinding of silicates, and dissolving the silicates into acidic or basic solutions (Oelkers et al. 2008). High temperature and grinding process contribute to high energy requirement during carbonation process. Ex-situ carbonation by using aqueous solution such as sodium and ammonium hydroxides was studied by Pérez-Moreno et al. (2015) which revealed that extraction by using alkaline solution assisted in speeding up the carbonation process. Since alkaline solutions have high potential to be used as absorbents for carbon dioxide removal, coupling carbon dioxide separation process with carbonation process is expected to lower down the cost for carbon dioxide capture technology through elimination of solvent regeneration process, while generating value-added carbonated products.

Inorganic carbonates, such as calcium carbonates and magnesium carbonates, have high industrial values, which are used as raw materials for the production of cement, building materials, paper, and optical glasses (Peters et al. 2011). Cyclic carbonates were obtained through subsequent utilization of carbon dioxide through the addition of epoxides (Xiao et al. 2014; Liu et al. 2016; Kumar et al. 2017). On top of that, carbon dioxide capture and subsequent utilization produced carbamates when reaction occurred in a non-aqueous media (Uma Maheswari and Palanivelu 2016; Hasib-ur-Rahman et al. 2012; Barbarossa et al. 2013; Uma Maheswari and Palanivelu 2014). Carbamates are useful precursors for drug synthesis and medicinal chemistry (Ghosh and Brindisi 2015). Carbamates can also be used to synthesis polymers for the production of plastics, coatings, adhesives and fibres (Ion et al. 2008). Carbon dioxide capture and subsequent utilization provides a more sustainable route for the synthesis of carbamate as compared to the conventional method for carbamate production, which involves the use of highly toxic materials, such as isocyanates, acyl azides, and nitroaromatic compounds (Ghosh and Brindisi 2015).

This chapter highlights the different types of solvent which were investigated as possible solvents for carbon capture and subsequent utilization of carbon dioxide, such as amines (Yoo et al. 2018; Sha et al. 2018; Kang et al. 2017; Guo et al. 2016; Arti et al. 2017; Park et al. 2016; Zhao et al. 2015; Kang et al. 2014; Dindi et al. 2014; Park et al. 2013), strong bases (Han and Wee 2016, 2017; Han et al. 2011), weak bases (Yeh et al. 2005; Kozak et al. 2009; Sutter et al. 2017), ionic liquids (Kumar et al. 2017; Liu et al. 2016), and amino acid salts (Shen et al. 2017). Depending on the medium used during carbon dioxide capture as well as reagents added, the main products that can be obtained through subsequent utilization are carbonates and carbamates.

2.2 Absorbents for Carbon Dioxide Capture and Subsequent Utilization

Carbon dioxide capture and subsequent utilization can be conducted by using a one-step transformation process (in-situ utilization) or two-step process, as shown in Figs. 2.1 and 2.2, respectively. In the one-step process, organic media, such as methanol (Sim et al. 2016; Han and Wee 2017), ethanol (Fonari et al. 2016), and vegetable oils (Uma Maheswari and Palanivelu 2014, 2016) were mixed with absorbents and used as solvents for carbon dioxide capture. The organic medium can be replaced with other non-aqueous solvents, namely deep eutectic solvent (Uma Maheswari and Palanivelu 2015) and ionic liquid (Hasib-ur-Rahman et al. 2012). During the carbon dioxide absorption process, the absorbents were directly converted into solid carbamate salts in a non-aqueous environment. Other than organic medium, inorganic carbonates can be instantaneously generated through a one-pot reaction when solvents, such as calcium hydroxide (Han et al. 2011) and aqueous ammonia (Sutter et al. 2017) are used as absorbents for carbon dioxide capture. Cyclic carbonates were also generated through the addition of epoxides into ionic liquids by using a one-step reaction (Kumar et al. 2017; Liu et al. 2016). Alternatively, a two-step reaction can also generate carbamate salts and carbonated products through the addition of reagents into a carbon dioxide-saturated absorbent (Mohamed

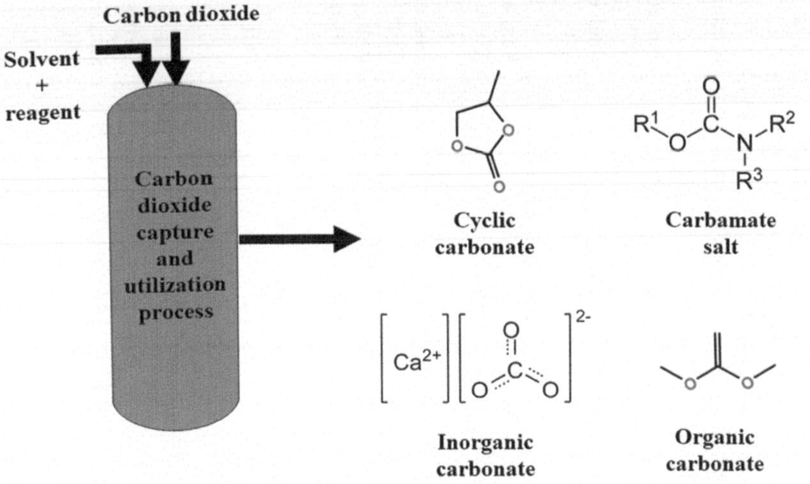

Fig. 2.1 One-step carbon dioxide capture and utilization process to generate carbonate and carbamate salts

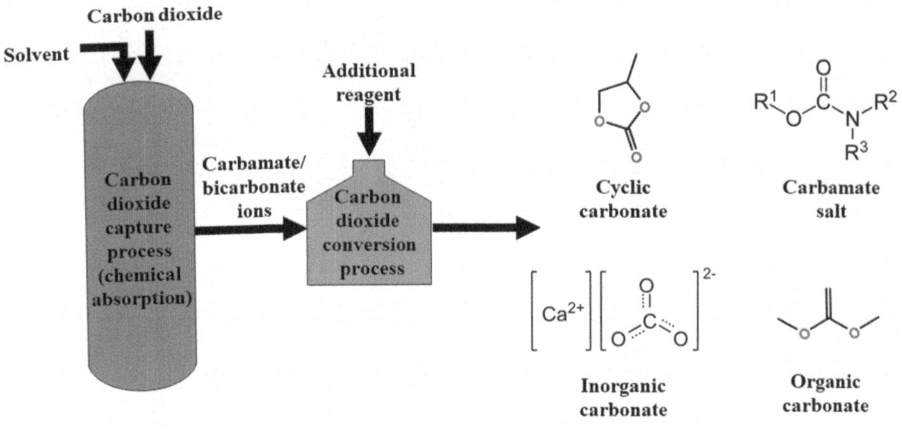

Fig. 2.2 Two-step carbon dioxide capture and utilization process to generate carbonate and carbamate salts

Mohsin et al. 2019; Kang et al. 2014; Park et al. 2016; Arti et al. 2017). Potential absorbents for carbon dioxide capture and subsequent utilization are listed in Table 2.1. The solvents were divided into a few categories, namely amine, strong base, ionic liquid, weak base, and amino acid salt. The types of medium used for carbon dioxide capture and the additional reagents added during carbon dioxide

Table 2.1 List of absorbents, media, and additional reagents used for carbon dioxide capture and utilization

Absorbents	Media for carbon dioxide capture	Additional reagents/ catalysts	Products of carbon dioxide conversion	Conditions during conversion process			Types of conversion		References
				Pressure (atmosphere)	Temperature (K)	Time (h)	One-step	Two-step	
Amine (aqueous phase)									
Monoethanolamine	Water	Aqueous calcium chloride,	Calcium carbonate	1	303.15	24	/		Park et al. (2013)
Monoethanolamine/diethanolamine/piperazine/2-amino-2methyl-1-propanol/N-methyldiethanolamine	Brine (aqueous sodium chloride)	Not applicable	Sodium bicarbonate	1	313.15	24	/		Dindi et al. (2014)
Monoethanolamine/diethanolamine/N-methyldiethanolamine	Water	Aqueous calcium oxide	Calcium carbonate	1	303.15	24		/	Kang et al. (2014)
Ethylenediamine	Polyethylene glycol	Aqueous calcium hydroxide	Calcium carbonate	1	363.15	2		/	Zhao et al. (2015)
Monoethanolamine/diethanolamine/N-methyldiethanolamine	Water	Aqueous barium chloride	Barium carbonate	1	303.15	2		/	Park et al. (2016)
Monoethanolamine/diethanolamine/N-methyldiethanolamine/2-amino-2methyl-1-propanol	Water	Anhydrous calcium chloride	Calcium carbonate	1	313.15	6		/	Arti et al. (2017)
Ethylenediamine	1,2-ethylene glycol	Aqueous calcium hydroxide	Calcium carbonate	1	373.15	1		/	Guo et al. (2016)
Monoethanolamine/diethanolamine/N-methyldiethanolamine	Water	Calcium ions (industrial wastewater)	Inorganic metal carbonates	1	298.15	3		/	Kang et al. (2017)

Monoethanolamine/ethylenediamine/triethanolamine/methyldiethanolamine/2-amino-2methyl-1-propanol/ammonium hydroxide	Water	Metal ions (Calcium ions/barium ions/strontium ion/cadmium ion/lead ion)	Metal carbonates	1–6 MPa	293.15–303.15	0.5–3		/	Sha et al. (2018)
Monoethanolamine	Water	Calcium hydroxide (industrial wastewater)	Calcium carbonates	1	298.15	3		/	Yoo et al. (2018)
3-Dimethylaminopropylamine mixed with glycine	Water	Ethanol	Carbamates	5	298.15	1		/	Mohamed Mohsin et al. (2019)
Amine (Non-aqueous phase)									
Diethanolamine (DEA)	Ionic liquid	Not applicable	DEA carbamate	1	298.15	6	/		Hasib-ur-Rahman et al. (2012)
2-amino-2methyl-1-propanol (AMP)	Ethylene glycol + ethanol/ethylene glycol + 1-propanol	Not applicable	AMP-carbamate, alcohol carbonates	1	293.15–313.15	24	/		Barbarossa et al. (2013)
Monoethanolamine/diethanolamine/2-amino-2methyl-1-propanol/triethanolamine/2-methylaminoethanol	Vegetable oil	Not applicable	Carbamates	1–8 bar	298.15–338.15	0.5–2.5	/		Uma Maheswari and Palanivelu (2014)
2-amino-2methyl-1-propanol (AMP)	Deep eutectic solvent	Not applicable	AMP carbamate	1–10 bar	298.15–338.15	0.5–2.5	/		Uma Maheswari and Palanivelu (2015)

(continued)

Table 2.1 (continued)

Absorbents	Media for carbon dioxide capture	Additional reagents/ catalysts	Products of carbon dioxide conversion	Conditions during conversion process			Types of conversion		References
				Pressure (atmosphere)	Temperature (K)	Time (h)	One-step	Two-step	
Ethylenediamine/diethylenetriamine/ triethylenetetramine	Vegetable oil	Not applicable	Carbamates	1.5, 10	298.15	0.5–1.5	/		Uma Maheswari and Palanivelu (2016)
Piperazine (PZ)	Water/ ethanol	Not applicable	PZ carboxamides	1	298.15	168	/		Fonari et al. (2016)
Piperazine (PZ)	Methanol	Not applicable	PZ-dicarbamate	1	298.15	Few hours	/		Sim et al. (2016)
Metal hydroxide solution (strong base)									
Calcium Hydroxide	Water	Not applicable	Calcium Carbonate	1	298.15	Not mentioned	/		Han et al. (2011)
Sodium hydroxide	Ethanol	Not applicable	Sodium ethyl carbonate	1	298.15	Not mentioned	/		Han and Wee (2016)
Sodium hydroxide	Methanol	Not applicable	Sodium methyl carbonate	1	298.15	0.5	/		Han and Wee (2017)
Ionic liquid									
Urea derivative-based ionic liquids	Not applicable	Propylene oxide	Propylene carbonate	0.5–2.5	373.15–413.15	0.5–3	/		Liu et al. (2016)
Bifunctionalized ionic liquid	Water	Styrene oxide	Styrene carbonate	1	373.15	12	/		Luo et al. (2016)
Histidine derived ionic liquid	Dimethyl carbonate	Epoxide	Cyclic carbonates	1	353.15	5	/		Kumar et al. (2017)

Weak base									
Chilled ammonia	Water	Not applicable	Ammonium carbonate and bicarbonate	Not mentioned	275.15–283.15	Not mentioned	/		Darde et al. (2009)
Aqueous ammonia	Water	Zinc sulfate heptahydrate/ zinc chloride/ barium chloride dehydrate	Metal carbonates	1	293	< 1	/		Gaur et al. (2010)
Amino acid salt									
Potassium Prolinate	Ethanol	Not applicable	Bircarbonate and proline salts	0.5	303.15	4	/		Shen et al. (2017)
Others									
Superbase (Diazabicyclo 5.4.0-undec-7-ene (DBU))	polyethylene glycol	n-butylamine	Dibutyl urea	1	373.15–403.15	12–24		/	Yang et al. (2011)
Weak acid (Ammonium nitrate + calcium oxide)	Water	Not applicable	Calcium carbonate	1	298.15	24	/		Lee et al. (2016)

conversion process will be discussed in the following sub-section. The experimental conditions, such as temperature, pressure, and reaction time are also summarized in Table 2.1.

2.2.1 Amine Solution

Amine solvents, such as monoethanolamine, diethanolamine, ethylenediamine, 2-amino-2-methyl-1-propanol, and piperazine are widely studied absorbent for carbon dioxide capture and subsequent utilization into value-added products (Table 2.1). The chemical reactions between amine and carbon dioxide are well understood based on the application of carbon dioxide removal from natural gas. Moreover, amine has high reactivity and it can be easily obtained at a reasonable cost (Liang et al. 2015). In theory, the reactions between carbon dioxide and aqueous amine solutions are divided into three steps. The first step involves the dissolution of carbon dioxide molecules into liquid phase, followed by the formation of bicarbonate ions, and finally the formation of carbamate ions (Mondal et al. 2012).

The ionic carbon dioxide molecules can be easily converted into metal carbonates through the addition of metal ions in an aqueous medium. This process was demonstrated through studies performed by various researchers (Yoo et al. 2018; Sha et al. 2018; Kang et al. 2017; Guo et al. 2016; Arti et al. 2017; Park et al. 2016; Zhao et al. 2015; Kang et al. 2014; Dindi et al. 2014; Park et al. 2013) who reported that metal carbonates, such as calcium, barium, and magnesium carbonates can be produced at low pressure and moderate temperature. However, the source of metal ions for the carbonation process is a crucial factor, which determines the commercial viability of this technology. In a study reported by Park et al. (2013) and Arti et al. (2017), calcium ions were obtained by dissolving solid calcium chloride in water. Meanwhile, Kang et al. (2014) used calcium oxide to provide metal ions for the conversion process. Similarly, other metal ions, such as barium and magnesium were also obtained from their respective metal oxides, which were subsequently converted to metal hydroxides (Sha et al. 2018; Guo et al. 2016; Zhao et al. 2015). The production of these metal oxides usually involves energy intensive processes, such as mining, conversion process, and transportation (Ejenstam 2010). Alternatively, the metal oxides can be replaced with metal ion waste available from desalination process. The concept of utilizing industrial waste as a source of metal ions was investigated by Dindi et al. (2014), Kang et al. (2017), and Yoo et al. (2018). Reject brine from the desalination process contained high amount of metal ions, which can be recycled instead of being released back into the seawater. This provides a more sustainable approach for the carbonation process. However, the use of reject brine is subjected to chlorine ions concentration in the waste solution, because high concentration of chlorine ions can cause equipment corrosions. Apart from inorganic carbonates, organic carbamate salts were also generated by using aquoues amine solution through addition of ethanol by using the two-step carbon dioxide capture and conversion process (Mohamed Mohsin et al. 2019).

Dilution of concentrated amine solutions by using water for the purpose of carbon dioxide capture was identified as one of the main causes of pipeline corrosion. According to Hasib-ur-Rahman et al. (2012), aqueous amine solution, which contained 15 weight percent (w/w %) diethanolamine showed high corrosion activity as compared to diethanolamine-ionic liquid emulsion, which indicated almost negligible corrosion activity. This was due to the absence of water, which prevented the formation of carbonic acid when water molecules reacted with carbon dioxide, and thus reducing its interaction with iron surfaces (Hasib-ur-Rahman et al. 2012). Moreover, the non-aqueous environment hindered the formation of bicarbonate and carbonate ions; hence, allowing the formation of solid carbamates. The study by Hasib-ur-Rahman et al. (2012) has prompted other authors to investigate the performances of carbon dioxide capture and subsequent utilization by using amine-based solvent mixed with organic solutions, such as methanol (Sim et al. 2016), ethanol (Fonari et al. 2016), ethylene glycol and ethanol mixtures (Barbarossa et al. 2013), as well as vegetable oils (Uma Maheswari and Palanivelu 2014, 2016). Deep eutectic solvents (Uma Maheswari and Palanivelu 2015) were also studied as potential non-aqueous media for carbon dioxide capture.

Apart from generation of solid carbamates, the non-aqueous medium may contribute to higher carbon dioxide loading capacity, depending on the nature of the solvent. Uma Maheswari and Palanivelu (2015), reported that the carbon dioxide loading capacity of amine mixed with deep eutectic solvent is higher as compared to aqueous alkanolamine solutions. This may be due to the high polar nature of deep eutectic solvents, which may attract more carbon dioxide molecules to dissolve in the solutions (Uma Maheswari and Palanivelu 2015). However, in the case of vegetable oils, the carbon dioxide absorption capacity is subjected to the oil's viscosity (Uma Maheswari and Palanivelu 2014, 2016). In general, vegetable oils, such as sunflower oil, gingelly oil, groundnut oil, and palm oil have high viscosities (Diamante and Lan 2014), which hinders the desorption of carbon dioxide molecules within the solution. Nonetheless, amine solution, especially 2-amino-2-methyl-1-propanol reported some improvements in carbon dioxide loading capacity when mixed with coconut oil (Uma Maheswari and Palanivelu 2014). This was due to the lower viscosity of coconut oil (Noureddini et al. 1992) as compared to other types of vegetable oil (Diamante and Lan 2014). This trend was also reported by Shamiri et al. (2016), where by mixing aqueous monoethanolamine with lower concentration of glycerol, showed higher carbon dioxide loading capacity as compared to monoethanolamine with higher glycerol concentration.

The addition of non-aqueous medium into pure amine solutions resulted in the formation of immiscible liquids, which are easier for product separation (Hasib-ur-Rahman et al. 2012). Two-phase flow system is a common phenomenon in many industrial processes, particularly in the oil and gas industry. However, despite advances in this technology, the mechanisms of carbon dioxide capture in a two-phase media is not fully understood. Moreover, characterization of two-phase flow requires complex mathematical calculations (Leclaire et al. 2013), which are both time-consuming and costly. The current study focuses on batch process (Uma Maheswari and Palanivelu 2014, 2016; Hasib-ur-Rahman et al. 2012), which is direct and straight forward. In contrast, the carbon dioxide capture process in the industry involves continuous flow and transportation of immiscible liquid will be more complicated. Therefore, despite the formation of value-added products, the use of non-aqueous solvent mixed with amine solution may not be suitable for large-scale commercial application. In contrary, the carbonation of carbon dioxide by using aqueous amine solution mixed with metal ions is regarded to bring a positive impact in the production of metal carbonates due to lower temperature and pressure requirement during conversion process.

2.2.2 *Metal Hydroxide Solution*

Metal hydroxide solution was also reported as a potential absorbent for carbon dioxide capture due to its high absorption capacity and fast capture rate (Yoo et al. 2013). Aqueous calcium hydroxide was previously studied for its application in carbon dioxide capture and subsequent utilization (Han et al. 2011). In an aqueous environment, calcium hydroxide dissolved into calcium and hydroxide ions. Calcium carbonates were formed instantaneously when calcium ions reacted with carbon dioxide in the aqueous phase. This process occurred at room temperature of 298.15 K and pressure of 1 atmosphere (Han et al. 2011). Sodium hydroxide is another type of strong base which was studied for carbon dioxide capture application. In the presence of carbon dioxide, aqueous sodium hydroxide produced sodium carbonate (Na_2CO_3), which were subsequently converted to sodium bicarbonate ($NaHCO_3$) (Yoo et al. 2013). In comparison with calcium hydroxide solution, the bicarbonate product obtained from sodium hydroxide solution was not recovered as solids due to its high solubility in water (Yoo et al. 2013).

Han and Wee (Han and Wee 2016, 2017) proposed the use of alcohol, such as ethanol (Han and Wee 2016) and methanol (Han and Wee 2017) as a medium for carbon dioxide capture and subsequent utilization. This method was successfully proven to generate sodium ethyl carbonate and sodium methyl carbonate when sodium hydroxide was dissolved in ethanol and methanol, respectively. The advantage of using alcohol is the presence of hydroxyl (OH) functional group, which participated in physical absorption of carbon dioxide (Han and Wee 2017). However, the use of organic medium was limited by its high viscosity, which may hinders carbon dioxide absorption process. This scenario was reported by Shamiri et al. (2016) such that concentration of organic medium added into the absorbent will directly influenced the carbon dioxide capture performance of monoethanolamine. At lower concentrations of organic medium, the carbon dioxide loading capacity of monoethanolamine-organic medium mixture were reported to be higher than the industrial standard of 30 w/w% aqueous monoethanolamine. However, as the concentration of organic medium increased, the carbon dioxide absorption capacity of monoethanolamine-organic medium was lower than the 30 w/w% aqueous monoethanolamine due to higher viscosity of the organic medium (Shamiri et al. 2016). Another drawback of using sodium hydroxide dissolved in ethanol as an absorbent is that higher concentration of sodium hydroxide is required for efficient removal of carbon dioxide (Han and Wee 2017). Similarly, the production of methyl carbonates only occurred at higher concentration of sodium hydroxide solution (Han and Wee 2017). Considering the fact that sodium hydroxide is produced from expensive metal ions, this process may not be economically viable. Industrial waste may be utilized as a source of metal ions, but in order to attain high purity products, the metal ions need to be segregated since industrial waste usually contains more than one types of metal ion (Kang et al. 2017). Moreover, metal hydroxide is also known to be highly corrosive (Ravanchi and Sahebdelfar 2014), and hence an unattractive choice of absorbent for carbon dioxide capture and subsequent utilization application.

2.2.3 *Ionic Liquid*

Ionic liquid is an emerging solvent for carbon dioxide capture which has gained significant attention mainly based on its properties, such as low volatility, high thermal stability, and tunable structures (Ma et al. 2017). On top of being an absorbent, ionic liquid can also be used as a catalyst for the production of cyclic carbonates. Cycloaddition of epoxides with carbon dioxide by using ionic liquids were investigated by a few authors, such as Kumar et al. (2017), Luo et al. (2016), and Liu et al. (2016). Experimental results revealed that epoxides were transformed into cyclic carbonates at atmospheric pressure. Moreover, the ionic liquid can be recycled up to six times with yields of more than 95%, as reported by Kumar et al. (2017).

In spite of the positive aspects of ionic liquid as an absorbent for carbon dioxide capture and utilization, the absorbent also has certain drawbacks. The synthesis of ionic liquid involves complex steps and occurs at high temperature, increasing its production cost (Huang et al. 2017). On top of that, the synthesis of ionic liquid as well as carbon dioxide conversion process occurs in the presence of organic media, such as toluene, acetonitrile (Liu et al. 2016), and dimethlylformamide (Kumar et al. 2017). These solvents are highly flammable and volatile, and thus release toxic fumes into the environment. Moreover, the production of ionic liquid generates organic waste product. Therefore, the development of ionic liquid for direct carbon dioxide capture remains a challenge mainly due to high cost and the use of toxic organic media.

2.2.4 *Weak Base*

The transformation of metal ions into metal carbonates through carbon dioxide capture and utilization can also be performed by using weak base as an absorbent. This process was studied by Gaur et al. (2010), in which metal carbonate was successfully synthesized when carbon dioxide reacted with a mixture containing metal salt and aqueous ammonia. In comparison to strong base, which produced carbonated products at higher concentration of absorbents (Han and Wee 2016, 2017), a smaller concentration of weak base was sufficient to transformed carbon dioxide into solid products (Gaur et al. 2010). The weak base can also be replaced by alkaline industrial waste, which produced metal carbonates when reacted with carbon dioxide and metal salts (Gaur et al. 2010).

Weak base can also generate precipitates directly during carbon dioxide capture process without the addition of reagent or the use of organic medium. Solid crystals, such as ammonium bicarbonate and ammonium carbonate can be easily generated by using aqueous ammonia as an absorbent, depending on pH of the solution (Yeh et al. 2005). However, pH of the solution was controlled to prevent the formation of by-products (Yeh et al. 2005). Nonetheless, the carbonated products have high commercial values as they are sources of fertilizers (Luis 2016). The disadvantage of using aqueous ammonia as an absorbent is the high volatility of the solvent, which leads to vapour loss into the environment. This problem can be resolved by decreasing the temperature of the absorbent through the chilled ammonia technology (Darde et al. 2009). In this process, ammonia was cooled down to a temperature between 2 and 10 °C before entering the absorption column, which prevented the evaporation of ammonia (Darde et al. 2009). Products which were recovered though chilled ammonia carbon dioxide capture process include ammonium bicarbonate, ammonium carbonate, and ammonium carbamate (Darde et al. 2009). However, the major drawback of chilled ammonia process is that the absorption and conversion process occur at very low temperature; thus, requires more energy for refrigeration and cooling processes.

2.2.5 Amino Acid Salt

Amino acid salt is another type of absorbent which was investigated for the purpose of carbon dioxide capture (Song et al. 2012; Sanchez Fernandez et al. 2013). The properties of amino acid salt include low oxidation degradation, high reactivity and low volatility, which makes it a good absorbent for carbon dioxide capture (Song et al. 2012). Similar to aqueous ammonia, pH of the absorbent plays an important role in the formation of precipitates during carbon dioxide capture, which leads to reduction in energy during solvent regeneration process through phase separation method (Sanchez Fernandez et al. 2013). However, these solids were regenerated to allow amino acid salt to be recycled back into the system. Currently, there are limited studies on carbon dioxide capture and subsequent utilization by using amino acid salts.

In a study conducted by Shen et al. (2017), amino acid salt was mixed with ethanol, which increased the rate for carbon dioxide capture as compared to aqueous amino acid salts. Although the study focused on phase change method through regeneration of precipitates, the formation of carbamate proline, bicarbonate, and ethyl carbonate salts proved that amino acid salt in organic medium has the capability to be used as an absorbent for carbon dioxide capture and subsequent utilization. However, the viscosity of the solution in ethanol was reported to be higher than in water (Shen et al. 2017), which may limits the dissolution of carbon dioxide in the absorbent. Nonetheless, as an environmentally friendly solvent, amino acid salt is an alternative absorbent which can potentially replace other types of absorbent for carbon dioxide capture and subsequent utilization application.

2.2.6 *Other Types of Absorbent*

Super base (Yang et al. 2011) and weak acid (Lee et al. 2016) were also reported as absorbents for carbon dioxide capture and subsequent utilization. Super base, namely diazabicyclo[5.4.0]-undec-7-ene (DBU) dissolved in polyethylene glycol solution was reacted with carbon dioxide to produce liquid amidinium carbonate salt, and subsequently converted into urea in the presence of amine solution. The reaction occurred at 1 bar and temperature of 100 °C to 130 °C (Yang et al. 2011) as compared to the conventional process, which occurred at temperature of 185 °C to 190 °C and pressure of 180 atmosphere to 200 atmosphere (Alper and Yuksel Orhan 2017). The use of weak acid as an absorbent was reported by Lee et al. (2016), where ammonium nitrate was used to extract metal ions from solid industrial waste. When ammonium nitrate was mixed with calcium oxide (solid industrial waste), ammonium hydroxide was produced, which increased the capacity of the mixture to absorb carbon dioxide. Simultaneously, calcium ions from industrial waste were directly converted to metal carbonates (Lee et al. 2016).

Super base has high carbon dioxide absorption capacity of approximately twice as compared to the conventional monoethanolamine solution (Yang et al. 2011). However, super base has high basicity, which can cause equipment and pipeline corrosions. Moreover, the use of organic medium increases the viscosity of the absorbent. Nonetheless, these few studies (Yang et al. 2011; Lee et al. 2016) suggested that apart from solvents listed in the previous categories, there are other types of absorbent which are also capable of capturing carbon dioxide and subsequently convert carbon dioxide into value-added products at moderate temperature and pressure conditions, such as the production of urea from super base and amine mixture.

2.2.7 *Overview*

The potential absorbents which can be used for carbon dioxide capture and subsequent utilization are as illustrated in Fig. 2.3. The solvents are divided into a few categories, namely amine solution, metal hydroxide solution, weak base, ionic liquid, amino acid salt, and others. Direct carbonation of carbon dioxide into metal carbonates were achieved through the addition of solutions containing metal ions. In the absence of water molecules, solid carbamates were generated. Certain absorbents such as calcium hydroxide and aqueous ammonia were able to transform carbon dioxide directly into solid products without the use of organic media or

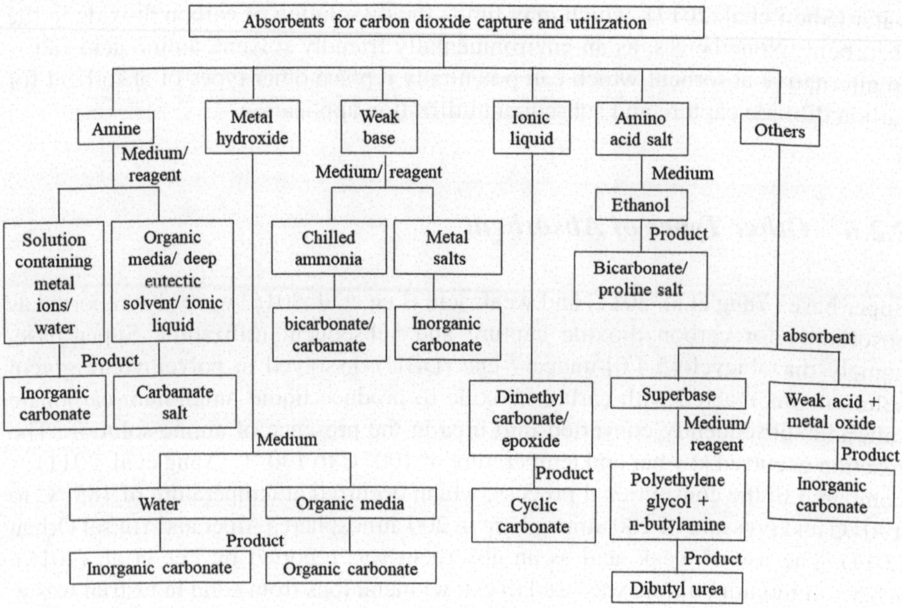

Fig. 2.3 Different types of absorbents for carbon dioxide capture, the media, and reagents used for conversion process, and products generated through the carbon capture and utilization process

additional reagent. Moreover, cyclic carbonates were generated without the use of expensive metal complexes by using ionic liquid as carbon dioxide absorbent and catalyst for the conversion process. However, the cycloaddition process may require the presence of an organic medium.

The advantages and disadvantages of each types of solvent are summarized in Table 2.2. Among all solvents, aqueous amine is the most mature solvent available for carbon dioxide capture. Moreover, the addition of metal ions was able to generate inorganic carbonates at low temperature and pressure. The expensive cost of metal ions was reduced by replacing metal ions with industrial waste and hence provides a more sustainable approach for the production of metal carbonates. However, the carbon dioxide capture capacity of amine solvent is much lower as compared to other types of solvents, such as ionic liquid and super base. The use of organic medium, such as methanol and ethanol is also attractive for the purpose of carbon dioxide capture and subsequent utilization. Moreover, the high viscosity of absorbent which may reduce the diffusivity of carbon dioxide can be overcome by using the two-step conversion method instead of one-step utilization process.

Table 2.2 Advantages and disadvantages of absorbents and reaction media/reagents for carbon capture and subsequent utilization

Types of absorbent	Reaction Media/ reagents	Advantages	Disadvantages	References
Amine solution	Metal ion solutions/ industrial waste	Mature solvent Generate metal carbonates at low temperature and atmospheric pressure	Energy intensive process for production of metal ions/metal salts High concentration of chlorine ions in industrial waste Metal ions in industrial waste need to be segregated to obtain high purity product	Sha et al. (2018); Kang et al. (2017); Guo et al. (2016); Arti et al. (2017); Park et al. (2016); Zhao et al. (2015); Kang et al. (2014); Dindi et al. (2014); Park et al. (2013)
	Organic media	Produce organic carbamate without the use of toxic chemicals One-step reaction – less capital cost for equipment	Organic media have high viscosity Formation of two-phase liquid	Hasib-ur-Rahman and Larachi (2012); Barbarossa et al. (2013); Uma Maheswari and Palanivelu (2014, 2015, 2016); Sim et al. (2016)
Metal hydroxide	Water	Generate metal carbonates at low temperature and atmospheric pressure One-step reaction – less capital cost for equipment	Energy intensive process for production of metal ions/metal salts Absorbents are highly corrosive	Han et al. (2011)
	Organic media	Produce organic carbonates in liquid form	Organic media have high viscosity Absorbents are highly corrosive	Han and Wee (2017); Han and Wee (2016)
Ionic liquid	Organic solvent/ epoxides	High carbon absorption capacity Transform epoxides into cyclic carbonates at low temperature and atmospheric pressure	The use highly toxic organic media (toluene, dimethylformamide) during synthesis of ionic liquid and during conversion process Ionic liquids are expensive Ionic liquids have high viscosity	Kumar et al. (2017); Liu et al. (2016); Luo et al. (2016)

(continued)

Table 2.2 (continued)

Types of absorbent	Reaction Media/ reagents	Advantages	Disadvantages	References
Weak base	Water	No additional media/ reagent is required for conversion process	High energy requirement for cooling process	Darde et al. (2009)
	Metal salts	Generate metal carbonates at low temperature and atmospheric pressure	Energy intensive process for production of metal ions/metal salts	Gaur et al. (2010)
			Aqueous ammonia has high volatility	
Amino acid salt	Organic media	Environmentally friendly absorbent	Low carbon dioxide loading capacity	Shen et al. (2017)
Others (super base and weak acid)	Polyethylene glycol (super base); metal oxide (weak acid)	Super base has high carbon dioxide absorption capacity	Organic media have high viscosity	Yang et al. (2011); Lee et al. (2016)
			High toxicity	
			Corrosive	

2.3 Conclusion

This chapter summarizes the different types of absorbent which were studied for carbon dioxide capture along with the reaction media and additional reagents which were used for carbon dioxide conversion. The carbon dioxide capture and subsequent utilization were divided into two types; one-step reaction or two-step reaction. One-step reaction occured mostly in the presence of organic and non-aqueous media, such as methanol, ethanol, vegetable oil, ethylene glycol, deep eutectic solvent, and ionic liquid. In the absence of water, the one-step process generated solid carbamates. However, some absorbents, such as calcium hydroxide and aqueous ammonia, produced precipitates through the one-step carbon dioxide capture process without the use of organic media. The two-step conversion process involved the addition of other materials, such as metal ions and epoxides, which generated metal carbonates and cyclic carbonates, respectively. Among all solvents, amine is regarded as a promising solvent for carbon dioxide capture and subsequent utilization due to its high reactivity and conversion activity. On top of that, as an environmentally friendly absorbent, amino acid salts also have the potential to be developed based on the precipitates formed during carbon dioxide capture. The integration between carbon dioxide capture and conversion process is expected to bring a positive impact to the industry, especially in reducing the energy requirement for carbon dioxide capture and generating value-added products which can bring revenue to investors. The current study focused on batch process mainly to test the concept of carbon dioxide capture and subsequent utilization. In the future, the study can be extended to include a continuous absorption process with extra attention on the solid separation process which is crucial to prevent equipment clogging.

Acknowledgement The authors would like to thank Carbon Dioxide Research Centre (CO2RES), Department of Chemical Engineering, Universiti Teknologi PETRONAS, and Information Resource Centre (IRC), Universiti Teknologi PETRONAS for providing resources and materials for this publication.

References

Alper E, Yuksel Orhan O (2017) Carbon dioxide utilization: developments in conversion processes. Petroleum 3(1):109–126. https://doi.org/10.1016/j.petlm.2016.11.003

Arti M, Youn MH, Park KT, Kim HJ, Kim YE, Jeong SK (2017) Single process for carbon dioxide capture and mineralization in various alkanolamines using calcium chloride. Energy Fuel 31(1):763–769. https://doi.org/10.1021/acs.energyfuels.6b02448

Barbarossa V, Barzagli F, Mani F, Lai S, Stoppioni P, Vanga G (2013) Efficient carbon dioxide capture by non-aqueous 2-amino-2-methyl-1-propanol (AMP) and low temperature solvent regeneration. RSC Adv 3(30):12349–12355. https://doi.org/10.1039/C3RA40933C

Basile A, Iulianelli A, Gallucci F, Morrone P (2010) 7 – advanced membrane separation processes and technology for carbon dioxide capture in power plants A2. In: Maroto-Valer MM (ed) Developments and innovation in carbon dioxide capture and storage technology, vol 1. Woodhead Publishing, Oxford, pp 203–242. https://doi.org/10.1533/9781845699574.2.203

Budzianowski WM (2015) Single solvents, solvent blends, and advanced solvent systems in carbon dioxide capture by absorption: a review. Int J Glob Warm 7(2):184–225. https://doi.org/10.1504/IJGW.2015.067749

Darde V, Thomsen K, Van Well WJ, Stenby EH (2009) Chilled ammonia process for carbon dioxide capture. Energy Procedia 1(1):1035–1042. https://doi.org/10.1016/j.egypro.2009.01.137

De Falco M, Iaquaniello G, Centi G (2013) Carbon dioxide: a valuable source of carbon. Springer, London

Diamante LM, Lan T (2014) Absolute viscosities of vegetable oils at different temperatures and shear rate range of 64.5 to 4835 per second. J Food Process 2014:6. https://doi.org/10.1155/2014/234583

Dindi A, Quang DV, Hadri NE, Rayer A, Abdulkadir A, Abu-Zahra MRM (2014) Potential for the simultaneous capture and utilization of carbon dioxide using desalination reject brine: amine solvent selection and evaluation. Energy Procedia 63:7947–7953. https://doi.org/10.1016/j.egypro.2014.11.829

Ejenstam J (2010) The lime industry, a potential business area for Kanthal. Dissertation, Uppsala Universitet

Fonari MS, Antal S, Castaneda R, Ordonez C, Timofeeva TV (2016) Crystalline products of carbon dioxide capture by piperazine aqueous solutions. CrystEngComm 18(33):6282–6289. https://doi.org/10.1039/C6CE01083K

Gaur A, Park J-W, Jang J-H (2010) Metal-carbonate formation from ammonia solution by addition of metal salts—an effective method for carbon dioxide capture from landfill gas (LFG). Fuel Process Technol 91(11):1500–1504. https://doi.org/10.1016/j.fuproc.2010.05.027

Ghosh AK, Brindisi M (2015) Organic carbamates in drug design and medicinal chemistry. J Med Chem 58(7):2895–2940. https://doi.org/10.1021/jm501371s

Gibson JA, Mangano E, Shiko E, Greenaway AG, Gromov AV, Lozinska MM, Friedrich D, Campbell EE, Wright PA, Brandani S (2016) Adsorption materials and processes for carbon capture from gas-fired power plants: AMPGas. Ind Eng Chem Res 55(13):3840–3851. https://doi.org/10.1021/acs.iecr.5b05015

Guo B, Zhao T, Sha F, Zhang F, Li Q, Zhang J (2016) A novel CCU approach of carbon dioxide by the system 1,2-ethylenediamine+1,2-ethylene glycol. Korean J Chem Eng 33(6):1883–1888. https://doi.org/10.1007/s11814-016-0023-8

Han S-J, Wee J-H (2016) Carbon dioxide fixation via the synthesis of sodium ethyl carbonate in NaOH-dissolved ethanol. Ind Eng Chem Res 55(46):12111–12118. https://doi.org/10.1021/acs.iecr.6b03250

Han S-J, Wee J-H (2017) Carbon dioxide fixation by combined method of physical absorption and carbonation in NaOH-dissolved methanol. Energy Fuel 31(2):1747–1755. https://doi.org/10.1021/acs.energyfuels.6b02709

Han S-J, Yoo M, Kim D-W, Wee J-H (2011) Carbon dioxide capture using calcium hydroxide aqueous solution as the absorbent. Energy Fuel 25(8):3825–3834. https://doi.org/10.1021/ef200415p

Hasib-ur-Rahman M, Siaj M, Larachi F (2012) Carbon dioxide capture in alkanolamine/room-temperature ionic liquid emulsions: a viable approach with carbamate crystallization and curbed corrosion behavior. Int J Greenhouse Gas Control 6:246–252. https://doi.org/10.1016/j.ijggc.2011.10.014

Huang Z, Jiang B, Yang H, Wang B, Zhang N, Dou H, Wei G, Sun Y, Zhang L (2017) Investigation of glycerol-derived binary and ternary systems in carbon dioxide capture process. Fuel 210:836–843. https://doi.org/10.1016/j.fuel.2017.08.043

Ion A, Van Doorslaer C, Parvulescu V, Jacobs P, De Vos D (2008) Green synthesis of carbamates from carbon dioxide, amines and alcohols. Green Chem 10(1):111–116. https://doi.org/10.1039/B711197E

Kang D, Park S, Jo H, Park J (2014) Carbon fixation using calcium oxide by an aqueous approach at moderate conditions. Chem Eng J 248:200–207. https://doi.org/10.1016/j.cej.2014.03.045

Kang D, Lee M-G, Jo H, Yoo Y, Lee S-Y, Park J (2017) Carbon capture and utilization using industrial wastewater under ambient conditions. Chem Eng J 308:1073–1080. https://doi.org/10.1016/j.cej.2016.09.120

Kozak F, Petig A, Morris E, Rhudy R, Thimsen D (2009) Chilled ammonia process for carbon dioxide capture. Energy Procedia 1(1):1419–1426. https://doi.org/10.1016/j.egypro.2009.01.186

Kumar P, Varyani M, Khatri PK, Paul S, Jain SL (2017) Post combustion capture and conversion of carbon dioxide using histidine derived ionic liquid at ambient conditions. J Ind Eng Chem 49:152–157. https://doi.org/10.1016/j.jiec.2017.01.022

Leclaire S, Reggio M, Trépanier J-Y (2013) Progress and investigation on lattice Boltzmann modeling of multiple immiscible fluids or components with variable density and viscosity ratios. J Comput Phys 246:318–342. https://doi.org/10.1016/j.jcp.2013.03.039

Lee M-G, Kang D, Yoo Y, Jo H, Song H-J, Park J (2016) Continuous and simultaneous carbon dioxide absorption, calcium extraction, and production of calcium carbonate using ammonium nitrate. Ind Eng Chem Res 55(45):11795–11800. https://doi.org/10.1021/acs.iecr.6b02880

Liang ZH, Rongwong W, Liu H, Fu K, Gao H, Cao F, Zhang R, Sema T, Henni A, Sumon K (2015) Recent progress and new developments in post-combustion carbon-capture technology with amine based solvents. Int J Greenhouse Gas Control 40:26–54. https://doi.org/10.1016/j.ijggc.2015.06.017

Liu M, Liang L, Li X, Gao X, Sun J (2016) Novel urea derivative-based ionic liquids with dual-functions: carbon dioxide capture and conversion under metal- and solvent-free conditions. Green Chem 18(9):2851–2863. https://doi.org/10.1039/C5GC02605A

Luis P (2016) Use of monoethanolamine (MEA) for carbon dioxide capture in a global scenario: consequences and alternatives. Desalination 380:93–99. https://doi.org/10.1016/j.desal.2015.08.004

Luo X, Chen K, Li H, Wang C (2016) The capture and simultaneous fixation of carbon dioxide in the simulation of fuel gas by bifunctionalized ionic liquids. Int J Hydrog Energy 41(21):9175–9182. https://doi.org/10.1016/j.ijhydene.2015.12.223

Ma T, Wang J, Du Z, Abdeltawab AA, Al-Enizi AM, Chen X, Yu G (2017) A process simulation study of carbon dioxide capture by ionic liquids. Int J Greenhouse Gas Control 58:223–231. https://doi.org/10.1016/j.ijggc.2017.01.017

Mazzotti M, Abanades JC, Allam R, Lackner KS, Meunier F, Rubin E, Sanchez JC, Yogo K, Zevenhoven R (2005) Mineral carbonation and industrial uses of carbon dioxide. IPCC special report on carbon dioxide capture and storage, pp 321–335

Metz B, Davidson O, De Coninck H, Loos M, Meyer L (2005) IPCC special report on carbon dioxide capture and storage. Intergovernmental panel on climate change, Geneva (Switzerland). Working group III

Mohamed Mohsin H, Mohd Shariff A, Johari K (2019) 3-Dimethylaminopropylamine (DMAPA) mixed with glycine (GLY) as an absorbent for carbon dioxide capture and subsequent utilization. Sep Purif Technol 222:297–308

Mondal MK, Balsora HK, Varshney P (2012) Progress and trends in carbon dioxide capture/separation technologies: a review. Energy 46(1):431–441. https://doi.org/10.1016/j.energy.2012.08.006

Noureddini H, Teoh B, Davis Clements L (1992) Viscosities of vegetable oils and fatty acids. J Am Oil Chem Soc 69(12):1189–1191. https://doi.org/10.1007/BF02637678

Oelkers EH, Gislason SR, Matter J (2008) Mineral carbonation of carbon dioxide. Elements 4(5):333–337. https://doi.org/10.2113/gselements.4.5.333

Park S, Lee M-G, Park J (2013) Carbon dioxide fixation by applying new chemical absorption-precipitation methods. Energy 59:737–742. https://doi.org/10.1016/j.energy.2013.07.057

Park S, Bang J-H, Song K, Jeon CW, Park J (2016) Barium carbonate precipitation as a method to fix and utilize carbon dioxide. Chem Eng J 284:1251–1258. https://doi.org/10.1016/j.cej.2015.09.059

Pérez-Moreno SM, Gázquez MJ, Bolívar JP (2015) Carbon dioxide sequestration by indirect carbonation of artificial gypsum generated in the manufacture of titanium dioxide pigments. Chem Eng J 262:737–746. https://doi.org/10.1016/j.cej.2014.10.023

Peters M, Köhler B, Kuckshinrichs W, Leitner W, Markewitz P, Müller TE (2011) Chemical technologies for exploiting and recycling carbon dioxide into the value chain. ChemSusChem 4(9):1216–1240. https://doi.org/10.1002/cssc.201000447

Ravanchi MT, Sahebdelfar S (2014) Carbon dioxide capture and utilization in petrochemical industry: potentials and challenges. Appl Petrochem Res 4(1):63–77. https://doi.org/10.1007/s13203-014-0050-5

Rochelle GT (2009) Amine scrubbing for carbon dioxide capture. Science 325(5948):1652–1654. https://doi.org/10.1126/science.1176731

Sanchez Fernandez E, Heffernan K, van der Ham LV, Linders MJ, Eggink E, Schrama FN, Brilman D, Goetheer EL, Vlugt TJ (2013) Conceptual design of a novel carbon dioxide capture process based on precipitating amino acid solvents. Ind Eng Chem Res 52(34):12223–12235. https://doi.org/10.1021/ie401228r

Sha F, Hong H, Zhu N, Qiao X, Zhao B, Ma L, Zhang J (2018) Direct non-biological carbon dioxide mineralization for carbon dioxide capture and utilization on the basis of amine-mediated chemistry. J Carbon Dioxide Util 24:407–418. https://doi.org/10.1016/j.jcou.2018.01.025

Shamiri A, Shafeeyan M, Tee H, Leo C, Aroua M, Aghamohammadi N (2016) Absorption of carbon dioxide into aqueous mixtures of glycerol and monoethanolamine. J Nat Gas Sci Eng 35:605–613. https://doi.org/10.1016/j.jngse.2016.08.072

Shen S, Bian Y, Zhao Y (2017) Energy-efficient carbon dioxide capture using potassium prolinate/ethanol solution as a phase-changing absorbent. Int J Greenhouse Gas Control 56:1–11. https://doi.org/10.1016/j.ijggc.2016.11.011

Sim J, Jo E, Jhon YH, Jang SG, Shim JG, Jang KR, Paek K, Kim J (2016) Isolation and crystal structure determination of piperazine dicarbamate obtained from a direct reaction between piperazine and carbon dioxide in methanol. Bull Korean Chem Soc 37(11):1854–1857. https://doi.org/10.1002/bkcs.10964

Song H-J, Park S, Kim H, Gaur A, Park J-W, Lee S-J (2012) Carbon dioxide absorption characteristics of aqueous amino acid salt solutions. Int J Greenhouse Gas Control 11:64–72. https://doi.org/10.1016/j.ijggc.2012.07.019

Sutter D, Gazzani M, Pérez-Calvo J-F, Leopold C, Milella F, Mazzotti M (2017) Solid formation in ammonia-based processes for carbon dioxide capture–turning a challenge into an opportunity. Energy Procedia 114:866–872. https://doi.org/10.1016/j.egypro.2017.03.1229

Teir S (2008) Fixation of carbon dioxide by producing carbonates from minerals and steelmaking slags. Helsinki University of Technology

Uma Maheswari A, Palanivelu K (2014) Absorption of carbon dioxide in alkanolamine and vegetable oil mixture and isolation of 2-amino-2-methyl-1-propanol carbamate. J Carbon Dioxide Util 6:45–52. https://doi.org/10.1016/j.jcou.2014.03.003

Uma Maheswari A, Palanivelu K (2015) Carbon dioxide capture and utilization by alkanolamines in deep eutectic solvent medium. Ind Eng Chem Res 54(45):11383–11392. https://doi.org/10.1021/acs.iecr.5b01818

Uma Maheswari A, Palanivelu K (2016) Alkyl amine and vegetable oil mixture—a viable candidate for carbon dioxide capture and utilization. Environ Sci Pollut Res:1–13. https://doi.org/10.1007/s11356-016-8306-5

Xiao L, Su D, Yue C, Wu W (2014) Protic ionic liquids: a highly efficient catalyst for synthesis of cyclic carbonate from carbon dioxide and epoxides. J Carbon Dioxide Util 6:1–6. https://doi.org/10.1016/j.jcou.2014.01.004

Xu G, Liang F, Yang Y, Hu Y, Zhang K, Liu W (2014) An improved carbon dioxide separation and purification system based on cryogenic separation and distillation theory. Energies 7(5):3484–3502. https://doi.org/10.3390/en7053484

Yang Z-Z, He L-N, Zhao Y-N, Li B, Yu B (2011) Carbon dioxide capture and activation by superbase/polyethylene glycol and its subsequent conversion. Energy Environ Sci 4(10):3971–3975. https://doi.org/10.1039/C1EE02156G

Yang Z-Z, He L-N, Gao J, Liu A-H, Yu B (2012) Carbon dioxide utilization with C–N bond formation: carbon dioxide capture and subsequent conversion. Energy Environ Sci 5(5):6602–6639. https://doi.org/10.1039/C2EE02774G

Yeh JT, Resnik KP, Rygle K, Pennline HW (2005) Semi-batch absorption and regeneration studies for carbon dioxide capture by aqueous ammonia. Fuel Process Technol 86(14–15):1533–1546. https://doi.org/10.1016/j.fuproc.2005.01.015

Yoo M, Han S-J, Wee J-H (2013) Carbon dioxide capture capacity of sodium hydroxide aqueous solution. J Environ Manag 114:512–519. https://doi.org/10.1016/j.jenvman.2012.10.061

Yoo Y, Kang D, Kim I, Park J (2018) Characteristics of metal cation carbonation and carbon dioxide utilization using seawater-based industrial wastewater. ChemistrySelect 3(32):9284–9292. https://doi.org/10.1002/slct.201800346

Zhao T, Guo B, Zhang F, Sha F, Li Q, Zhang J (2015) Morphology control in the synthesis of $CaCO_3$ microspheres with a novel carbon dioxide-storage material. ACS Appl Mater Interfaces 7(29):15918–15927. https://doi.org/10.1021/acsami.5b03568

Chapter 3
Metal Oxides for Carbon Dioxide Capture

Lakshminarayana Kudinalli Gopalakrishna Bhatta, Umananda Manjunatha Bhatta, and Krishna Venkatesh

Abstract Anthropogenic emission of greenhouse gases, particularly carbon dioxide, is believed to be the main cause for current climate change. This has compelled the humankind to explore numerous technological options to reduce carbon dioxide emission. Among these, the carbon dioxide capture and storage technology is the only cost-effective option to reduce carbon dioxide emission from large point sources.

The development low-cost efficient carbon dioxide capture material is of paramount importance for economical viability of carbon dioxide capture and storage. In this regard, a significant progress has been made in the last one and half decade. The present chapter provides an overview of recent trends in the development of metal oxide based materials for carbon dioxide capture. They find applications in carbon dioxide capture technologies such as adsorption, calcium looping, and chemical looping. The basic principles of adsorption and chemical looping are provided in the beginning of chapter. Subsequent sections discuss the different aspects of capture materials based on magnesium oxide, layered double oxides, calcium oxide, and transition metal oxides. Magnesium oxide is an attractive candidate for pre-combustion carbon dioxide capture owing to its favorable temperature window (200–500 °C). In general, layered double oxides exhibit excellent hydrothermal stability, high selectivity, and high regeneration capability. Calcium oxide based adsorbents operate in the temperature range of 500–900 °C. Oxides of nickel, iron, copper, manganese, and cobalt are the major oxygen carrier materials for chemical looping systems. The insightful observations with respect to carbon dioxide capture performance of these metal oxide systems are presented in the concluding section.

Keywords Carbon dioxide capture · Magnesium oxide · Calcium oxide · Strategy · Mixed metal oxide · Transition metal oxide · Adsorption · Oxygen carrier · Chemical looping · Calcium looping

---

L. K. G. Bhatta (✉) · U. M. Bhatta · K. Venkatesh
Centre for Incubation, Innovation, Research and Consultancy (CIIRC), Jyothy Institute of Technology, Bengaluru, Karnataka, India

Inamuddin et al. (eds.), *Sustainable Agriculture Reviews 38*, Sustainable Agriculture Reviews 38, https://doi.org/10.1007/978-3-030-29337-6_3

3.1 Introduction

Anthropogenic emission of greenhouse gases, particularly carbon dioxide, is believed to be the main cause for current climate change. Fossil fuels are the main sources of carbon dioxide emission. Further, several industries such as steel, cement and chemicals manufacturing are also responsible for considerable amount of carbon dioxide emission. Hence, global efforts for carbon mitigation are underway with a mission target of keeping the increase in the global average temperature to well below 2 °C above pre-industrial levels. In this regard, several clean energy technologies such as solar, wind, modern hydrothermal, geothermal, and biomass have been developed to phase out fossil fuels. Nevertheless, fossil fuels will remain major source of energy in near future (Shafiee and Topal 2009) and carbon-intensive industries will continue to play their major role in world's economy. In this context, the carbon dioxide capture and storage technology is the only cost-effective option to reduce carbon dioxide emission from these sources (Clarke et al. 2014).

In 2005, Intergovernmental Panel on Climate Change published a special report (IPCC 2005) on carbon dioxide capture and storage, identifying the gaps in knowledge. A significant progress has been made ever since with a huge expert community around the world working in different domains of carbon dioxide capture technology. Today, about 21 large-scale and 100 small-scale carbon dioxide capture and storage projects are either being functional or in pipeline (Abanades et al. 2015). The current research is focused on dramatic reduction in cost of capture, resource consumption, resource footprint, and efficiency penalties.

The carbon dioxide capture technologies are classified mainly in two ways. Depending on the stage at which carbon dioxide is being separated in fossil fuel power plants, they are classified into post-combustion, pre-combustion, and oxy fuel combustion. On the other hand, based on gas separation principle, they are classified into following types: (1) absorption, (2) adsorption, (3) membranes, (4) cryogenic distillation, and (5) High temperature solid looping.

Metal oxides and their composites are promising candidates for adsorption and high temperature solid looping systems. The Sects. 3.2 and 3.3 provide salient features of adsorption and high temperature solid looping systems. Various aspects of metal oxide based adsorbents are discussed in subsequent sections.

3.2 Adsorption Technology for Carbon Dioxide Capture

Among separation technologies mentioned in Sect. 3.1, adsorption is considered an alternative to benchmark absorption technology (Younas et al. 2016). The adsorption technique involves mainly two steps: (1) adsorption, and (2) regeneration of adsorbent. In first step, the gas mixture is passed through a bed of solid adsorbent under specified conditions. During this step, preferential adsorption of a component takes place. In second step, the selectively adsorbed component is removed from the adsorbent (desorption). Thus, the adsorption and desorption operate in cycles to achieve the desirable gas separation.

A variety of process configurations are used for effective gas separation. Among these, fixed bed, fluidized bed, and moving bed are most common. Regeneration of adsorbent can be achieved by different techniques. These include, pressure swing adsorption, temperature swing adsorption, vacuum swing adsorption, and other hybrid techniques. All these techniques are essentially cyclic processes involving adsorption desorption, and auxiliary steps (Ben-Mansour et al. 2016). The choice of a particular technique depends upon several technical as well as economical factors.

A variety of solid adsorbents are used for carbon dioxide capture. On basis of adsorption/desorption temperatures, these are grouped into low, intermediate, and high temperature adsorbents. The temperature range corresponds to below 200 °C, between 200 and 400 °C, and above 400 °C, respectively. The selection of suitable adsorbent depends on several factors such as type of application, desired process performance and configuration as well as capture stage along the gas path. The characteristics of an ideal adsorbent are described in the Table 3.1 (Balasubramanian and Chowdhury 2015; Bhatta et al. 2015a).

Table 3.1 Essential characteristics of an ideal adsorbent. In reality there exists a trade-off among these characteristics

Property	Description
High working capacity	Reduces the mass of adsorbent required for a given throughput, lowers the volume of adsorbent bed and minimizes the related equipment size and cost
High selectivity	Results in the production of highly pure CO_2, reducing the cost of subsequent CO_2 transportation and storage/utilization
Dimensionless breakthrough time	Useful metric for PSA that combines both selectivity and working capacity
Low regeneration energy	Reduces energy penalty. However, this property depends on many factors including heat of adsorption and CO_2 loading
Fast reaction kinetics	Reduces reactor volume, capital costs, energy penalty and cycle times. However, trade-off between kinetics and regeneration should be taken into account
High thermal/ chemical stability	Reduces adsorbent make-up rate, leading to cost reduction
Adequate mechanical strength	Allows mechanical processing of adsorbent to maximize the dense packing of bed and bulk handling
High tolerance to impurities	Avoids pre-treatment steps, leading to cost reduction.
Cost	Should be inexpensive for obvious economical reasons.
High cyclic stability	Reduces sorbent replacement rate, leading to cost reduction.
Compatibility with catalyst	Useful for sorption enhanced reaction process (SERP) applications
Low risk for health and safety	For obvious reasons

3.3 High Temperature Solid Looping

High temperature solid looping is an emerging technology. It involves a fast reversible reaction between adsorbent and gas at very high temperature. The main advantage of this technology is that it offers effective integration with the existing steam cycle concepts for power generation. High temperature solid looping is classified into two types: (1) chemical looping involving O_2 and oxygen-carrier materials, (2) calcium looping involving carbon dioxide and calcium oxide.

3.3.1 Chemical Looping

The chemical looping is a very flexible concept, applicable to wide range of carbon dioxide capture schemes. It can be further classified into mainly three types: (1) chemical looping combustion, (2) chemical looping reforming for syngas production, and (3) chemical looping for H_2 production. The detailed account of all these technologies can be found elsewhere (Mattisson et al. 2018; Adanez et al. 2018; Lyngfelt and Linderholm 2017; Spallina et al. 2017; Luo et al. 2018; Fan 2010; Fennell and Anthony 2015). The fundamental principle governing these technologies is described briefly as follows.

The schematic diagram of chemical looping combustion is shown in Fig. 3.1. It consists of two reactors typically: a fuel reactor containing a metal oxide, and an air reactor. The fuel enters into the fuel reactor and reacts with metal oxide as shown in Eq. (3.1)

$$(2n+m)M_xO_y + C_nH_{2m} \rightarrow (2n+m)M_xO_{y\ 1} + mH_2O + nCO_2 \tag{3.1}$$

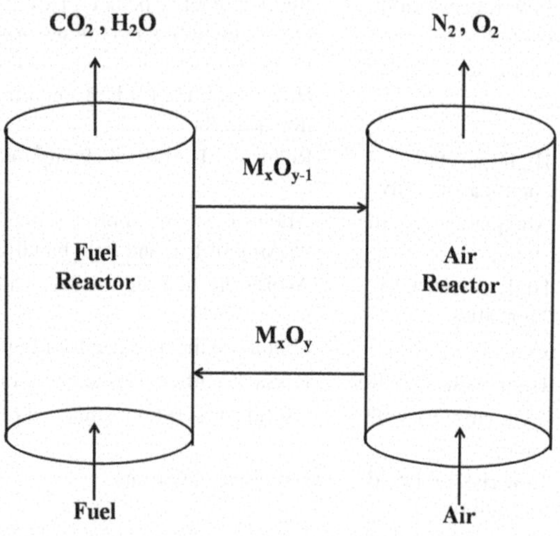

Fig. 3.1 Schematic chemical looping combustion process. The fuel reacts with metal oxide in fuel reactor. The spent carrier is then circulated into air reactor and regenerated

The resulting stream consists of carbon dioxide and steam. The steam is condensed out to get pure carbon dioxide stream for storage. The spent carrier (M_xO_{y-1}) is then circulated into air reactor and regenerated according to Eq. (3.2).

$$M_xO_{y\ 1} + \frac{1}{2}O_2 \rightarrow M_xO_y \tag{3.2}$$

The outlet stream from air reactor contains nitrogen and unutilized oxygen. The net quantity of heat liberated is almost equal to that of conventional combustion. However, carbon dioxide is separated from other flue gas components inherently in this scheme and thus reducing the energy penalty.

3.4 Magnesium Oxide Based Adsorbents

Magnesium oxide is an attractive candidate for pre-combustion carbon dioxide capture owing to its favorable temperature window (200–500 °C). Other advantages include easy of regeneration using temperature swing adsorption, the availability of low cost precursors and moderate selectivity in presence of steam. However, virgin magnesium oxide obtained from abundant natural minerals has several limitations such as low adsorption capacity, slow reaction rate, and poor thermo-mechanical stability (Wang et al. 2011a). These limitations hindered the candidature of magnesium oxide as carbon dioxide capture material for many years. However, numerous studies on the first principles in the last two decades have revived the research interest in developing effective magnesium oxide based adsorbents. These aspects are discussed in the following sub-sections.

3.4.1 Adsorption Mechanism

The carbon dioxide adsorption on magnesium oxide is a surface phenomenon, involving acid – base type of reaction. The acidic carbon dioxide molecules react with basic O^{2-} sites to form different types of complexes (Eq. 3.3).

$$Mg - O + CO_2 \rightarrow Mg - O \cdots CO_{2(ad)} \tag{3.3}$$

Depending on the conditions and environment of capture, the complexes formed include carbonates (monodentate, bidentate, and bridging), carbonate ions, and bicarbonates. In general, monodentate carbonates are formed at low temperature due to rapid physical adsorption. The bidentate carbonates are found predominantly at higher temperature due to relatively slow chemical adsorption (Choi et al. 2009).

The strength of basic sites depends on surface geometry at atomic level. Oxygen atoms at edges and corners of magnesium oxide crystal possess low coordination numbers. Such atoms are more basic than those present in basal planes. Hence, carbon dioxide will be adsorbed more effectively at these low coordinated sites (Wang 2011a). In general, bidentate complexes are formed on corner sites and monodentate complexes are formed on edge sites (Jensen et al. 2005).

The carbonate layer formed on the surface during the initial stage of adsorption hinders further adsorption of carbon dioxide molecules. This results in low performance of pure magnesium oxide. The low surface area pure magnesium oxide is another reason. Hence, the recent research is focused on developing magnesium oxide based adsorbents with enhanced performance using different strategies. These strategies include: (1) Synthesis of mesoporous magnesium oxide, (2) use of support matrix for better dispersion of magnesium oxide nanoparticles, (3) synthesis of molten salts – modified magnesium oxide, and (4) use of mixed metal oxides. These are discussed in the following section.

3.4.2 *Performance Enhancement Strategies*

High surface area and appropriate porosity enhance carbon dioxide capture properties such as adsorption capacity, cyclic stability, and selectivity. Accordingly, mesoporous magnesium oxide shows superior carbon dioxide capture performance than commercial magnesium oxide over a broad range of temperature and pressure (Bian et al. 2010; Ruminski et al. 2011). The rate of adsorption in case of such porous material is controlled mainly by film diffusion and intraparticle diffusion (Song et al. 2016).

The superior carbon dioxide capture properties can also be obtained by dispersing magnesium oxide nanoparticles onto a porous support such as mesoporous carbon. This improves the accessibility of basic sites to carbon dioxide molecules (Kim et al. 2014; Bhagiyalakshmi et al. 2011). Depending on the synthesis route, the support can also act as a source of basic sites (Liu et al. 2013).

The porosity, basic sites, and stability of pure magnesium oxide can be reinforced by combining magnesium oxide with other metal oxides such as aluminium oxide and titanium dioxide (Han et al. 2014). Such mixed metal oxides exhibit different structural properties than those from the corresponding metal oxides, accounting for enhanced performance. For instance, magnesium oxide – titanium dioxide mixed oxide exhibited higher adsorption capacity than those of pure magnesium oxide or titanium dioxide owing to increased surface area, pore volume, and generation of bimodal pores (Jeon et al. 2012).

Another effective method for performance enhancement is the modification of magnesium oxide with molten salts of various combinations (Gao et al. 2017). The molten salts such as nitrates and carbonates of lithium, potassium, sodium, and caesium are employed widely. Several synthesis routes such as wet impregnation, aerogel method, supercritical drying method, precipitation, solvothermal, and wet mixing are used to obtain molten salts – modified magnesium oxide. In comparison with pure magnesium oxide, these novel adsorbents exhibit better carbon dioxide capture properties. For example, the adsorption capacity of mesoporous magnesium oxide – sodium carbonate – sodium nitrate composite obtained by supercritical drying method was found to be 12.7 mmol g^{-1} at 325 °C in pure carbon dioxide (Vu et al. 2016). The commercial magnesium oxide sample doped with 10 mol% ($Li_{0.3}$ $Na_{0.6}$ $K_{0.1}$) NO_3 demonstrated highest adsorption capacity ever reported in literature (16.8 mmol g^{-1} at 300 °C) (Qiao et al. 2017). The actual mechanism of such alkali promotion, leading to superior performance, has not yet been completely realized. However, studies indicated that the formation of mixed phase with either determined or undetermined stochiometry is the reason for improved capture properties compared with pure magnesium oxide.

3.5 Layered Double Oxides

Layered double hydroxides (hydrotalcite like compounds) are brucite-like two-dimensional anionic basic clay materials with general formula $\left(M_{1-x}^{2+}M_x^{3+}(OH)_2\right)^{x+}\left(A_{x/n}^{n-} \bullet mH_2O\right)^{x-}$. Here, M^{2+} represents the divalent cation (Mg^{2+}, Cu^{2+}, Zn^{2+}, Ni^{2+}, or Mn^{2+}), M^{3+} represents the trivalent cation (Al^{3+}, Cr^{3+}, or Fe^{3+}), and A^{n-} represents the anion (CO_3^{2-}, SO_4^{2-}, NO_3^-, Cl^-, or OH^-). In general, the value of x is in the range 0.1–0.5.

The pristine layered double hydroxides are not considered for carbon dioxide adsorption owing to poor basic strength and presence of charge compensating anions that hinder carbon dioxide adsorption. However, they undergo structural transformation during thermal treatment (calcination) to yield three dimensional network of nearly amorphous metastable mixed metal oxides. Such mixed metal oxides derived from layered double hydroxides are called layered double oxides. These are capable of adsorbing carbon dioxide in the temperature range of 350–500 °C. Hence, they find potential applications in sorption enhanced water-gas shift (SEWGS) process of an integrated gasification combined cycle (IGCC) or natural gas combined cycle (NGCC) plant to maximize hydrogen production. They are also considered for the post-combustion carbon dioxide capture under hot conditions and biomass reforming processes.

In general, layered double oxides exhibit excellent hydrothermal stability, high selectivity, and high regeneration capability. However, they present several challenges such as poor adsorption capacity, arguably slow kinetics, expensiveness compared with conventional alkali-metal-based oxides, and poor mechanical stability over multi-cyclic operations.

Among several types of layered double oxides, the one derived from $(Mg)_{1-x}$. $(Al)_x$ $(OH)_2$ $(CO_3)_{x/2}$ m H_2O has been studied intensively for carbon dioxide adsorption. This is abbreviated generally as Mg-Al-CO_3 layered double oxide. The crystal structure of derived layered double oxide is highly dependent on calcination temperature. The optimal calcination temperature for Mg-Al-CO_3 layered double oxide was found to be 400 °C (Wang et al. 2011b; Bhatta et al. 2015a).

Several factors such as chemical composition, heat treatment, operating conditions and synthesis routes influence the carbon dioxide capture performance of layered double oxides. Hence, previous studies were focused on the optimization of these influencing factors. In spite of intensive research, the reported adsorption capacity was usually small (<15 weight %).

The research on layered double oxides is being revived in recent times due to their potential applicability in sorption enhanced reaction processes. The recent strategies to enhance the performance are as follows: (1) insertion of organic anions into Mg–Al CO_3 layered double oxide, (2) synthesis of composite or hybrid materials, (3) development of novel methods to impregnate alkali metal carbonate, (4) control of particle size, and (5) in-depth study on carbon dioxide adsorption mechanism (Wang et al. 2014). These are discussed in the following sub-section.

3.5.1 *Methodologies for Improving Capture Characteristics*

It is well known that structural properties and thermal stability of layered double hydroxides are influenced greatly by the type of anion used for obtaining the layered double hydroxides. The inter-layer distance of Mg–Al CO_3 layered double oxide can be increased drastically by intercalating organic anions. Organic anions such as carboxylates and dicarboxylates are used for intercalation. These anions undergo decomposition at low temperature and their length can also be tailored easily. When such an organo-layered double hydroxide is subjected to thermal treatment, the layered structures will breakdown to more disordered microstructure due to evolution of huge amount of gaseous products (carbon dioxide and water). This will result in enhanced surface basic sites and low degree of crystallinity with more stable quasi-amorphous structure, leading to improved carbon dioxide capture properties. For example, stearate intercalated Mg-Al layered double oxide showed 2.5 times higher carbon dioxide adsorption capacity (1.25 mmol g^{-1}) than the conventional Mg-Al CO_3 layered double oxide (0.5 mmol g^{-1}) at 300 °C (Wang et al. 2012). The carbon dioxide capture capacity also depends on the length of carbon chain in organic anion. Longer the length of carbon chain, higher is the adsorption capacity. The quasi-amorphous structure of resulting layered double oxide becomes more stable with increase in the length of carbon chain (Qin et al. 2017).

The carbon dioxide capture properties can also be enhanced by employing layered double oxides based hybrid or composite materials. Generally, porous carbon, graphene oxide, carbon nanotubes, and activated carbon are used for obtaining hybrid materials (Wang et al. 2016; Zhu et al. 2015; Wang et al. 2015a; Marco et al. 2017; Bhatta et al. 2016; Bhatta et al. 2015b). These materials, being highly porous, ensure better dispersion and stabilization of exfoliated layers. The resulting composite materials exhibit higher surface area with increased surface basic sites, well developed porosity, and better thermo-mechanical stability. Such hybrid materials show superior carbon dioxide capture properties compared with conventional layered double oxides.

The impregnation of alkali metal carbonate is another effective methodology to achieve better performance. Earlier studies focused mainly on potassium carbonate impregnation by different methods using water as a solvent. However, high degree of impregnation (more than 20%) will block active pores due to poor dispersion, leading to low adsorption capacity. Hence, the current research is focused on novel methods for better impregnation of potassium carbonate on all accessible layered surfaces to form dense dispersed basic sites, which react with carbon dioxide after calcination (Li et al. 2015; Jang et al. 2014). There are also few efforts to use other alkali metal carbonates such as lithium carbonate, sodium carbonate, rubidium carbonate, and caesium carbonate (Min et al. 2014). Among these, lithium carbonate showed the better promotion effect compared to all other alkali metal carbonates (Wang et al. 2011b).

The particle size of precursor affects the structural characteristics of resulting layered double oxide. Layered double hydroxides show typical sand rose or platelet-like morphologies irrespective of synthesis techniques, with particularly water as a solvent. The layered double oxides derived from such largely aggregated precursor possess low carbon dioxide adsorption capacity. Further, the synthesized oxides revert to hydroxide structure in the presence of water and carbonate ions. Hence, numerous efforts are being made to address this problem by controlling the particle size of layered double hydroxides using novel solvents, techniques, and protocols. Layered double hydroxides with different morphologies such as nanospheres, nanoflowers, and nanosheets are obtained by adopting novel synthesis routes and by adding supporting matrix (Kou et al. 2018; Wang et al. 2013; Sidek et al. 2016). The resulting layered double oxides from such precursors were found to exhibit better carbon dioxide capture properties due to increased number of surface basic sites. Exfoliation of layered double hydroxides into nanosheets can also be achieved by modifying the synthesis protocol of co-precipitation method, which is most widely used technique. During the synthesis, organic solvents like methanol and acetone are used to remove the water present in inter layers. These solvents evaporate easily at later stage of synthesis. Layered double hydroxides plates can now slide easily on each other to form nanosheets. This synthesis technique is known as aqueous miscible organic solvent (AMOS) treatment (Zhu et al. 2019).

Earlier studies revealed that layered double oxides exhibiting metastable quasi-amorphous structure, which is achieved at minimum possible temperature during heat treatment, show highest capture capacity. However, actual mechanism of adsorption is more complex as it depends on many factors including capture conditions and environment. Recent studies attempted to provide more insights on actual mechanism of adsorption. According to one such study, the formation of basic sites (active Mg-O species) at optimal thermal treatment involves two mechanisms (Gao et al. 2013). In the first mechanism, Al^{3+} ions would replace some of Mg^{2+} ions present in the periclase magnesium oxide lattice. Insertion of each Al^{3+} ion into the periclase magnesium oxide lattice would result in the formation of two active sites. In the second mechanism, some of these Al^{3+} ions will diffuse out of the octahedral sites subsequently and occupy interlayer tetrahedral sites. Consequently, such octahedral site occupied previously by Al^{3+} becomes vacant, producing three active sites. Another study established an asymmetrical "volcano" type relationship between carbon dioxide capture properties and the relative crystallinity of layered double oxides by both experiment and molecular simulation with density functional theory (Kwon et al. 2017).

3.6 Calcium Oxide Based Adsorbents

Calcium oxide reacts reversibly with carbon dioxide as shown in Eqs. 3.4 and 3.5.

Calcium oxide adsorbs carbon dioxide in the temperature range of 500 °C to 800 °C. This exothermic reaction is called carbonation. The carbon dioxide desorption in the temperature range of 800 °C to 950 °C is an endothermic reaction. This is called decarbonation or calcination.

$$CaO_{(S)} + CO_{2(g)} \leftrightarrow CaCO_{3(S)} \quad \Delta H = -175 kJ\,mol^{-1} \tag{3.4}$$

$$CaCO_{3(S)} \leftrightarrow CaO_{(S)} + CO_{2(g)} \quad \Delta H = +179.2 kJ\,mol^{-1} \tag{3.5}$$

This reversible reaction between calcium oxide and carbon dioxide forms the basis of the concept called calcium looping. Several other terminologies are also used in the literature for this concept, including carbonate looping, regenerative calcium cycle, Ca-looping, and CaL.

Calcium looping offers potential advantageous application in almost every type of carbon dioxide removal pathways. Major advantages of calcium looping are: (1) it reduces efficiency penalty in thermal power plants, (2) it finds easy application in industry-scale circulating fluidized beds, which is energy efficient mature technology, (3) it can be integrated economically with other carbon intensive industries (for example, cement), and (4) it involves extremely cost-effective, low health risk adsorbent (limestone or dolomite). Accordingly, calcium looping concept can be applied to broad range of carbon dioxide capture schemes with a varying degree of technological readiness levels. More details on such schemes can be found elsewhere (Abanades et al. 2015; Sun et al. 2018: Salaudeen et al. 2018; Blamey et al. 2010). Considering the scope of this chapter, the general aspects of calcium oxide based adsorbents are covered here.

From Eqs. 3.4 and 3.5, it is clear that the stoichiometric adsorption capacity is very high, corresponding to 78.6 wt% carbon dioxide. However, actual capacity is comparatively low due to several factors, including nature of precursors, adsorbent particle size and morphology. Though the natural adsorbents such as limestone and dolomite are very cheap, they have severe attrition problem. Synthetic calcium oxide based adsorbents exhibit much higher attrition resistance, but are not cost-effective as the natural ones. Further, adsorption capacity is affected negatively in the presence of sulfur dioxide due to cumulative sulphonation process (Manovic and Anthony 2010).

Above all, the biggest problem is the rapid decline in carbonation conversion capacity of the adsorbent during continuous carbonation-calcination cycle. Carbonation is believed to occur in two steps. The rapid chemical reaction takes place in the first step forming a product layer on the surface of particles. In the second step, carbon dioxide molecules diffuse through this product layer into inner layers slowly. As the thickness of product layer grows with time, further diffusion becomes more difficult. This suggests that thickness of product layer at grain boundary, the particle size, surface morphology, and porosity are critical parameters for total carbon dioxide uptake and reaction rate. Sintering of particles during multiple cycles is the main cause for observed decay in reactivity (Ives et al. 2008; Ramkumar and Fan 2010; Acharya et al. 2012). The sintering leads to blockage of pores, causing undesirable changes in adsorbent geometry. Sintering occurs due to various reasons, including the growth of calcium oxide sub-grain, high carbon dioxide concentration in the feed gas, presence of impurities in fuel or adsorbent, and extreme decarbonation conditions with respect to time and temperature (Gonzalez et al. 2008; Manovic et al. 2009). Further, attrition and fragmentation of particles also contribute to loss in reactivity.

The decay in reactivity reduces the performance efficiency of adsorbents, ultimately leading to energy penalty. Hence, several strategies with respect to adsorbent performance and process optimization have been researched for sustainable reactivity. For the sake of simplicity, these strategies can be classified as follows with respect to adsorbent performance: (1) development of robust synthetic calcium oxide, (2) modification of adsorbent with additives and supports, (3) modification of precursor, and (4) reactivation of spent adsorbent (Bhatta et al. 2015a). These are discussed in the following sub-section.

3.6.1 *Strategies for Sustainable Reactivity*

Robust effective synthetic calcium oxide can be obtained by two techniques: (1) use of novel precursors, and (2) modification of existing synthesis methods or use of novel synthesis routes. Different precursors including nano sized calcium carbonate, calcium acetate monohydrate, calcium formate, calcium D-gluconate monohydrate, calcium nitrate tetrahydrate, and calcium L-lactate hydrate are used to obtain pure calcium oxide (Liu et al. 2009; Florin and Harris 2009; Wu et al. 2008). The resulting adsorbent exhibits better carbonation conversion due to high surface area and appropriate porosity. Such better structural properties can also be obtained by novel synthesis techniques such as wet precipitation, flame spray pyrolysis, carbon dioxide bubbling, and sol-gel (Yang et al. 2009; Karami and Mahinpey 2012; Santos et al. 2012; Lu et al. 2009).

Enhanced sustainable reactivity can also be achieved by mixing calcium oxide particles with highly porous inert materials, which possess higher Tamman temperature than calcium oxide. These inert materials act as physical barriers and thus prevent the sintering of calcium oxide particles by forming fine and homogeneous dispersion. They can be grouped into two types on the basis of their final form. First type of supports reacts with a portion of calcium oxide to produce final inert support matrix. Examples are mayenite ($Ca_{12}Al_{14}O_{33}$), tricalcium dialuminate ($Ca_9Al_6O_{18}$), calcium titanate ($CaTiO_3$), and calcium zirconium oxide ($CaZrO_3$) (Broda et al. 2012; Broda and Muller 2012; Xu et al. 2013; Wu and Zhu 2010; Koirala et al. 2011). However, such supports are inert to carbon dioxide. Other type of supports does not react either with calcium oxide or carbon dioxide. Examples are magnesium oxide, yttrium oxide, ytterbium oxide, and cerium oxide (Li et al. 2009; Broda et al. 2014a; Zhang et al. 2014; Hu et al. 2016a; Wang et al. 2015b). It should be noted that employing huge amount inert material is undesirable in large scale industrial operation as it might lead to increased capital and operational costs.

Though these strategies enhance carbon dioxide uptake and cyclic stability, they increase the production cost of adsorbent. Besides, these strategies compromise with the major advantage of calcium looping, namely, use of low cost naturally available precursors such as dolomite and limestone. Hence, recent studies re-addressed the inherent sintering problem of calcium oxide obtained from these precursors. These precursors are treated with organic acids such as acetic, formic, tartaric, propionic, and oxalic acids to obtain corresponding organic salts. These salts decompose to $CaCO_3$ and other compounds on heating. The type of other compounds depends upon the type of organic acid used. The calcium oxide obtained from such treated $CaCO_3$ will possess improved structural properties compared to untreated material. This acidification technique is employed by using variety of organic acids (Ridha et al. 2013; Hu et al. 2016b; Radfarnia and Iliuta 2013; Sun et al. 2013; Li et al. 2011; Al-Jeboori et al. 2013). One major problem with such organic acid treatment is the toxicity and flammability of the secondary products, for example, formaldehyde and acetone. Hence, some ensuing studies have employed mineral acids for treatment (Gonzalez et al. 2016).

Another strategy is the application of animal shells as precursor materials. Animal shells are considered bio-based waste. However, they are porous, rich in calcium carbonate, and cost-effective. Hence, use of animal shells not only reduces the pollution, but also enhances circular economy. Numerous types of shells are being investigated, including egg shell, oyster shell, cockle shell, mussel shell, clam shell, and scallop shell (Castilho et al. 2013; Rashidi et al. 2011; Witoon 2011; Soisuwan et al. 2014). Calcinations of such shells produce calcium oxide along with minor quantities of oxides of magnesium, potassium, sodium, phosphorous, and silicon. As these other oxides are present in variable quantities in different samples, there is an inconsistency in the reported literature about their physico-chemical properties and carbon dioxide capture capacity. Moreover, most of shell derived calcium oxide adsorbents suffer from decay in reactivity over multiple cycles.

Considering the cost reduction, another attractive option is the reactivation of spent adsorbent. Studies have explored different techniques for reactivation. Much of research has been focused on following techniques: (1) hydration, (2) re-pelletization, and (3) re-carbonation. Hydration of spent adsorbent results in swelling and fragmentation particles due to the formation of calcium hydroxide. Having larger molar volume than calcium oxide, calcium hydroxide particles undergo disintegration upon dehydration to produce adsorbent with better porosity and surface area. In re-pelletization technique, use of water to rebind the spent pellets combines hydration and pelletization techniques, leading to better porosity and resistance to attrition. In re-carbonation technique, adsorbent is partially carbonized in the regular carbonator under the specified conditions in the first stage. This is followed by second carbonation in re-carbonator under severe conditions. Upon calcination, carbon dioxide escapes out from the inner layers, generating the more pores. Hence, re-carbonation increases the residual conversion without the need for additional reagents. The recent literature reports various aspects of all these techniques (Arias et al. 2010; Coppola et al. 2014; Manovic and Anthony 2007; Rong et al. 2013; Manovic and Anthony 2007, 2009; Broda et al. 2014b; Arias et al. 2012).

3.7 Metal Oxides for Oxygen Transfer in Chemical Looping Technology

The current research on chemical looping systems is focused on several technical aspects including reaction kinetics, system efficiency, reactor design, and oxygen carrier development and prototype testing. Development of robust oxygen carrier materials is most fundamental to the advancement of chemical looping technology. The important criteria for an ideal oxygen carrier are given in the Table 3.2 (Lyngfelt et al. 2008; Fang et al. 2009). Today more than nine hundred oxygen carrier materials are being developed at laboratory scale. Some of them are also tested for large scale industry applications (Abanades et al. 2015). The general considerations of oxygen carrier materials are discussed in the following paragraphs.

Majority of oxygen carrier materials are based on few transition metal oxides, namely, nickel, iron, copper, manganese, and cobalt. The active metal oxides are

Table 3.2 Important criteria for ideal oxygen carrier material. No single oxygen carrier material can meet all these essential characteristics

Property	Description
High reactivity with fuel and air	Allows smaller reactor size, lowering capital expenditure
Low fragmentation and attrition	Reduces make-up rate and energy penalty, leading to reduction in operating expenditures
Low tendency for agglomeration	Maintains the initial reactivity
High stability under operating conditions	Improves cycle efficiency
Oxygen transport capacity	Lowers solid circulation flow rate, allows smaller reactor size and less material production cost
Low cost	Lowers production cost reduces overall CO_2 capture cost
Low risk for health and safety	Meets stringent environmental regulations and makes public approval easy

mixed generally with inert material. The advantages of adding inert material are as follows: (1) it acts as a porous support and reduces sintering of oxygen carrier particles, (2) it enhances desirable characteristics of oxygen carrier particles with its binder properties, and (3) it increases the ion permeability in the solid particles. However, the ratio of free oxygen in a particle decreases with the addition of inert material. Most commonly used inert materials are aluminum oxide, titanium dioxide, yttria-stabilized zirconia, silica, and zirconium dioxide.

Nickel oxide based oxygen carriers have received much attention mainly due to high oxygen transfer capacity and their applicability at high temperature chemical looping combustion (900–1100 °C). However, apart from expensiveness, their fuel conversion efficiency is limited to 99–99.5% owing to thermodynamic restriction. Off late, the research on nickel oxide based oxygen carriers is being discouraged as they are potential human carcinogens.

Cobalt oxide based oxygen carriers also face same kind of challenges. The maximum achievable conversion for cobalt metal oxide system at 1000 °C is 93.0% only due to thermodynamic limitation. Further, it is quite expensive and associated with health as well as safety issues.

Compared to nickel and cobalt oxide based oxygen carriers, copper oxide based oxygen carriers present many advantages: (1) they possess high oxygen transport capacity, (2) both reactions are exothermic. It means there is no need to supply heat to the reduction reactor, (3) reduction reaction is thermodynamically favourable, (4) they are cost-effective (Fang et al. 2009). However, low melting temperature is the main disadvantage of copper oxide based oxygen carriers.

Iron based oxides are best oxygen carriers in terms of availability, cost, health risk, and environmental safety. However, they suffer from very low oxygen transport capacity and low reactivity. For the same reasons, manganese based oxygen carriers are also explored to lesser extent.

It is clear that each metal oxide system has its own advantages and challenges. Several factors including thermodynamic feasibility, structural characteristics, chemical properties, and mechanical properties influence the performance of oxygen carrier materials (Hossain and Lasa 2008). Evidently, there is huge scope for development of oxygen carriers with excellent reactivity and stability.

3.8 Conclusions

Tremendous progress has been made in carbon dioxide capture and storage technology since Intergovernmental Panel on Climate Change published a special report on carbon dioxide capture and storage in 2005. Several technologies have been realized to different levels of technology readiness. Among these, adsorption, calcium looping, and chemical looping have received much attention owing to their potential capability to capture carbon dioxide at lower energy penalties and cost. Several types of metal oxide based materials find applications in these technologies. Magnesium oxide based adsorbents and layered double oxides have been investigated for carbon dioxide capture at intermediate temperature. Calcium looping involving calcium oxide based adsorbents is best applied for carbon dioxide capture at high temperature. Transition metal oxides have been explored for oxygen transfer in high temperature chemical looping technology.

This chapter gave an overview of recent progress made in the development of metal oxide based materials for these applications. The essential characteristics of an ideal carbon dioxide capture material were detailed in Tables 3.1 and 3.2. The insightful observations made from the careful review of plethora of publications could be summarized as follows.

There are several merits and challenges with each type of metal oxide and its derivatives. The choice of a particular type depends on application type, process type, capture conditions and capture environment. However, several general observations can be made with respect to materials' properties.

The carbon dioxide adsorption capacity of metal oxide based adsorbents depends mainly on number of active basic sites accessible to carbon dioxide molecules. The rate of reaction could be related largely to carbon dioxide diffusion into the inner layer or pores. The selectivity depends on pore characteristics and chemical affinity. The energy requirement for regeneration is associated with heats of adsorption. The poor cyclic capacity could be related to thermo-mechanical instability and drastic changes in the structural characteristics of material during multi-cycle operation. Hence, the physical as well as chemical properties of material such as surface area, pore volume, pore size distribution, chemical composition, particle size, surface geometry at atomic scale, and stability are very critical for better carbon dioxide capture characteristics.

In general, optimization of these properties is the underlying principle of different kinds of property enhancement strategies. Often, there would be trade-off among desirable properties while developing a strategy. For example, use of inert support material leads to improved carbon dioxide capture properties. Meanwhile, it might increase the reactor size leading to higher operational expenditure. Moreover, the production cost, human health risk, and environmental safety need to be considered while developing novel strategies. Hence, there should be more emphasis on systematic studies focusing on rational design of adsorbents.

With regard to oxygen carrier development for chemical looping, there is a huge scope for research to develop oxygen carrier materials with excellent stability and reactivity. The future research should focus on novel oxygen carrier materials, mixed oxides of tested materials, novel supports, varying the proportion of active material and support, different synthesis routes and conditions, and novel precursors.

In summary, numerous possibilities might result in multitude of possible materials with varying properties. The optimization of carbon dioxide capture properties for a particular material would be a daunting task considering the fact it depends upon the application also. Hence, material and process development should progress together for successful deployment of carbon dioxide capture and storage technology at very large scale in near future.

References

Abanades JC, Arias B, Lyngfelt A, Mattisson T, Wiley DE, Li H, Brandani S (2015) Emerging CO_2 capture systems. Int J Greenhouse Gas Control 40:126–166. https://doi.org/10.1016/j.ijggc.2015.04.018

Acharya B, Dutta A, Basu P (2012) Circulating-fluidized-bed-based calcium-looping gasifier: experimental studies on the calcination–carbonation cycle. Ind Eng Chem Res 51(25):8652–8660. https://doi.org/10.1021/ie300629a

Adánez J, Abad A, Mendiara T, Gayán P, de Diego LF, García-Labiano F (2018) Chemical looping combustion of solid fuels. Prog Energy Combust Sci 65:6–66. https://doi.org/10.1016/j.pecs.2017.07.005

Al-Jeboori MJ, Nguyen M, Dean C, Fennell PS (2013) Improvement of limestone-based CO_2 sorbents for Ca looping by HBr and other mineral acids. Ind Eng Chem Res 52(4):1426–1433. https://doi.org/10.1021/ie302198g

Arias B, Grasa GS, Abanades JC (2010) Effect of sorbent hydration on the average activity of CaO in a Ca-looping system. Chem Eng J 163(3):324–330. https://doi.org/10.1016/j.cej.2010.08.009

Arias B, Grasa GS, Alonso M, Abanades JC (2012) Post-combustion calcium looping process with a highly stable sorbent activity by recarbonation. Energy Environ Sci 5(6):7353–7359. https://doi.org/10.1039/c2ee03008j

Balasubramanian R, Chowdhury S (2015) Recent advances and progress in the development of graphene-based adsorbents for CO_2 capture. J Mater Chem A 3(44):21968–21989. https://doi.org/10.1039/c5ta04822b

Ben-Mansour R, Habib MA, Bamidele OE, Basha M, Qasem NAA, Peedikakkal A, Ali M (2016) Carbon capture by physical adsorption: materials, experimental investigations and numerical modeling and simulations- a review. Appl Energy 161:225–255. https://doi.org/10.1016/j.apenergy.2015.10.011

Bhagiyalakshmi M, Hemalatha P, Ganesh M, Mei PM, Jang HT (2011) A direct synthesis of mesoporous carbon supported MgO sorbent for CO_2 capture. Fuel 90(4):1662–1667. https://doi.org/10.1016/j.fuel.2010.10.050

Bhatta LKG, Subramanyam S, Chengala MD, Olivera S, Venkatesh K (2015a) Progress in hydrotalcite like compounds and metal-based oxides for CO_2 capture: a review. J Clean Prod 103:171–196. https://doi.org/10.1016/j.jclepro.2014.12.059

Bhatta LKG, Subramanyam S, Chengala MD, Bhatta UM, Venkatesh K (2015b) Enhancement in CO_2 adsorption on hydrotalcite-based material by novel carbon support combined with K_2CO_3 impregnation. Ind Eng Chem Res 54(43):10876–10884. https://doi.org/10.1021/acs.iecr.5b02020

Bhatta LKG, Subramanyam S, Chengala MD, Bhatta UM, Guha P, Dinakar RPH, Venkatesh K (2016) Layered double hydroxides/multiwalled carbon nanotubes-based composite for high-temperature CO_2 adsorption. Energy Fuel 30(5):4244–4250. https://doi.org/10.1021/acs.energyfuels.6b00141

Bian SW, Baltrusaitis J, Galhotra P, Grassian VH (2010) A template-free, thermal decomposition method to synthesize mesoporous MgO with a nanocrystalline framework and its application in carbon dioxide adsorption. J Mater Chem 20(39):8705–8710. https://doi.org/10.1039/c0jm01261k

Blamey J, Anthony EJ, Wang J, Fennell PS (2010) The calcium looping cycle for large-scale CO_2 capture. Prog Energy Combust Sci 36(2):260–279. https://doi.org/10.1016/j.pecs.2009.10.001

Broda M, Müller CR (2012) Synthesis of highly efficient, Ca-based, Al_2O_3-stabilized, carbon gel-templated CO_2 Sorbents. Adv Mater 24(22):3059–3064. https://doi.org/10.1002/adma.201104787

Broda M, Kierzkowska AM, Müller CR (2012) Influence of the calcination and carbonation conditions on the CO_2 uptake of synthetic Ca-based CO_2 sorbents. Environ Sci Technol 46(19):10849–10856. https://doi.org/10.1021/es302757e

Broda M, Kierzkowska AM, Müller CR (2014a) Development of highly effective CaO-based, MgO-stabilized CO_2 sorbents via a scalable "one-pot" recrystallization technique. Adv Funct Mater 24(36):5753–5761. https://doi.org/10.1002/adfm.201400862

Broda M, Manovic V, Anthony EJ, Müller CR (2014b) Effect of pelletization and addition of steam on the cyclic performance of carbon-templated, CaO-based CO_2 sorbents. Environ Sci Technol 48(9):5322–5328. https://doi.org/10.1021/es405668f

Castilho S, Kiennemann A, Pereira MFC, Dias APS (2013) Sorbents for CO_2 capture from biogenesis calcium wastes. Chem Eng J 226:146–153. https://doi.org/10.1016/j.cej.2013.04.017

Choi S, Drese JH, Jones CW (2009) Adsorbent materials for carbon dioxide capture from large anthropogenic point sources. ChemSusChem 2(9):796–854. https://doi.org/10.1002/cssc.200900036

Clarke L et al (2014) Assessing transformation pathways in: climate change 2014: mitigation of climate change. Contribution of working group III to the fifth assessment report of the intergovernmental panel on climate change. Edenhofer O et al. (Eds.) Cambridge University Press. Available online at https://www.ipcc.ch/site/assets/uploads/2018/02/ipcc_wg3_ar5_ chapter6.pdf. Accessed Feb 2019

Coppola A, Salatino P, Montagnaro F, Scala F (2014) Reactivation by water hydration of the CO_2 capture capacity of a calcium looping sorbent. Fuel 127:109–115. https://doi.org/10.1016/j.fuel.2013.09.059

DeMarco M, Menzel R, Bawaked SM, Mokhtar M, Obaid AY, Basahel SN, Shaffer MS (2017) Hybrid effects in graphene oxide/carbon nanotube-supported layered double hydroxides: enhancing the CO_2 sorption properties. Carbon 123:616–627. https://doi.org/10.1016/j.carbon.2017.07.094

Fan L (2010) Chemical looping systems for fossil energy conversions. Wiley, Hoboken

Fang H, Haibin L, Zengli Z (2009) Advancements in development of chemical-looping combustion: a review. Int J Chem Eng 2009:1–16. https://doi.org/10.1155/2009/710515

Fennell P, Anthony B (eds) (2015) Calcium and chemical looping technology for power generation and carbon dioxide (CO_2) capture. Elsevier Ltd, Amsterdam

Florin NH, Harris AT (2009) Reactivity of CaO derived from nano-sized $CaCO_3$ particles through multiple CO_2 capture-and-release cycles. Chem Eng Sci 64(2):187–191. https://doi.org/10.1016/j.ces.2008.10.021

Gao Y, Zhang Z, Wu J, Yi X, Zheng A, Umar A, Wang Q (2013) Comprehensive investigation of CO_2 adsorption on Mg–Al–CO_3 LDH-derived mixed metal oxides. J Mater Chem A 1(41):12782–12790. https://doi.org/10.1039/c3ta13039h

Gao W, Zhou T, Gao Y, Louis B, O'Hare D, Wang Q (2017) Molten salts-modified MgO-based adsorbents for intermediate-temperature CO_2 capture: a review. J Energy Chem 26(5):830–838. https://doi.org/10.1016/j.jechem.2017.06.005

González B, Grasa GS, Alonso M, Abanades JC (2008) Modeling of the deactivation of CaO in a carbonate loop at high temperatures of calcination. Ind Eng Chem Res 47(23):9256–9262. https://doi.org/10.1021/ie8009318

González B, Blamey J, Al-Jeboori MJ, Florin NH, Clough PT, Fennell PS (2016) Additive effects of steam addition and HBr doping for CaO-based sorbents for CO_2 capture. Chem Eng Process:Process Intensif 103:21–26. https://doi.org/10.1016/j.cep.2015.09.019

Han SJ, Bang Y, Kwon HJ, Lee HC, Hiremath V, Song IK, Seo JG (2014) Elevated temperature CO_2 capture on nano-structured MgO–Al_2O_3 aerogel: effect of Mg/Al molar ratio. Chem Eng J 242:357–363. https://doi.org/10.1016/j.cej.2013.12.092

Hossain MM, de Lasa HI (2008) Chemical-looping combustion (CLC) for inherent CO_2 separations—a review. Chem Eng Sci 63(18):4433–4451. https://doi.org/10.1016/j.ces.2008.05.028

Hu Y, Liu W, Sun J, Yang X, Zhou Z, Zhang Y, Xu M (2016a) High temperature CO_2 capture on novel Yb_2O_3-supported CaO-based sorbents. Energy Fuel 30(8):6606–6613. https://doi.org/10.1021/acs.energyfuels.6b01185

Hu Y, Liu W, Sun J, Li M, Yang X, Zhang Y, Xu M (2016b) Structurally improved CaO-based sorbent by organic acids for high temperature CO_2 capture. Fuel 167:17–24. https://doi.org/10.1016/j.fuel.2015.11.048

IPCC et al (2005) Special report on carbon dioxide capture and storage. In: Metz B (Ed.) Cambridge University Press. Available online at https://www.ipcc.ch/report/carbon-dioxide-capture-and-storage Accessed Feb 2019

Ives M, Mundy RC, Fennell PS, Davidson JF, Dennis JS, Hayhurst AN (2008) Comparison of different natural sorbents for removing CO_2 from combustion gases, as studied in a bench-scale fluidized bed. Energy Fuel 22(6):3852–3857. https://doi.org/10.1021/ef800417v

Jang HJ, Lee CH, Kim S, Kim SH, Lee KB (2014) Hydrothermal synthesis of K_2CO_3-promoted hydrotalcite from hydroxide-form precursors for novel high-temperature CO_2 sorbent. ACS Appl Mater Interfaces 6(9):6914–6919. https://doi.org/10.1021/am500720f

Jensen MB, Pettersson LG, Swang O, Olsbye U (2005) CO_2 sorption on MgO and CaO surfaces: a comparative quantum chemical cluster study. J Phys Chem B 109(35):16774–16781. https://doi.org/10.1021/jp052037h

Jeon H, Min YJ, Ahn SH, Hong SM, Shin JS, Kim JH, Lee KB (2012) Graft copolymer templated synthesis of mesoporous MgO/TiO_2 mixed oxide nanoparticles and their CO_2 adsorption capacities. Colloids Surf A Physicochem Eng Asp 414:75–81. https://doi.org/10.1016/j.colsurfa.2012.08.009

Karami D, Mahinpey N (2012) Highly active CaO-based sorbents for CO_2 capture using the precipitation method: preparation and characterization of the sorbent powder. Ind Eng Chem Res 51(12):4567–4572. https://doi.org/10.1021/ie2024257

Kim TK, Lee KJ, Yuh J, Kwak SK, Moon HR (2014) Multi-core MgO NPs@ C core–shell nanospheres for selective CO_2 capture under mild conditions. New J Chem 38(4):1606–1610. https://doi.org/10.1039/C4NJ00067F

Koirala R, Gunugunuri KR, Pratsinis SE, Smirniotis PG (2011) Effect of zirconia doping on the structure and stability of CaO-based sorbents for CO_2 capture during extended operating cycles. J Phys Chem C 115(50):24804–24812. https://doi.org/10.1021/jp207625c

Kou X, Guo H, Ayele EG, Li S, Zhao Y, Wang S, Ma X (2018) Adsorption of CO_2 on MgAl-CO_3 LDHs-derived sorbents with 3D nanoflower-like structure. Energy Fuel 32(4):5313–5320. https://doi.org/10.1021/acs.energyfuels.8b00024

Kwon HJ, Kwon S, Seo JG, Jung IS, Son YH, Lee CH, Lee HC (2017) Predictive guide for collective CO_2 adsorption properties of Mg– Al mixed oxides. ChemSusChem 10(8):1701–1709. https://doi.org/10.1002/cssc.201601581

Li L, King DL, Nie Z, Howard C (2009) Magnesia-stabilized calcium oxide absorbents with improved durability for high temperature CO_2 capture. Ind Eng Chem Res 48(23):10604–10613. https://doi.org/10.1021/ie901166b

Li Y, Sun R, Liu H, Lu C (2011) Cyclic CO_2 capture behavior of limestone modified with pyroligneous acid (PA) during calcium looping cycles. Ind Eng Chem Res 50(17):10222–10228. https://doi.org/10.1021/ie2007455

Li S, Ribeiro AM, Shi Y, Moreira MN, Cai N, Rodrigues AE (2015) Synthesis, pelleting, and performance evaluation of a novel K-promoted γ-alumina/MgAl-layered double oxide composite adsorbent for warm gas H_2/CO_2 separation. Ind Eng Chem Res 54(28):7154–7163. https://doi.org/10.1021/acs.iecr.5b01342

Liu W, Low NW, Feng B, Wang G, Diniz da Costa JC (2009) Calcium precursors for the production of CaO sorbents for multicycle CO_2 capture. Environ Sci Technol 44(2):841–847. https://doi.org/10.1021/es902426n

Liu WJ, Jiang H, Tian K, Ding YW, Yu HQ (2013) Mesoporous carbon stabilized MgO nanoparticles synthesized by pyrolysis of $MgCl_2$ preloaded waste biomass for highly efficient CO_2 capture. Environ Sci Technol 47(16):9397–9403. https://doi.org/10.1021/es401286p

Lu H, Smirniotis PG, Ernst FO, Pratsinis SE (2009) Nanostructured Ca-based sorbents with high CO_2 uptake efficiency. Chem Eng Sci 64(9):1936–1943. https://doi.org/10.1016/j.ces.2008.12.038

Luo M, Yi Y, Wang S, Wang Z, Du M, Pan J, Wang Q (2018) Review of hydrogen production using chemical-looping technology. Renew Sust Energ Rev 81:3186–3214. https://doi.org/10.1016/j.rser.2017.07.007

Lyngfelt A, Linderholm C (2017) Chemical-looping combustion of solid fuels–status and recent progress. Energy Procedia 114:371–386. https://doi.org/10.1016/j.egypro.2017.03.1179

Lyngfelt A, Johansson M, Mattisson T (2008) Chemical-looping combustion-status of development. In 9th international conference on circulating fluidized beds 13:16.05 Chalmers University of Technology

Manovic V, Anthony EJ (2007) Steam reactivation of spent CaO-based sorbent for multiple CO_2 capture cycles. Environ Sci Technol 41(4):1420–1425. https://doi.org/10.1021/es0621344

Manovic V, Anthony EJ (2009) Long-term behavior of CaO-based pellets supported by calcium aluminate cements in a long series of CO_2 capture cycles. Ind Eng Chem Res 48(19):8906–8912. https://doi.org/10.1021/ie9011529

Manovic V, Anthony EJ (2010) Competition of sulphation and carbonation reactions during ooping cycles for CO_2 capture by CaO-based sorbents. J Phys Chem A 114(11):3997–4002. https://doi.org/10.1021/jp910536w

Manovic V, Charland JP, Blamey J, Fennell PS, Lu DY, Anthony EJ (2009) Influence of calcination conditions on carrying capacity of CaO-based sorbent in CO_2 looping cycles. Fuel 88(10):1893–1900. https://doi.org/10.1016/j.fuel.2009.04.012

Mattisson T, Keller M, Linderholm C, Moldenhauer P, Ryden M, Leion H, Lyngfelt A (2018) Chemical-looping technologies using circulating fluidized bed systems: status of development. Fuel Process Technol 172:1–12. https://doi.org/10.1016/j.fuproc.2017.11.016

Min YJ, Hong SM, Kim SH, Lee KB, Jeon SG (2014) High-temperature CO_2 sorption on Na_2CO_3-impregnated layered double hydroxides. Korean J Chem Eng 31(9):1668–1673. https://doi.org/10.1007/s11814-014-0116-1

Mohd Sidek HB, Jo YK, Kim IY, Hwang SJ (2016) Stabilization of layered double oxide in hybrid matrix of graphene and layered metal oxide nanosheets: an effective way to explore efficient CO_2 adsorbent. J Phys Chem C 120(41):23421–23429. https://doi.org/10.1021/acs.jpcc.6b08065

Qiao Y, Wang J, Zhang Y, Gao W, Harada T, Huang L, Wang Q (2017) Alkali nitrates molten salt modified commercial MgO for intermediate-temperature CO_2 capture: optimization of the Li/Na/K ratio. Ind Eng Chem Res 56(6):1509–1517. https://doi.org/10.1021/acs.iecr.6b04793

Qin Q, Wang J, Zhou T, Zheng Q, Huang L, Zhang Y, Wang Q (2017) Impact of organic interlayer anions on the CO_2 adsorption performance of mg-Al layered double hydroxides derived mixed oxides. J Energy Chem 26(3):346–353. https://doi.org/10.1016/j.jechem.2017.01.003

Radfarnia HR, Iliuta MC (2013) Limestone acidification using citric acid coupled with two-step calcination for improving the CO_2 sorbent activity. Ind Eng Chem Res 52(21):7002–7013. https://doi.org/10.1021/ie400277q

Ramkumar S, Fan LS (2010) Thermodynamic and experimental analyses of the three-stage calcium looping process. Ind Eng Chem Res 49(16):7563–7573. https://doi.org/10.1021/ie100846u

Rashidi NA, Mohamed M, Yusup S (2011) A study of calcination and carbonation of cockle shell. World Acad Sci Eng Technol 60:818–823

Ridha FN, Manovic V, Macchi A, Anthony MA, Anthony EJ (2013) Assessment of limestone treatment with organic acids for CO_2 capture in Ca-looping cycles. Fuel Process Technol 116:284–291. https://doi.org/10.1016/j.fuproc.2013.07.007

Rong N, Wang Q, Fang M, Cheng L, Luo Z, Cen K (2013) Steam hydration reactivation of CaO-based sorbent in cyclic carbonation/calcination for CO_2 capture. Energy Fuel 27(9):5332–5340. https://doi.org/10.1021/ef4007214

Ruminski AM, Jeon KJ, Urban JJ (2011) Size-dependent CO_2 capture in chemically synthesized magnesium oxide nanocrystals. J Mater Chem 21(31):11486–11491. https://doi.org/10.1039/c1jm11784j

Salaudeen SA, Acharya B, Dutta A (2018) CaO-based CO_2 sorbents: a review on screening, enhancement, cyclic stability, regeneration and kinetics modelling. J CO2 Util 23:179–199. https://doi.org/10.1016/j.jcou.2017.11.012

Santos ET, Alfonsín C, Chambel AJS, Fernandes A, Dias AS, Pinheiro CIC, Ribeiro MF (2012) Investigation of a stable synthetic sol–gel CaO sorbent for CO_2 capture. Fuel 94:624–628. https://doi.org/10.1016/j.fuel.2011.10.011

Shafiee S, Topal E (2009) When will fossil fuel reserves be diminished? Energy Policy 37(1):181–189. https://doi.org/10.1016/j.enpol.2008.08.016

Soisuwan S, Phommachant J, Wisaijorn W, Praserthdam P (2014) The characteristics of green calcium oxide derived from aquatic materials. Procedia Chem 9:53–61. https://doi.org/10.1016/j.proche.2014.05.007

Song G, Zhu X, Chen R, Liao Q, Ding YD, Chen L (2016) An investigation of CO_2 adsorption kinetics on porous magnesium oxide. Chem Eng J 283:175–183. https://doi.org/10.1016/j.cej.2015.07.055

Spallina V, Shams A, Battistella A, Gallucci F, van Sint AM (2017) Chemical looping technologies for H_2 production with CO_2 capture: thermodynamic assessment and economic comparison. Energy Procedia 114:419–428. https://doi.org/10.1016/j.egypro.2017.03.1184

Sun R, Li Y, Wu S, Liu C, Liu H, Lu C (2013) Enhancement of CO_2 capture capacity by modifying limestone with propionic acid. Powder Technol 233:8–14. https://doi.org/10.1016/j.powtec.2012.08.011

Sun H, Wu C, Shen B, Zhang X, Zhang Y, Huang J (2018) Progress in the development and application of CaO-based adsorbents for CO_2 capture—a review. Mater Today Sustain 1:1–27. https://doi.org/10.1016/j.mtsust.2018.08.001

Vu AT, Ho K, Jin S, Lee CH (2016) Double sodium salt-promoted mesoporous MgO sorbent with high CO_2 sorption capacity at intermediate temperatures under dry and wet conditions. Chem Eng J 291:161–173. https://doi.org/10.1016/j.cej.2016.01.080

Wang S, Yan S, Ma X, Gong J (2011a) Recent advances in capture of carbon dioxide using alkali-metal-based oxides. Energy Environ Sci 4(10):3805–3819. https://doi.org/10.1039/c1ee01116b

Wang Q, Tay HH, Chen L, Liu Y, Chang J, Zhong Z, Borgna A (2011b) Preparation and CO_2 capture capacity of alkali metal carbonates promoted hydrotalcite. J Nanoeng Nanomanuf 1(3):298–303. https://doi.org/10.1166/jnan.2011.1034

Wang Q, Tay HH, Zhong Z, Luo J, Borgna A (2012) Synthesis of high-temperature CO_2 adsorbents from organo-layered double hydroxides with markedly improved CO_2 capture capacity. Energy Environ Sci 5(6):7526–7530. https://doi.org/10.1039/c2ee21409a

Wang Q, Gao Y, Luo J, Zhong Z, Borgna A, Guo Z, O'Hare D (2013) Synthesis of nano-sized spherical $Mg_3Al–CO_3$ layered double hydroxide as a high-temperature CO_2 adsorbent. RSC Adv 3(10):3414–3420. https://doi.org/10.1039/c2ra22607c

Wang J, Huang L, Yang R, Zhang Z, Wu J, Gao Y, Wang Q, O'Hareb D, Zhong Z (2014) Recent advances in solid sorbents for CO_2 capture and new development trends. Energy Environ Sci 7:3478–3518. https://doi.org/10.1039/C4EE01647E

Wang J, Mei X, Huang L, Zheng Q, Qiao Y, Zang K, Guo Z (2015a) Synthesis of layered double hydroxides/graphene oxide nanocomposite as a novel high-temperature CO_2 adsorbent. J Energy Chem 24(2):127–137. https://doi.org/10.1016/S2095-4956(15)60293-5

Wang S, Fan S, Fan L, Zhao Y, Ma X (2015b) Effect of cerium oxide doping on the performance of CaO-based sorbents during calcium looping cycles. Environ Sci Technol 49(8):5021–5027. https://doi.org/10.1021/es5052843

Wang J, Huang L, Zheng Q, Qiao Y, Wang Q (2016) Layered double hydroxides/oxidized carbon nanotube nanocomposites for CO_2 capture. J Ind Eng Chem 36:255–262. https://doi.org/10.1016/j.jiec.2016.02.010

Witoon T (2011) Characterization of calcium oxide derived from waste eggshell and its application as CO_2 sorbent. Ceram Int 37(8):3291–3298. https://doi.org/10.1016/j.ceramint.2011.05.125

Wu SF, Zhu YQ (2010) Behavior of $CaTiO_3$/nano-CaO as a CO_2 reactive adsorbent. Ind Eng Chem Res 49(6):2701–2706. https://doi.org/10.1021/ie900900r

Wu SF, Li QH, Kim JN, Yi KB (2008) Properties of a nano CaO/Al_2O_3 CO_2 sorbent. Ind Eng Chem Res 47(1):180–184. https://doi.org/10.1021/ie0704748

Xu P, Xie M, Cheng Z, Zhou Z (2013) CO_2 capture performance of CaO-based sorbents prepared by a sol–gel method. Ind Eng Chem Res 52(34):12161–12169. https://doi.org/10.1021/ie401600e

Yang Z, Zhao M, Florin NH, Harris AT (2009) Synthesis and characterization of CaO nanopods for high temperature CO_2 capture. Ind Eng Chem Res 48(24):10765–10770. https://doi.org/10.1021/ie901137s

Younas M, Sohail M, Leong LK, Bashir MJ, Sumathi S (2016) Feasibility of CO_2 adsorption by solid adsorbents: a review on low-temperature systems. Int J Environ Sci Technol 13(7):1839–1860. https://doi.org/10.1007/s13762-016-1008-1

Zhang X, Li Z, Peng Y, Su W, Sun X, Li J (2014) Investigation on a novel $CaO–Y_2O_3$ sorbent for efficient CO_2 mitigation. Chem Eng J 243:297–304. https://doi.org/10.1016/j.cej.2014.01.017

Zhu X, Wang Q, Shi Y, Cai N (2015) Layered double oxide/activated carbon-based composite adsorbent for elevated temperature H_2/CO_2 separation. Int J Hydrog Energy 40(30):9244–9253. https://doi.org/10.1016/j.ijhydene.2015.05.116

Zhu X, Chen C, Suo H, Wang Q, Shi Y, O'Hare D, Cai N (2019) Synthesis of elevated temperature CO_2 adsorbents from aqueous miscible organic-layered double hydroxides. Energy 167:960–969. https://doi.org/10.1016/j.energy.2018.11.009

Chapter 4
Hybrid Membranes for Carbon Capture

Masumeh Momeni, Mohammad Mesbah, Ebrahim Soroush, and Shohreh Shahsavari

Abstract Advances in separation technology, the continuing development of material science and increased rate of carbon emission have provided an important opportunity to find novel materials for separation of different species from eachother. Membrane-based gas separation technology have been used for capturing CO_2 and offered many advantages due to their inherent attributes such as energy-saving and continuous operation compared to traditional methods like absorption and adsorption processes. To fabricate the membranes, various organic and inorganic materials can be utilized. Mixed matrix membranes, a family of the hybrid organic-inorganic membranes, are known as the innovative membranes for separation of CO_2 from gas stream. Zeolites, silica, carbon nano tube (CNT), carbon molecular sieve (CMS), metal organic framework (MOF) and graphene are the most popular filler particles which have been utilized for adding into polymer matrix. Each has its own advantages and drawbacks when they are incorporated within the polymer. The performance of the mixed matrix membranes fabricated with different types of filler particles based on existing literature were reviewed with special focus on post- and pre-combustion carbon capture, for separation of CO_2/N_2 and H_2/CO_2, respectively. In most of cases, the CO_2 permeability and selectivity are increased after embedding of fillers within polymer matrix compared to the neat polymers alone. Mixed matrix membranes preparation methods and formation of interfacial defects were also discussed briefly.

M. Momeni
Department of Chemical Engineering, Sahand University of Technology, Tabriz, Iran

M. Mesbah (✉)
Young Researchers and Elites Club, Science and Research Branch, Islamic Azad University, Tehran, Iran
e-mail: m_mesbah@alum.sharif.edu

E. Soroush
Young Researchers and Elites Club, Ahvaz Branch, Islamic Azad University, Ahvaz, Iran

S. Shahsavari
Young Researchers and Elite Club, Shiraz Branch, Islamic Azad University, Shiraz, Iran

Inamuddin et al. (eds.), *Sustainable Agriculture Reviews 38*, Sustainable Agriculture Reviews 38, https://doi.org/10.1007/978-3-030-29337-6_4

Keywords Mixed matrix membrane · Zeolite · Silica · Carbon nanotube · Carbon molecular sieve · Graphene · Metal organic framework · CO_2 · Carbon capture · Hybrid membrane

Abbreviations

6FDA	2,20 bis(3,40-dicarboxyphenyl)hexafluoropropane dianhydride
ALPO	Aluminophosphate
CA	celluloseacetate
CMS	Carbon molecular sieve
CNT	Carbon nano tube
CuTPA	Copper and terephthalic acid
GO	Graphene oxide
MIL	Materials institute lavoisier
MMM	Mixed matrix membrane
MOF	Metal organic framework
MWCNT	Multi-walled carbon nanotube
PBI	Polybenzimidazole
PDMS	Polydimethylsiloxane
PEBA	Poly(ether-block-amide)
PEEK-WC	Polyetheretherketone
PES	Polyethersulfone
PI	Polyimide
PIM-1	Tetrafluoroterephtalonitrile
PMDA-ODA	Poly[4,4′-oxydiphenylenepyromellitimide]
PMMA	Poly(methyl methacrylate)
PPO	Poly (phenylene oxide)
PPZ	Polyphosphazene
PSF	Polysulfone
PVAc	poly (vinyl acetate)
SAPO	silico alumino phosphate
SEM	Scaning electron microscope
SWCNT	Single-walled carbon nanotube
TMPDA	2,4,6-trimethyl-1,3-phenylenediamine
XLPEO	Cross-linked poly(ethylene oxide)
ZIF	Zeolitic imidazolate frameworks

4.1 Introduction

Through the growth of CO_2 emission into the atmosphere, special focus has been put on finding viable ways to reduce the concentration of atmospheric carbon dioxide. In recent decades, membranes have been known as the potential candidate that can effectively capture CO_2 from stationary point sources like power plants. Due to high stability, high efficiency, low operating and capital costs, environmentally friendliness, compact design, ease of operation and low energy requirement, membranes have attracted remarkable attention for separation processes (Dalane et al. 2017). Membranes can be used as an alternative to liquid amine absorption in order to remove CO_2 from other species and save large amounts of energy. It is expected that membrane technology will act as a valuable tool for replacing or repowering the old separation units. They have the great potential to be used in any of pre-, post- and oxyfuel combustion systems. In this chapter, after a brief introduction on membrane process, a potential of hybrid membranes as the new generation membranes for CO_2 capture will be discussed based on the existing literature. Thus, the contribution of this chapter is to cover the fabricated hybrid membranes which have been applied for CO_2/N_2 (post-combustion) and H_2/CO_2 separations (pre-combustion). In addition, the performance of synthesized hybrid membranes for O_2/N_2 (oxy-combustion) and CO_2/CH_4 separations are also discussed.

4.2 Background

The membrane is a thin, semipermeable barrier which selectively separates the species. In other words, it allows some species to pass through and stops others. A feed consisting of a mixture of two or more components is firstly introduced to the membrane. One or more species of feed mixture move faster than another or other species within the membrane matrix and appears at the "permeate" side. The driving force for separation is often pressure or concentration gradient across the membrane. However, electric potential and temperature gradient may be other driving forces for separation. A section of feed which is retained and does not pass through the membrane is called "retentate".

Permeability and selectivity are two critical factors that can determine the performance of a membrane. To be effective for separating a mixture of components, the membrane should possess high permeability and selectivity. The term "Permeability" is the ability of permeants to pass through the membrane material which limits flow rates. Selectivity is the quality of a membrane of being selective and is defined as the ratio of permeability of two species. Selectivity is used to compare the separating capacity of a membrane for two or more species. For practical separation, membranes with both high permeability and selectivity are required. Permeability can be written as:

$$P_i = \frac{LN_i}{\Delta P} \tag{4.1}$$

Where, L is the membrane thickness (cm), N_i represents the flow rate of gas permeated through the film ($cm^3/cm^2.s$) and ΔP refers to pressure difference across the membrane (cmHg). Permeability unit is often given as Barrer (1 Barrer = 10^{-6} cm^3(STP).cm/cm^2.s.cmHg). The ideal selectivity, or separation factor, of a membrane can be determined by this relationship as Eq. 4.2. In this equation, P_i and P_j are the permeabilities of component i and component j, respectively.

$$\alpha_{i/j} = \frac{P_i}{P_j} \tag{4.2}$$

Membranes can be made from different materials like organic polymers or even inorganic materials such as ceramics, pyrolyzed carbon and metals. Polymers are commonly utilized for fabrication of membranes due to their low cost and ease of preparation. However, they have several restrictions due to their low selectivity, swelling, thermal instability, decomposition in organic solvents and subject to mechanical damage (Vinoba et al. 2017). Polymeric membrane performance has shown a trade-off behavior between membrane permeability and selectivity. Indeed, polymeric membranes with high permeability are generally less selective and vice versa. These drawbacks have convinced membranologists to develop inorganic materials in place of polymers. Inorganic membranes have shown higher permeability and selectivity in comparison with polymeric membranes. They are advantageous due to their long-term stability at high temperatures, resistance to harsh environment and high pressure drops, high chemical stability, high mechanical strength and longer lifespan. Despite the favorable characteristics of inorganic membranes, their high capital costs, poor processability, low membrane surface area per module volume and brittleness are the challenges to utilize them effectively in gas separation industries (Vinoba et al. 2017). Selection of appropriate materials for fabrication can strongly affect the performance of the resulted membrane. The need for better membranes and solving the permeability/selectivity trade-off has led to extensive study in this field. Researchers are still trying to synthesize the more potent materials with better performance, which capable to exceed the Robeson upper bound (Robeson 1991). In recent years, organic-inorganic hybrid membranes, consisting of mixed matrix membranes and organic-inorganic composite membranes, have being developed to take the advantageous of individual polymeric and inorganic membranes simultaneously and also enhance the gas separation performance. Organic-inorganic composite membranes can be prepared by modification of the inorganic support with organic modifier agent, or the organic porous support can be modified by inorganic agent (Luebke et al. 2006; Lin et al. 2018). However, mixed matrix membranes can be prepared by blending of the inorganic particles in organic matrix. To date, most organic-inorganic hybrid membranes have been fabricated in mixed matrix membranes type. In this chapter, the attempts have been focused on introducing the different type of mixed matrix membranes and their separation performance for carbon dioxide capture.

4.3 Mixed Matrix Membranes

Mixed matrix membranes are membrane materials in which the inorganic fillers are embedded in polymer matrix. Incorporation of inorganic fillers into the polymer phase enables such membranes to promote permeability and/or selectivity compared to pristine polymeric membranes. The synergism between the properties of organic and inorganic materials yields excellent properties of mixed matrix membranes. Both materials, polymer and inorganic filler, should be chosen appropriately to successful implementation of mixed matrix membranes. The study on such membranes have been increased since 1980s (Buck II et al. 2009). mixed matrix membranes can be prepared for wide ranges of applications such as tissue engineering, electro dialysis, natural gas sweetening, O_2/N_2 separation, reverse osmosis, pervaporative separation, H_2 separation and even CO_2 removal (Peng et al. 2005; Ismail et al. 2009). Due to good potential of mixed matrix membranes for separation of CO_2, researchers have intensified their efforts to improve these types of membranes for carbon capture.

Numerous types of porous and nonporous fillers can be utilized for preparation of mixed matrix membranes. Characteristics of final mixed matrix membrane can be improved by fillers due to chemical interaction between filler and host material. To make CO_2 transport through the mixed matrix membranes easier, the filler should have high selectivity and good compatibility with polymer matrix. A variety of fillers have been used in different shape, size and chemical structures to increase the modulus and strength of continuous polymeric phase. Most of the porous filler particles that are used in CO_2-selective mixed matrix membranes, have the size of 0.34–2.6 nm. In porous fillers, the gases separate by their sizes or shapes (Wang et al. 2017). Fillers such as metal oxide, zeolite, silica, carbon molecular sieve, graphene oxide, carbon nanotube have been used for incorporating into polymer matrix (Nafisi and Hägg 2014). They can be circular, plate like, fibrous or irregular in shapes. The performance of mixed matrix membranes with different fillers will be discussed in the subsequent section in details.

4.3.1 Zeolites

Zeolite is a porous aluminosilicate with a relatively open three-dimensional (3D) crystal structure. The cage-like structure of zeolites makes them favorable for gas separation. They possess shape selective properties, high porosity, appropriate pore size, excellent thermal and chemical stability. Contrary to polymers, zeolites do not undergo thermal decomposition or deformation when dealing with high-temperature conditions. Because of their impressive separation properties, zeolites can be utilized as potent filler particles for preparation of mixed matrix membranes. Jusoh et al. (2017) fabricated the mixed matrix membranes using zeolite-T as an inorganic filler and 4,4′-(Hexafluoroiso propylidene)diphthalic anhydride (6FDA)-durene polyimide as the polymeric phase. After synthesis of zeolite-T, these fillers were incorporated into polymer matrix at different loadings of zeolite up to 15 wt.%. Pure gas permeation test indicated that the maximum CO_2 permeability of 843.6 Barrer was obtained at 1 wt.% of zeolite loading. CO_2/CH_4 mixed gas permeation test also performed for mixed matrix membrane containing 1 wt.% of zeolite-T. The results revealed that the mixed gas permeability for CO_2 reduced to 511.1 Barrer. This reduction is reported to be due to the CH_4 inhibition effect, gas polarization effect and non-ideality gas phase bahavior. It was also observed that fabricated mixed matrix membranes had higher gas separation performance than that of pristine 6FDA-durene polymer. Zarshenas et al. (2016) studied the effects of zeolite particle size and zeolite loading on the gas separation performance of zeolite NaX/poly (ether-b-amide) Pebax mixed matrix membrane. Zeolite NaX particle were embedded within Pebax-1657 copolymer to form a selective layer on the porous polyether sulfone support. The results indicated that the gas permeability increased in order of $CO_2 > O_2 > N_2$ for both pristine Pebax membrane and Pebax/zeolite NaX mixed matrix membrane. This behavior is because of the higher CO_2 condensability and its smaller kinetic diameter compared to the O_2 and N_2, which leads to the higher solubility and diffusivity of the CO_2 molecule through Pebax polymeric matrix. It was observed that the embedding of the zeolite within Pebax leads to the reduction of the permeability of all gas species, while the higher CO_2/N_2 selectivity was achieved. The pore blockage and chain rigidification of polymer were reported to be the cause of this behavior. However, the partial pore blockage had the major role. They also observed that the mixed matrix membranes containing nano-sized zeolite had better gas separation performance than micro-sized one. Bryan et al. (2014) fabricated zeolite-based mixed matrix membrane using zeolite 13X to investigate the CO_2 separation performance for post-combustion carbon capture. The gas permeation properties were studied for different 13X filler loading of 5, 10 and 15 wt.% in Pebax polymeric phase. It was observed that the CO_2 permeability increased with the increase of the zeolite loading. MFI-zeolite particles were embedded into polyether sulfone membrane to investigate the CO_2 separation performance of the resultant membrane (Yu et al. 2013). The Polyethersulfone/10% MFI zeolite mixed matrix membrane showed the CO_2/N_2 selectivity of 35. Scaning electron microscope (SEM) images of the fabricated mixed matrix membrane and the neat polyethersulfone membrane is also shown in Fig. 4.1. As can be seen, the surface of the neat polyethersulfone is uniform and no obvious defects can be seen in this magnification (Fig. 4.1a). While, the poor adhesion between filler particles and polymer phase results in the formation of interfacial voids around the zeolite particles (Fig. 4.1b). These results showed that the morphology of the mixed matrix membrane membranes depends on the distribution of the particles within the continuous phase. Detailed descriptions about interfacial defects will be discussed in Sect. 4.4.

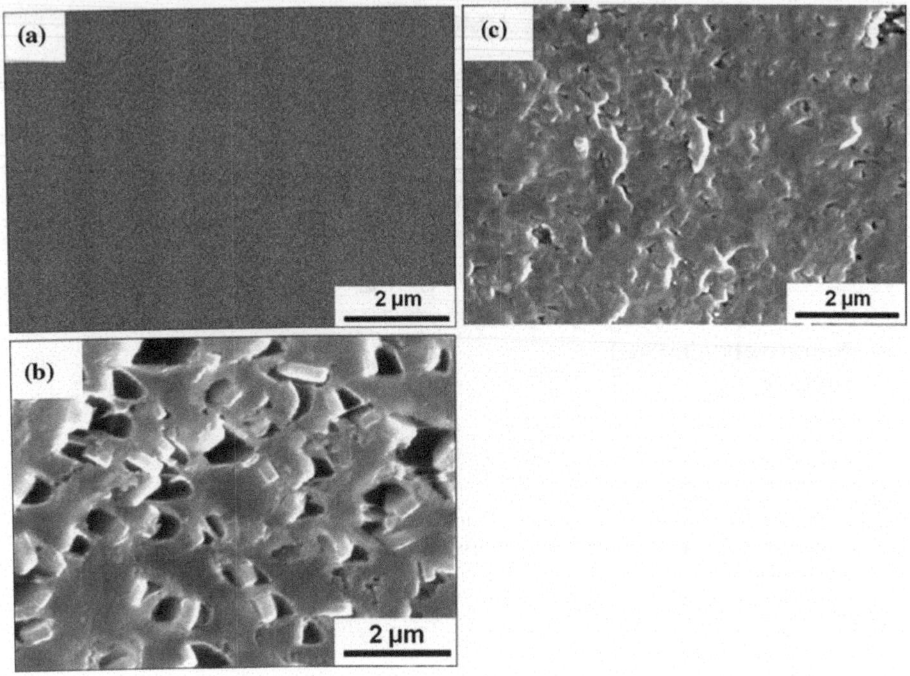

Fig. 4.1 Surface images of (**a**) the neat polyethersulfone membrane which has no obvious defects, (**b**) the zeolite-based mixed matrix membranes which has interfacial voids due to poor adhesion between polymer and zeolite particles and (**c**) the zeolite-based mixed matrix membranes which has good compatibility between polymer and filler particles. (Reprinted with permission of [An approach to prepare defect-free PES/MFI-type zeolite mixed matrix membranes for CO2/N2 separation, Yu, J., L. Li, N. Liu and R. Lee, Springer]' from (Yu et al. 2013))

Modification of zeolite surface with CO_2-philic material is an effective strategy to enhance the separation properties of such materials. The introduction of the interactive functional groups on zeolite surface has been suggested to improve the compatibility of polymer and separation performance. The influence of silico alumino phosphate (SAPO)-34 zeolite modification using ionic liquid has been studied by Ahmed et al. (2017). Owing to CO_2-affinitive moieties, CO_2 are highly soluble in ionic liquids. 1-ethyl-3-methyl imidazolium bis(tri-fluoromethylsulfonyl) amide ([emim][TF_2N]) ionic liquid was utilized to modify surface of the zeolite. The schematic view of the modified zeolite is shown in Fig. 4.2. The mixed matrix membranes were prepared by incorporating of ionic liquid-modified SAPO-34 into polysulfone. Enhanced CO_2 permeance of 29% and CO_2/N_2 selectivity of 206% were observed compared to those of the pristine polysulfone membrane. This results showed that the ionic liquid improves polymer/filler interphase morphology and provides the pathways for CO_2 transport through the membrane. Table 4.1 shows the ability of zeolite-based mixed matrix membranes to separate different types of gases. As can be seen, zeolites such as Na-A,X,Y, ZSM-5, MFI, and silicalites have been used for CO_2 separation from H_2 and N_2, and even O_2/N_2 separation.

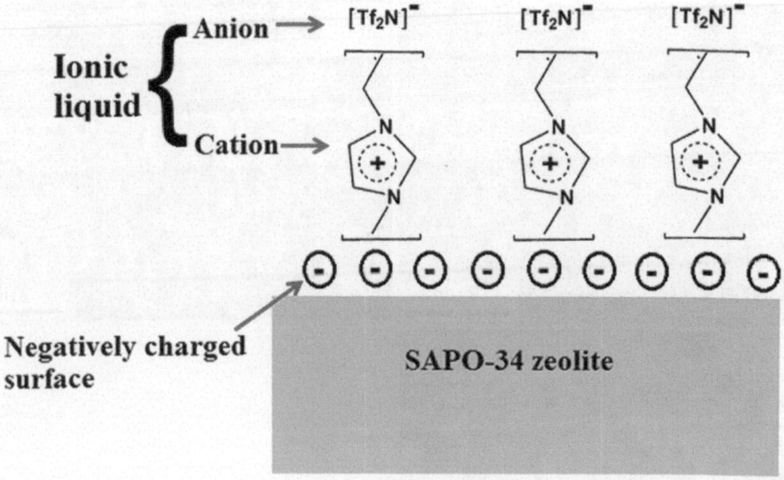

Fig. 4.2 The schematic view for adsorption of ionic liquid on the modified zeolite surface (*SAPO: Silicoaluminophosphate*), (Reprinted with permission of [Modification of gas selective SAPO zeolites using imidazolium ionic liquid to develop polysulfone mixed matrix membrane for CO_2 gas separation, Ahmad, N., C. Leo, A. Mohammad and A. Ahmad, Elsevier]' from (Ahmad et al. 2017))

4.3.2 Silica

The silica-based fillers are class of conventional filler particles which have been used for mixed matrix membranes' fabrication. Due to their stability, low toxicity, relatively low cost, excellent molecular sieving properties and ability to be functionalized with a range of molecules and polymers, these fillers are applied in membrane-based gas and liquid separations (Elma et al. 2012). Zhuang et al. (2014) investigated the performance of the fabricated poly (2,6-dimethyl-1,4-phenylene oxide) poly (phenylene oxide)-silica mixed matrix membranes for H_2/CO_2 separation. In order to obtain defect free mixed matrix membrane, the silica filler particles were synthesized via in-situ sol-gel method. The effect of acid- and base-catalyzed silica on the performance of mixed matrix membrane was also studied. The obtained results showed that H_2 and CO_2 permeabilities in acid-catalyzed silica were 10 times faster than those of base-catalyzed. It is attributed to the better dispersion of acid-catalyzed clusters in the poly (phenylene oxide) continuous phase, as shown in Fig. 4.3. Smaller particles provide higher surface area, thereby enhancing the gas permeability. It was also shown that the increase of silica loading increased the permeability of H_2 from 82.17 to 548.7 Barrer and CO_2 permeability from 48.81 to 154.05 Barrer. The mixed matrix membrane also showed the enhanced selectivity of H_2/CO_2 from 1.68 to 3.56.

Carbon-silica nanocomposite and Mobil Composition of Matter No. 41 (MCM-41) silica fillers were utilized by Anjum et al. (2015) to fabricate the mixed matrix membranes. Carbon-silica nanocomposite s were prepared by treating the MCM-41

Table 4.1 The performance of different zeolite-based mixed matrix membranes in gas separation

Mixed matrix membrane composition		Permeability(Barrer)				Selectivity			Reference
Zeolite	Polymer	CO_2	N_2	H_2	O_2	CO_2/N_2	H_2/CO_2	O_2/N_2	
Silicalite	PDMS	2000–5000	180–350	–	370–1000	10–14	–	–	Tantekin-Ersolmaz et al. (2000)
ZSM-5	PEBA	100–450	7–450	–	10.5–400	1.03–17.69	–	–	Friess et al. (2011)
4A	PEEK-WC	–	0.16–90	–	–	0.87–21	–	–	Clarizia et al. (2008)
4A	PES	1.56–2.32	–	4.94–8.3	0.363–0.583	–	3.17–3.58	–	Huang et al. (2006)
SAPO-34	PES	1.53–5.12	–	7.06–12.57	–	–	2.45–4.64	–	Karatay et al. (2010)
4A, 5A, 13X, NaY	PI	0.185–33.4	0.00181–1.35	–	0.033–6.58	20.6–102	–	–	Yong et al. (2001)
ZSM-5	PI	8.27–15.41	–	19.56–36.31	1.35–3.17	–	2.30–2.57	–	Zhang et al. (2008a)
Silicalite-1	PI/PES	7.2–18.7	–	15.4–38.4	2.3–2.8	39.8–41.7	–	–	Zornoza et al. (2011a)
SBA-15	PPO	55–80	2.5–5.5	100–152	–	17–20	1.5–2	–	Weng et al. (2010)
SAPO-34, ALPO	PPZ	17–64	–	–	–	55	–	–	Jha and Way (2008)
Silicalite-1	Teflon AF 1600	176–3720	29.8–1140	1450–5590	32.2–2050	3.1–5.9	–	–	Golemme et al. (2006)
3A	PSf	–	–	18.2		–	13	–	Guiver et al. (2002)
Silicalite	CA	–	–	–	–	–	–	4.3	Kulprathipanja et al. (1988)
4A	PVAc	–	–	–	0.28–0.35	–	–	9.7–10.4	Mahajan and Koros (2000)

(continued)

Table 4.1 (continued)

Mixed matrix membrane composition		Permeability(Barrer)				Selectivity			Reference
Zeolite	Polymer	CO_2	N_2	H_2	O_2	CO_2/N_2	H_2/CO_2	O_2/N_2	
4A	PSf	–	–	–	1.8	–	–	7.7	Wang et al. (2002)
4A	PEI	–	–	–	0.28	–	–	12.9	Mahajan and Koros (2002)
5A	PES	–	–	–	0.7	–	–	7.4	Li et al. (2005)
NaY-NH2	CA	4.1	–	–	–	26	–	–	Sanaeepur et al. (2014)

PDMS polydimethylsiloxane, *PEBA* poly(ether-block-amide), *PES* polyethersulfone, *SAPO* silico alumino phosphate, *PI* polyimide, *ALPO* aluminophosphate, *PPO* poly (phenylene oxide), *PSf* polysulfone, *CA* celluloseacetate, *PVAc* poly (vinyl acetate), *PEEK-WC* polyetheretherketone, *PPZ* polyphosphazene, *PEI* polyetherimide

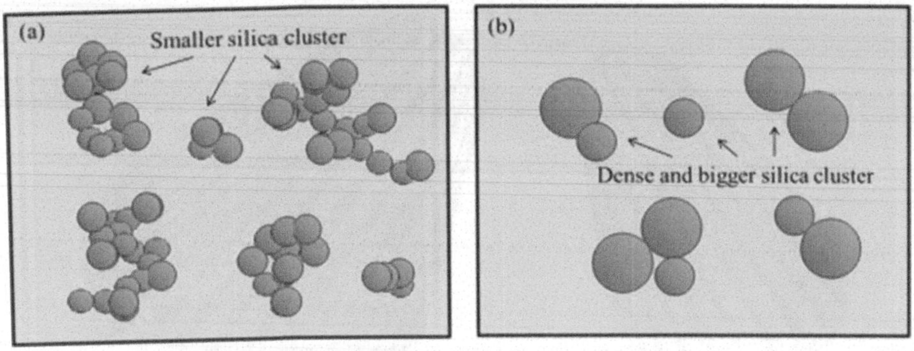

Fig. 4.3 Schematic view of (**a**) acid-catalyzed silica and (**b**) base-catalyzed silica. The smaller silica cluster of acid-catalyzed silica enhance gas permeability. (Reprinted with permission of [Preparation of poly (phenylene oxide)-silica mixed matrix membranes by in-situ sol–gel method for H_2/CO_2 separation, Zhuang, G.-L., H.-H. Tseng and M.-Y. Wey, Elsevier]' from (Zhuang et al. 2014))

fillers with a carbon precursor. Both fillers were incorporated into Matrimid matrix. The observation demonstrated that the presence of a carbon phase inside the MCM-41 silica particles provide the better interaction with CO_2 gas molecules. Embedding the MCM-41within Matrimid polymer increased the selectivity of CO_2/N_2 from 22.8 to 37.8. In comparison with the neat Matrimid, enhanced CO_2/N_2 selectivity of 67% was also observed after incorporating of carbon-silica nanocomposite filler into polymer matrix. Ghadimi et al. (2014) prepared silicon dioxide (SiO_2)/poly (ether-block-amide) mixed matrix membranes for separation of CO_2. Cis-9- octa-decenoic acid were utilized as modifier agent on silica spheres in order to restrict their agglomeration within the polymeric matrix. The results showed that after incorporating of the modified SiO_2 particles into Pebax, the selectivity of CO_2/N_2 and CO_2/H_2 increased from 61 and 9 to 137 and 17, respectively. Chemical modification of particles reduced the –OH groups on surface of fillers, hence eliminated the agglomeration of the particle within polymer continuous phase.

4.3.3 Carbon Nano Tube

Carbon nanotubes (CNTs) are recognized as the interesting fillers in mixed matrix membranes preparation. They possess unique properties such as rapid transport for gases and mechanical strength (Ismail et al. 2009). They can improve mechanical strength of the polymers even at low concentration in the polymer matrix. Due to the extraordinary inherent smoothness of carbon nanotubes walls, these fillers show higher gas transport behavior compared to other inorganic filler particles. Carbon nanotubes are composed of sp^2-bonded carbon atoms which cause them to have high stiffness. There are two main types of carbon nanotubes: single-walled carbon nanotube (SWCNT) and multi-walled carbon nanotube (MWCNT), as can be seen

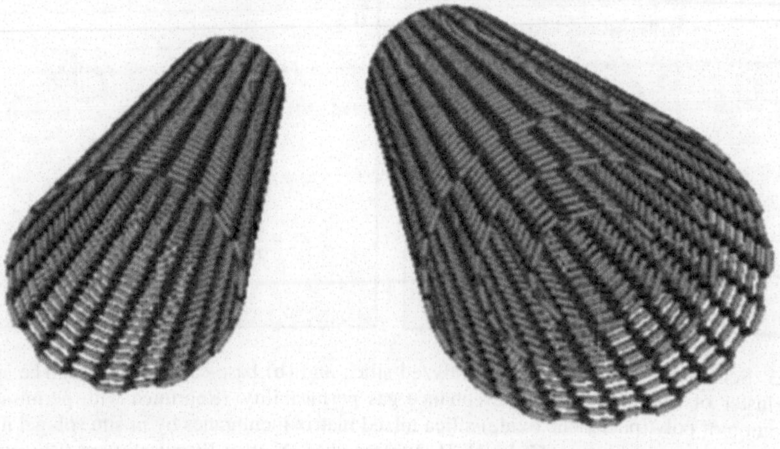

Fig. 4.4 Sketch of the various size structures of carbon nanotubes (**a**) single-walled carbon nanotube (SWCNT) (**b**) multi-walled carbon nanotubes (MWCNT), Reprinted with permission of [Transport and separation properties of carbon nanotube-mixed matrix membrane, Ismail, A., P. Goh, S. Sanip and M. Aziz, Elsevier]' from (Ismail et al. 2009)

in Fig. 4.4 (Ismail et al. 2009). Multi-walled carbon nanotubes have multiple rolled layers of graphene.

Carbon nanotubes have been utilized to improve the polymer properties and enhance the permeability of different species through polymer matrix. A comprehensive review of carbon nanotube mixed marix membranes was reported by Ismail et al. (2009). Their report reviewed the development of carbon nanotubes application in the fabrication of mixed matrix membranes for gas and liquid separations. Zhao et al. (2017) reported that three different carbon nanotubes were mixed with Pebax to prepare high performance mixed matrix membranes. The effect of CO_2 content was varied, up to 35 wt.%. They observed that CO_2 permeabilities of the prepared membranes increase with the enhancement of carbon nanotubes content. Their results also showed that permeability of CO_2 were about 350 Barrer for Pebax/multi-walled carbon nanotubes-NH_2 membrane at the highest carbon nanotubes content. The Glycerol triacetate was also used as an additive to enhance the CO_2 solubility of Pebax/ carbon nanotubes membranes. The Fourier-transform infrared spectroscopy indicated a good mixing between glycerol triacetate and Pebax/ multi-walled carbon nanotubes -NH_2. CO_2 permeability of the resultant membrane was about 1408 Barrer, which was higher than that of Pebax/ multi-walled carbon nanotubes-NH_2. This increment in CO_2 permeability of Glycerol triacetate containing mixed matrix membranes is because of the existence of CO_2-philic acetate groups. Further, Wang et al. (2014) fabricated the flat-sheet hybrid membrane containing multi-walled carbon nanotube, Pebax and poly (ethylene glycol)- based polymer. The membrane was prepared by dispersing multi-walled carbon nanotube into polymer solution. They found that CO_2 diffusivity and solubility increase with incorporation of multi-walled carbon nanotube because of the enhanced chain mobility. The gas permeation result indicated that for resultant membranes, the permeability of CO_2 and selectivity of CO_2/N_2 increased 49.28% and 21.24% compared to their polymeric counterparts, respectively. Increased selectivity of CO_2/N_2 and permeability of CO_2 were observed with increasing the multi-walled carbon nanotube loading up to 5 wt.%. Mixed gas permeation performance of the fabricated membranes was also determined at room temperature, and the CO_2 permeability of membrane was about 743 Barrer and its CO_2/N_2 selectivity was almost 108. Zhao et al. (2014) reported that the multi-walled carbon nanotubes were incorporated into the crosslinked Polyvinyl alcohol-poly (siloxane) matrix containing amines. The resultant mixed matrix membranes showed the very high CO_2 permeability of about 3400 Barrer, and CO_2/H_2 selectivity of 179 at 380.15 K and pressure 0.2 MPa. In this mixed matrix membrane, amines facilitate the transport of CO_2 molecules through membrane and help to promote CO_2 permeability and selectivity of membrane against gases.

Despite good properties of carbon nanotubes, the dispersion of pristine carbon nanotube into the polymer matrix can cause agglomeration of carbon nanotubes and poor interfacial interaction between carbon nanotubes and polymer matrix. To tackle this problem, functionalization of the surface of carbon nanotubes is suggested as an effective solution. There are three different types of functionalization including defect functionalization, covalent functionalization and noncovalent functionalization (Sahoo et al. 2010). Sanip et al. (2011) functionalized multi-walled carbon nanotube with beta-cyclodextrin and then, fabricated the mixed matrix membrane by mixing of the functionalized multi-walled carbon nanotube with polyimide. They observed that functionalization treatment of multi-walled carbon nanotube with beta-cyclodextrin had the great influence on the solubility and homogenous dispersion of the carbon nanotubes into polymer.

Pure polyimide membrane, untreated and acid-treated multi-walled carbon nanotube/ polyimide membranes were fabricated by solution-casting method (Sun et al. 2017). Figure 4.5 shows the schematic view of the acid-treated mixed matrix membrane. The gas permeation results showed that acid-treated membranes had better performance in comparison with pristine multi-walled carbon nanotube-containing membrane. CO_2/N_2 selectivity of polyimide /1 wt.% untreated multi-walled carbon nanotube and polyimide 1 wt.% acid-treated multi-walled carbon nanotube were reported as 18.44 and 26.61, respectively. In this case, polar functional groups on the surface of acid-treated multi-walled carbon nanotubes provide channels to transport gas molecules more easily through the membrane matrix. CO_2/N_2 separation factor of pure polyimide membrane and PI/3 wt.% acid-treated multi-walled carbon nanotube mixed matrix membranes were 15.40 and 37.74, respectively. Due to the strong interaction of polar carboxylic (-COOH) and hydroxyl (-OH) group with polar gases like CO_2, the fabricated mixed matrix membranes indicated higher selectivity than the pristine PI membrane.

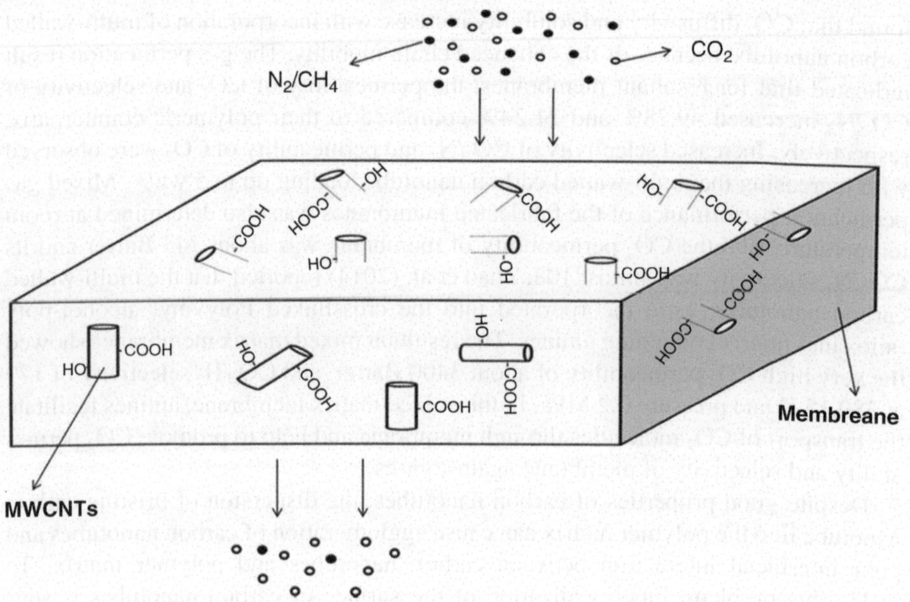

Fig. 4.5 The acid treated multi-walled carbon nanotubes within mixed matrix membrane. Gas transport mechanism in prepared mixed matrix membranes in which vertical short-cut multi-walled carbon nanotubes disperse in different direction and act as channels in the gas transport mechanism and cause increasing in permeability coefficient, Reprinted with permission of [Fabrication of polyimide and functionalized multi-walled carbon nanotubes mixed matrix membranes by in-situ polymerization for CO_2 separation, Sun, H., T. Wang, Y. Xu, W. Gao, P. Li and Q. J. Niu, Elsevier]' from (Sun et al. 2017)

4.3.4 Carbon Molecular Sieve

Carbon molecular sieves (CMSs) are identified as attractive carbonaceous materials, which have been widely utilized in industrial applications. They produce from the pyrolysis of polymers at high temperatures (Shin et al. 2019). Due to their constricted apertures, carbon molecular sieves can separate the gas molecules with very similar size. Tunable nature of such materials allows them to be used in membrane-based gas separation applications. Recently, Joglekar et al. (2019) developed carbon molecular sieve hollow fiber membranes for flue gas carbon capture at both ambient and sub-ambient temperatures. Cho et al. (2018) modified the surface of the carbon molecular sieve with H_2O_2 to investigate the CO_2 uptake capacity. They introduced various functional groups onto the carbon molecular sieve surface under different reaction conditions. It was observed that the CO_2 adsorption capacity increased as the carboxylic group concentration increased on the carbon molecular sieve surface. Because of superior adsorptivity of carbon molecular sieves, they have been used as filler particles for mixed matrix membranes' preparation. Vu et al. (2003) embedded carbon molecular sieve fillers into Matrimid® 5218 and Ultem® 1000 to examine the CO_2 separation performance of the resultant membranes. The membrane performance was evaluated at different loadings of carbon molecular sieves up to 35 wt.% and also compared with the pure polymers. The results showed that the incorporating of the carbon molecular sieves into the both polymer matrices enhanced the CO_2 permeability. In addition, the enhancements of 45% and 40% in CO_2/CH_4 selectivity were observed for carbon molecular sieve dispersed in Matrimid and carbon molecular sieve dispersed in Ultem phases, respectively.

4.3.5 *Metal Organic Framework*

During recent years, metal organic frameworks (MOFs), as a class of crystalline compounds, have been utilized in gas storage, separation, ion exchange, molecular separation and heterogeneous catalysts (Adams et al. 2010; D'Alessandro et al. 2010). They possess unique structural and chemical properties including: high void volume, low density, high surface area-to-weight ratio, controlled porosity, high thermal and chemical stability (Goh et al. 2011). These microporous materials consist of metal-based nodes which are connected to organic bridging ligands to form a 3D extended network. The pore dimensions and chemical properties of the metal organic frameworks can be manipulated by adding the different functional groups to facilitate the adsorption of target gases (Wong-Foy et al. 2006). They possess a lower density and a higher pore volume than the zeolites. Therefore, at the same amount of filler, they can affect the membrane properties more than zeolites (Seoane et al. 2015).

Several types of the metal organic frameworks have been used as the fillers for preparation of the mixed matrix membranes, including copper-based metal organic frameworks (Cu-based metal organic framework), zirconium 1,4-dicarboxybenzene (UiO-66), zeolitic imidazolate frameworks (ZIFs), materials institute lavoisier metal organic frameworks (MILs) etc. (Ahmadi et al. 2018). Adams et al. (2010) were dispersed the two-dimensionally coordinated metal organic framework particles of copper and terephthalic acid in continuous polymeric phase of poly (vinyl acetate) to fabricate mixed matrix membranes. Due to high affinity of Cu-based metal organic frameworks with polar molecules, they are known as CO_2-philic metal organic frameworks (Ahmadi et al. 2018). An optical micrograph of copper and terephthalic acid particles has shown that these type of metal organic framework particles have the crystalline appearance (Fig. 4.6). It was reported that the CO_2 permability of fabricated mixed matrix membranes and neat poly (vinyl acetate) membrane were about 3.26 and 2.44 Barrer, respectively. Also, the comparative results with pristine poly (vinyl acetate) indicated that CO_2/N_2 selectivities were about 35.4 for mixed matrix membranes and 32.1 for poly (vinyl acetate) membrane.

Perez et al. (2009) fabricated MOF-5/Matrimid mixed matrix membranes for CO_2 separation. Pure gas permeability test showed that the permeability of CO_2, H_2, N_2, O_2 and CH_4 with the increment of MOF-5 loading up to 30 wt.%. The enhanced permeability of the resultant membrane was reported due to increase in diffusivity

Fig. 4.6 An optical micrograph of the synthesized terephthalic acid particles, the particles have a crystalline structure. (Reprinted with permission of [Metal organic framework mixed matrix membranes for gas separations, Adams, R., C. Carson, J. Ward, R. Tannenbaum and W. Koros, Elsevier]' from (Adams et al. 2010))

in the membrane owing to the porosity of the MOF-5 nanocrystals. A class of zirconium-based metal organic framework (MOF) UiO-66 has gained much attention as a potential filler for preparation of mixed matrix membranes. This filler shows high affinity towards CO_2. The top surface of the UiO-66 can be functionalized with hydroxyl (-OH), nitro (-NO_2), amino (-NH_2) and methoxy (-OMe) groups (Moghaddam et al. 2018; Rada et al. 2018). UiO-66 and UiO-66-NH_2 were added to the Pebax matrix for mixed matrix membrane preparation (Shen et al. 2016). The CO_2 permeability and CO_2/N_2 selectivity were also investigated at different metal organic frameworks loading for both membranes. It was observed that for metal organic framework loadings up to 20 wt.%, CO_2 permeability increased for both UiO-66/Pebax and UiO-66-NH_2/Pebax membrane. At higher metal organic frameworks loading, CO_2/N_2 selectivity showed the decreased trend for both fabricated membranes. This reduction was assigned to the agglomeration of filler particles and formation of non-selective interface defects at higher filler loading. The effect of temperature was also investigated and the results were compared to that of pure poly(ether-block-amide) membrane. The results showed when the temperature rose from 20 to 80 °C, the gas permeability increased, while the selectivity of CO_2/N_2 reduced. Further, the effect of humidity was also studied to see how the fabricated mixed matrix membranes perform in the presence of water vapor. For both mixed matrix membranes, permeability of CO_2 and CO_2/N_2 selectivity increased after humidifying. This behavior attributes to the hydrophilic nature of the utilized fillers. Permeability of CO_2 changed from 96.3 to 139.7 Barrer and selectivity of CO_2/N_2 increased from 56.6 to 61.7 for UiO-66/Pebax membrane. For UiO-66-NH_2/Pebax membrane, permeability of CO_2 increased from 87.0 to 130.2 Barrer for dry and humid state, respectively. In addition, CO_2/N_2 selectivity altered from 66.1 to 72.2 after humidifying.

Zeolitic imidazolate frameworks (ZIFs), as a new class of metal organic frameworks, have been known as a potential material for separation of CO_2. These type of materials possessed zeolite topology and their pore window size can be tuned for separation of specific gases. Several researchers have utilized zeolitic imidazolate frameworks including ZIF-8, ZIF-90 and ZIF-7 as porous fillers (Bae et al. 2010; Liu et al. 2011; Yang et al. 2011; Zhang et al. 2012).

Imidazole linkers can interact with CO_2 and enhance the selective sorption and diffusion of this small gas in zeolitic imidazolate framework crystal. In addition to this, large cavities and tight pore apertures of zeolitic imidazolate frameworks make them attractive for CO_2 capture. In pre-combustion capture, where separation of H_2/CO_2 is desired, zeolitic imidazolate frameworks with pore size of about 3A show the precise size-selective molecular sieving ability and good separation performance (Goh et al. 2011). Yang and Chung (2013a, b) examined the H_2/CO_2 separation performance of ZIF-8/polybenzimidazole mixed matrix membranes for applications such as integrated gasification combined cycle power plant and syngas processing. Two types of membranes were prepared with different loadings of ZIF-8 (38.2 and 63.6 vol.%). The ZIF-8 (38.2 vol.%)/polybenzimidazole showed H_2 permeability of 470.5 Barrer, and that of ZIF-8 (63.6 vol.%)/polybenzimidazole membrane was high about 2014.8 Barrer. H_2/CO_2 selectivies are reported as 26.3 and 12.3 for ZIF-8 (38.2 vol.%)-containing and ZIF-8 (63.6 vol.%)-containing mixed matrix membranes, respectively.

Li et al. (2013) prepared a CO_2 selective composite membrane which contains ZIF-7 nanoparticles. The CO_2 permeability of their fabricated membrane was about 145 Barrer and the CO_2/N_2 selectivity was around 97. Nafisi and Hägg (2014) developed the flat sheet-type membranes of Pebax-2533 as the main continuous matrix and ZIF-8 as a filler. The fabricated membrane consisted of two different layers; polymer-based layer and inorganic-based layer. The cause of these two deferent layers' formation is reported to be due to the solvent evaporation and agglomeration of the filler particles during preparation procedure. Scanning electron microscope (SEM) images revealed that as the filler concentration increased up to 50 wt.%, the thickness of the inorganic layer increased (Fig. 4.7). CO_2, N_2, O_2 and CH_4 single gas permeation as well as CO_2/N_2 binary gas permeation tests were also performed with a constant volume method. Single gas permeation test indicated that permeability of CO_2 within Pebax-2533/ZIF-8 (15% loading of ZIF-8) was about 574 Barrer. In addition, selectivities of CO_2/N_2 and CO_2/CH_4 were almost 30.3 and 10.4, respectively. Compared to neat Pebax-2533, Pebax-2533/ZIF-8 (35% loading of ZIF-8) exhibited CO_2 permeability of about 3.66-fold larger than that of pristine Pebax-2533. The binary gas permeation test (10% CO_2 and 90% N_2) was approximately similar to single gas permeation result.

Materials Institute Lavoisier (MILs) are another well-known sub-family of metal organic frameworks. Perez et al. (2017) examined VTEC™ PI-1388/(Al) NH_2-MIL-53 mixed matrix membrane for H_2/CO_2 separation at various pressures (5–30 bar) and temperatures (35–300 °C). Figure 4.8 depicts the H_2/CO_2 selectivity

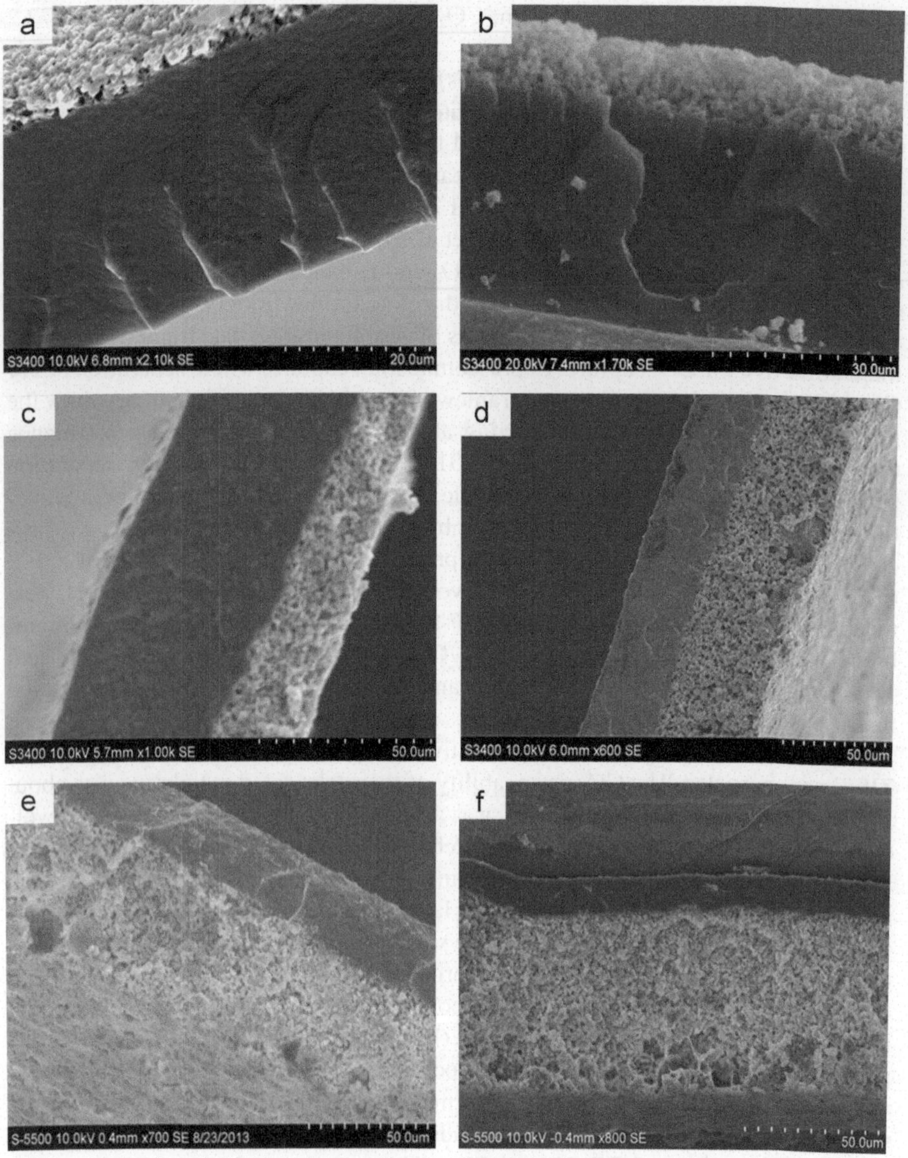

Fig. 4.7 The cross-section scanning electron microscope (SEM) images of the fabricated ZIF-8/Pebax mixed matrix membranes at different zeolitic imidazolate framework loadings, The thickness of the inorganic layer increases by increasing zeolitic imidazolate framework loadings in mixed matrix membrane. (Reprinted with permission of [Development of dual layer of ZIF-8/PEBAX-2533 mixed matrix membrane for CO2 capture, Nafisi, V. and M.-B. Hägg, Elsevier]' from (Nafisi and Hägg 2014))

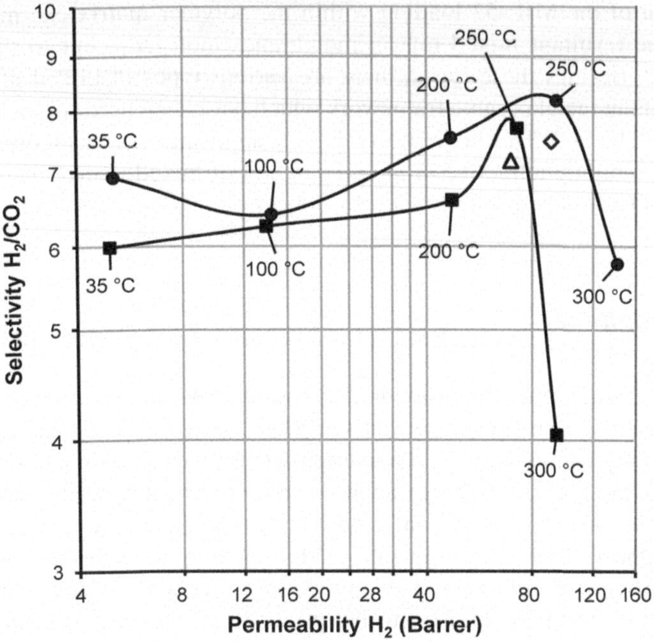

Fig. 4.8 The performance of the neat VTEC membrane (■) and NH_2-MIL-53/VTEC mixed matrix membrane (●) at different temperatures, due to the strong affinity of (Al) NH_2-MIL-53 for VTEC™, mixed matrix membrane has the higher performance than VTEC™ membrane. (Reprinted with permission of [Amine-functionalized (Al) MIL-53/VTEC™ mixed-matrix membranes for H_2/CO_2 mixture separations at high pressure and high temperature, Perez, E. V., G. J. Kalaw, J. P. Ferraris, K. J. Balkus and I. H. Musselman, Elsevier]' from (Perez et al. 2017))

versus H_2 permeability of the prepared VTEC™/(Al) NH_2-MIL-53 mixed matrix membrane, which is compared to the performance of the VTEC™ membrane. As can be seen, both VTEC™ and mixed matrix membrane had the highest H_2 permeability and H_2/CO_2 selectivity at temperature of 250 °C. It was also shown that the mixed matrix membrane had the better performance than VTEC™ membrane. This higher performance is attributed to the strong affinity of (Al) NH_2-MIL-53 for VTEC™ PI-1388. Dorosti et al. (2014) studied the performance of mixed matrix membranes which contain different weight percentage of MIL-53. MIL-53 as the disperse phase was added to Matrimid 5218 continuous polymeric phase. As the MIL-53 loading increased from 0 to 20 wt.%, CO_2 permeability increased from 6.2 to 14.52 Barrer. The enhanced CO_2 permeability is due to the quadrupole moment of CO_2 molecules, which makes strong affinity of CO_2 with hydroxyl groups in MIL-53 structure. The maximum CO_2/CH_4 selectivity of 50 was also obtained at 15 wt.% loading of MIL-53. In addition, the performance of the fabricated mixed matrix membranes was compared with the Robeson curve. It was observed that the enhancement of the MIL-53 loading within the polymer matrix can make performance of the resultant mixed matrix membranes more adjacent to the Robeson upper bound. Besides these types, there are various types of mixed matrix membrane containing metal organic framework which have been used for gas separation, as shown in Table 4.2. This table shows that the application of metal organic framework fillers is highly attractive in the preparation of mixed matrix membranes for separation of CO_2.

4.3.6 *Graphene*

Graphene is thinnest two-dimensional (2D) carbon material, in which carbon atoms arranged in a regular hexagonal pattern. Graphene possesses high tensile strength of 130 GPa, Young's module of 1 TPa and remarkably large specific surface area of 2600 m^2/g (Ding and Li 2018). In addition to its mechanical properties, it is also highly flexible and has high thermal conductivity (Huang et al. 2018). The oxidized form of graphene, known as graphene oxide is widely used in the fabrication of membranes. Because of its remarkable properties, graphene oxide is the subject of intense research in this field. Due to the presence of oxygen functionalities, graphene oxide is easily dispersed in water and other organic solvents, as well as in different matrixes (Konios et al. 2014).

Dong et al. (2016) synthesized porous reduced graphene oxide (PRG) through a dehydration reaction and then mixed them into the Pebax polymer to prepare the porous graphene oxide -based gas separation membrane. The obtained results indicated that, with the increase of porous reduced graphene oxide concentration, the maximum CO_2 permeability of about 120 Barrer and CO_2/N_2 selectivity of about 104 were achieved at the porous reduced graphene oxide concentration of 5 wt.%. The presence of graphene oxide fillers in the polymer matrix led to increase in permeability of CO_2 and CO_2/N_2 selectivity of 100% and 86.02%, respectively. The increment in permeability of CO_2 and selectivity has been reported to be attributed to the well-constructed pathways of the porous reduced graphene oxide.

Li et al. (2015b) fabricated the mixed matrix membranes by embedding carbon nanotube and graphene oxide into a Matrimid polymeric phase. Due to the favorable properties of these two fillers, the combination of both materials can significantly improve the CO_2 separation performance of the membrane. The nanofillers were dispersed into the appropriate solvent, sonicated until form a homogeneous suspension. After that, the nonofiller-containing suspension was added into polymeric solution. A good nanofiller-polymer contact can be observed from scanning electron microscope image (Fig. 4.9).

In order to increase the transport properties of the graphene oxide -based mixed matrix membranes, graphene oxide is often modified with different functional groups such as amino acids, polyethylene glycol monomethyl ether and polyethyleneimine (Li et al. 2015a; Xin et al. 2015). Jia et al. (2018) fabricated hybrid membranes by incorporation UiO-66-NH_2@graphene oxide into polyimide matrix to

Table 4.2 The gas separation performance of the metal organic framework (MOF) -based mixed matrix membranes

Mixed matrix membrane composition		Permeability(Barrer)			Selectivity		Reference
Metal organic frameworks (MOFs)	Polymer	CO_2	N_2	H_2	CO_2/N_2	H_2/CO_2	
Cu-BPY-HFS	Matrimid	10.4	0.3	20.3	33.2	2	Zhang et al. (2008b)
CuTPA	PVAc	3.3	0.1	–	35.7	–	Adams et al. (2010)
HKUST-1	PI	37.2	7.8	1270	4.8	34.14	Hu et al. (2010)
HKUST-1	PDMS	3053	341	740	8.9	–	Hu et al. (2010)
HKUST-1	PSf	7.9	0.9	14.9	8.8	–	Car et al. (2006)
Mn(HCOO)2	PSf	6.8	0.2	10.2	29.5	1.43	Car et al. (2006)
IRMOF-1	Matrimid	38.8	–	114.9	–	2.96	Liu et al. (2009)
MOF-5	Matrimid	13.8	0.4	38.3	34.5	2.78	Perez et al. (2009)
ZIF-8	Matrimid	14.2	0.6	47.2	24.1	3.33	Ordoñez et al. (2010)
ZIF-8	Matrimid	16.6	0.9	63.5	18.9	3.33	Song et al. (2012)
ZIF-8	Ultem	1.8	0.04	–	44	–	Dai et al. (2012)
Cu 1,4-BDC	PVAc	2.4	–	–	32.1	–	Adams et al. (2010)
HKUST-1	PMDA-ODA	306.6	–	3066	8	10	Hu et al. (2010)
ZIF-7	PBI	–	–	3.7	–	8.7	Yang et al. (2011)
ZIF-8 + S1C	PSf	5.9	–	–	24.6	–	Zornoza et al. (2011b)
MIL-53(Al)	Matrimid	11.5	–	–	18	–	Zornoza et al. (2011b)
ZIF-7	Pebax	72	–	–	34	–	Li et al. (2013)
ZIF-8	PIM-1	4390	–	1630	24.4	0.4	Bushell et al. (2013)
ZIF-90	PBI	–	–	4.1		8.9	Yang and Chung (2013a, b)
HKUST-1	PPO	68.7	–	75	16	1.1	Ge et al. (2013)
ZIF-8	6FDA-durene	468.5	–	518.5	13.4	1.1	Wijenayake et al. (2013)
NH2-CAU-1	PMMA	–	–	5000	–	3	Cao et al. (2013)
CPO-27(mg)	XLPEO	380	–	–	22	–	Bae and Long (2013)
CPO-27(mg)	6FDA-TMPDA	650	–	–	14	–	Bae and Long (2013)
CPO-27(mg)	PDMS	3100	–	–	9.5	–	Bae and Long (2013)

(continued)

Table 4.2 (continued)

Mixed matrix membrane composition		Permeability(Barrer)			Selectivity		
Metal organic frameworks (MOFs)	Polymer	CO_2	N_2	H_2	CO_2/N_2	H_2/CO_2	Reference
ZIF-8	Pebax	351	–	–	33.8	–	Nafisi and Hägg (2014)
ZIF-8	PBI-BuI	2.3	–	6.2	26.8	2.7	Bhaskar et al. (2014)
ZIF-8	DMPBI-BuI	3.8	–	12.8	21.7	3.4	Bhaskar et al. (2014)
ZIF-8	DBzPBI-BuI	25.8	–	61.4	12.9	2.4	Bhaskar et al. (2014)
c-MOF-5	PEI	1.7	–	10.1	16.8	6	Arjmandi and Pakizeh (2014)
ZIF-71	6FDA-durene	959	–	756	14.7	0.8	Japip et al. (2014)
UiO-66 -NH2	Matrimid 5218	24	–	–	36	–	Venna et al. (2015)
UiO-66-$(COOH)_2$	PIM-1	5190	–	–	20.4	–	Khdhayyer et al. (2017)
MOP-18	Matrimid 5218	14	–	–	22.9	–	Perez et al. (2014)
P-MIL-53	PDMS	36	–	–	25.8	–	Zhu et al. (2017)
Fe(BTC)	Matrimid 5218	217.9	–	–	23.1	–	Nabais et al. (2018)
Ti-UiO-66	PIM-1	2.54	–	–	1.05	–	Carter et al. (2017)

PVAc poly (vinyl acetate), *PDMS* polydimethylsiloxane, *PI* polyimide, *PSf* polysulfone, *PPO* poly phenylene oxide, *PMDA-ODA* Poly[4,4′-oxydiphenylenepyromellitimide], *PBI* polybenzimidazole, *PMMA* Poly(methyl methacrylate), *XLPEO* cross-linked poly(ethylene oxide), 6FDA: 2,20 bis(3,40-dicarboxyphenyl)hexafluoropropane dianhydride, *TMPDA* 2,4,6-trimethyl-1,3-phenylenediamine, *PDMS* polydimethylsiloxane, *PEI* polyetherimide, *PIM-1* tetrafluoroterephtalonitrile, *MOF* metal organic framework, *CuTPA* copper and terephthalic acid, *ZIF* zeolitic imidazolate frameworks

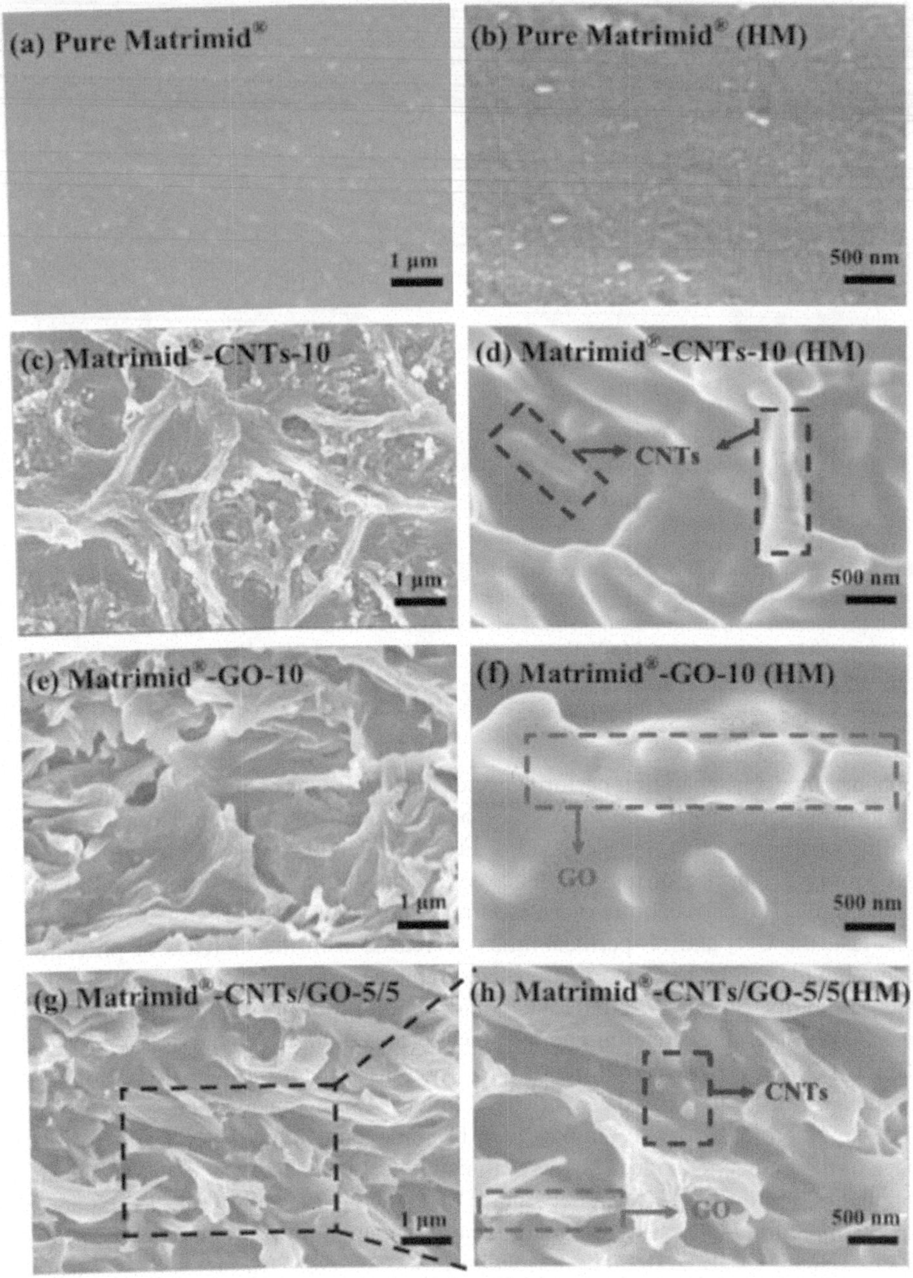

Fig. 4.9 The scanning electron microscope (SEM) images of the (**a**, **b**) Pure Matrimid membrane, (**c**, **d**) Carbon nanotube -based mixed matrix membrane, (**e**, **f**) graphene oxide(GO)-based mixed matrix membrane and (**g**, **h**) Carbon nanotube/graphene oxide-based mixed matrix membrane. (Reprinted with permission of [Synergistic effect of combining carbon nanotubes and graphene oxide in mixed matrix membranes for efficient CO2 separation, Li, X., L. Ma, H. Zhang, S. Wang, Z. Jiang, R. Guo, H. Wu, X. Cao, J. Yang and B. Wang, Elsevier]' from (Li et al. 2015b))

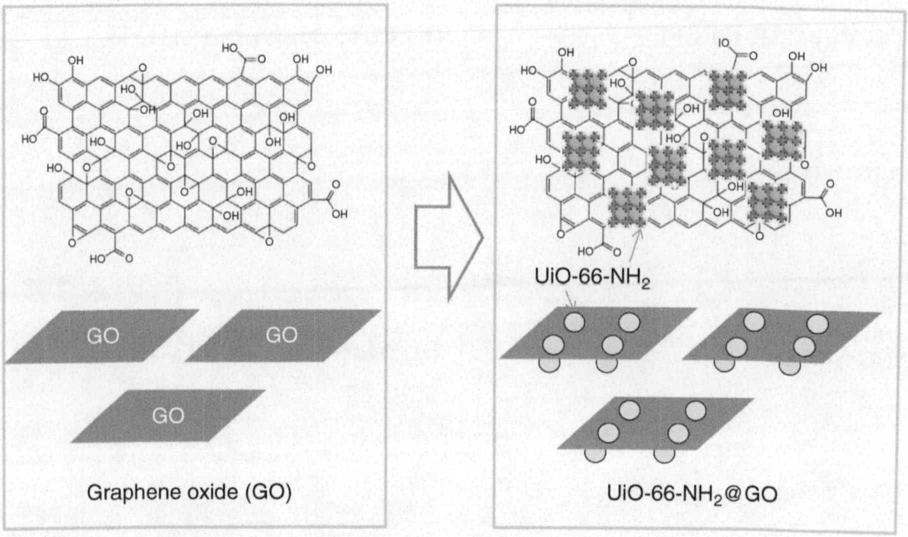

Fig. 4.10 Schematic view of the functionalized graphene oxide (GO), the grown of UiO-66-NH_2 nanocrystals on the graphene oxide nanosheets surface to form Uio-66-NH2@GO. (Reprinted with permission of [Amine-functionalized MOFs@ GO as filler in mixed matrix membrane for selective CO_2 separation, Jia, M., Y. Feng, J. Qiu, X. Zhang and J. Yao, Elsevier]' from (Jia et al. 2018))

enhance the CO_2 separation performance of the membrane. They studied the effect of growing UiO-66-NH_2 on the surface of graphene oxide to improve adsorption ability of fillers to CO_2. As depicted in Fig. 4.10, UiO-66-NH_2 nanocrystals were grown on the surface of the graphene oxide nanosheets. The resultant membrane showed the CO_2 permeability of 7.28 Barrer and CO_2/N_2 selectivity of 52. For pure gas test, the CO_2 and N_2 permeabilities decreased with the increase of pressure. Ionic liquids can be used to functionalize graphene oxide surface and enhance the CO_2 solubility. Among different ionic liquids, imidazolium-based ionic liquids have gained considerable attention due to their great CO_2-philic properties (Noble and Gin 2011). It was shown that the CO_2/N_2 separation factor of Pebax hybrid membranes was increased with covalent functionalization of the 1-(3-aminopropyl)-3-methylimidazolium-bromide ionic liquid (IL-NH_2) on the surface of the graphene oxide. The gas permeation test indicated that an enhancement of about 90% in CO_2/N_2 selectivity and 50% in CO_2 permeability for the graphene oxide–ionic liquid based hybrid membranes compared to the pristine Pebax membrane (Huang et al. 2018). The formation of strong hydrogen bonds between the hard segment of the polymer matrix and the fillers can enhance the interfacial compatibility, which results in better separation performance.

4.4 Preparation of Mixed Matrix Membranes

For fabrication of mixed matrix membranes, organic or inorganic filler with inherent superior separation characteristic is added into the bulk polymeric phase. Material selection and membrane preparation method are two important factors in the fabrication of high performance mixed matrix membranes. The following methods are described by Aroon et al. for dope solution preparation (Aroon et al. 2010):

1. The filler particles are firstly added into the desired solvent. After mixing, polymer is added to this homogeneous suspension to obtain the dope solution.
2. As shown in Fig. 4.11b, polymer pellets are first dissolved in solvent while stirring to get a polymer solution. Afterward, fillers are slowly added into the polymer solution and dispersed for several hours.
3. In this case, a homogeneous suspension containing inorganic fillers and a polymeric solution are prepared separately and then, they both mix together.

The selection of membrane configuration is greatly dependent on the application. Mixed matrix membranes can be made into hollow-fiber and flat sheet configurations as depicted in Fig. 4.12. Flat sheet membranes can be prepared into dense and thin-film composite membranes. In thin-film composite mixed matrix membranes, a thin, dense skin layer that contains highly selective fillers is formed on the thicker porous support by various techniques. The skin layer performs the gas separation function, while the porous support only provides mechanical strength for membrane and has no resistance against transportation of gas molecules through membrane matrix. In contrast to flat sheet membranes, a hollow fiber offers a more practical configuration. Hollow fiber membranes have very high membrane surface area per unit volume. These types of membrane have microporous structure as support layer and a dense selective layer on inside or outside surface of the fiber (Strathmann 2005).

Distribution of inorganic particles into the polymeric solution is of great importance because the particles would agglomerate during solution preparation. Among these fabrication methods, methods (a) and (c) can be used for better distribution of filler particles and preventing of agglomeration (Aroon et al. 2010). Figure 4.13 shows the agglomerated fillers in polymer matrix.

Agglomeration of particles is resulted from sedimentation or surface pattern (migration to the surface). Different physical properties and difference in density between particles and polymers cause the particles to precipitate during mixed matrix membrane preparation. Strong tendency to agglomeration and also weak contact of particles in the polymer phase might form some defects at the polymer-particle interface and enhance non-ideality of the mixed matrix membrane. An ideal mixed matrix membrane is the mixed matrix that has no defects at the interface between polymer and particles. Defects can potentially affect the performance of membrane and push the overall performance below the upper bound

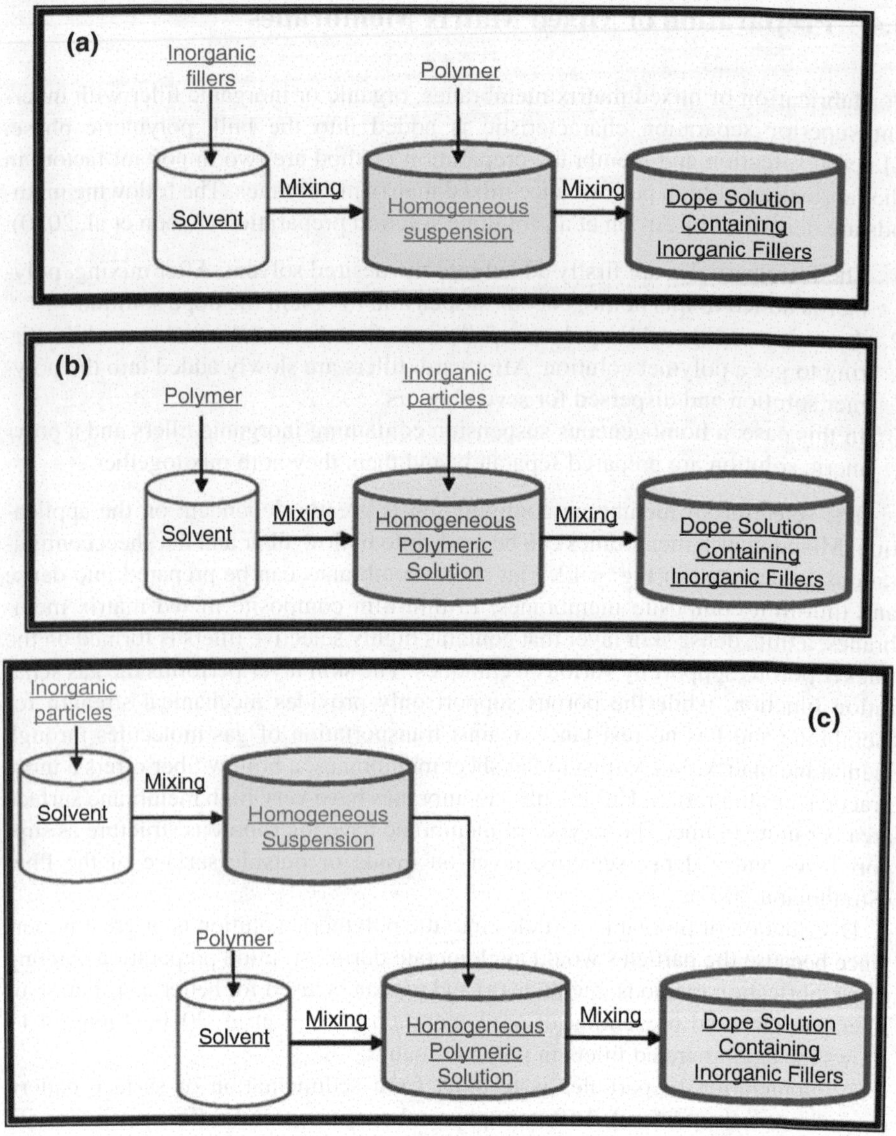

Fig. 4.11 Different procedures for mixed matrix membrane preparation. (Reprinted with permission of [Performance studies of mixed matrix membranes for gas separation: a review, Aroon, M., A. Ismail, T. Matsuura and M. Montazer-Rahmati, Elsevier]' from (Aroon et al. 2010))

trade-off curve. These defects can allow the gases to flow non-selectively around the solid particles. These interfacial defects are schematically depicted in Fig. 4.14. Case 1 shows an ideal mixed matrix membrane with ideal interfacial morphology that has no defects in its structure. The interfacial defects can be existed in various

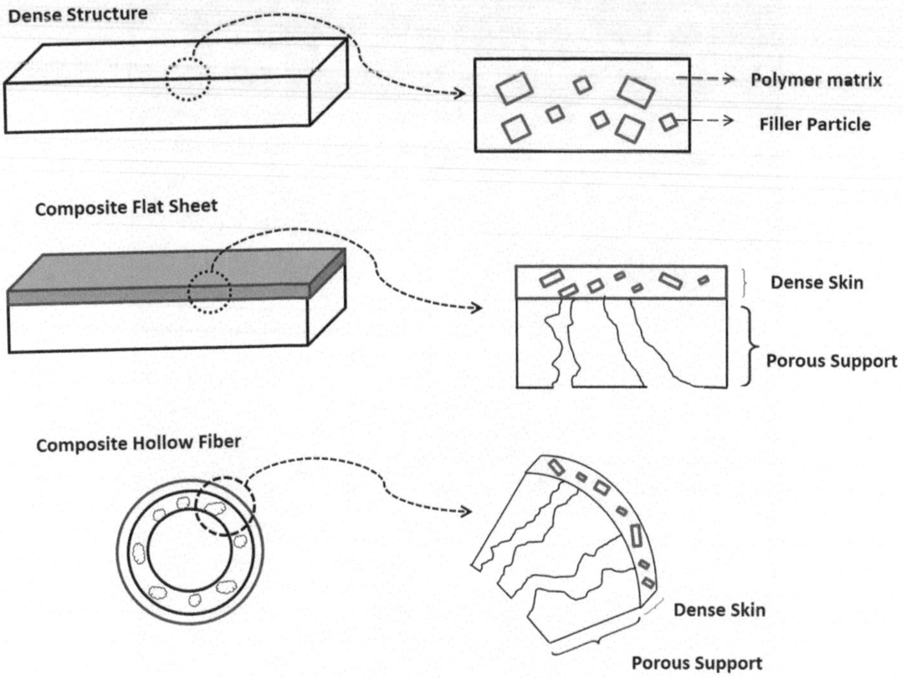

Fig. 4.12 Different configurations of mixed matrix membranes

structures between particle-polymer interface and classified into 3 major categories: (1) interface voids or sieves-in-a-cage, (2) rigidified polymer layer, (3) particle pore blockage (Bastani et al. 2013). Figure 4.15 shows scanning electron microscope image of sieve-in-a-cage defect. Membranes containing both porous and nonporous fillers might experience two first interfacial defects but, pore blockage, either partial or total, is formed in the porous fillers. The pores of fillers may be clogged or plugged by polymer chain, solvent or even contaminant in a feed gas during or after membrane preparation. When the pores of filler particle plug completely, the particle acts like an impermeable filler and gas molecules cannot pass through these pores.

When solvent evaporates during membrane preparation procedure, the stress arises around the particles that results in formation of interface voids and rigidified polymer layer near the filler particle (Moore and Koros 2005). Poor adhesion between polymer and particle, disruption of polymer in the vicinity of the inorganic phase are another reasons for interface voids formation (Aroon et al. 2010). Due to mentioned difficulties, fabrication of ideal mixed matrix membranes seems challenging and complicated. Despite the difficulties of fabricating the mixed matrix

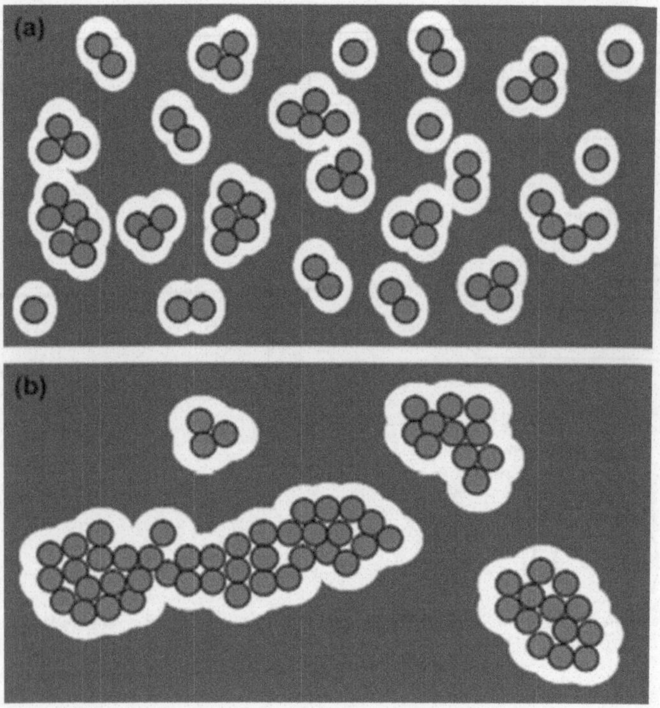

Fig. 4.13 The possible distribution of the filler particles within polymer matrix: (**a**) the dispersed particles and (**b**) the agglomerated particles which is resulted from sedimentation or surface pattern and the cause of precipitate formation during mixed matrix membrane preparation is difference in density between particles and polymers. (Reprinted with permission of [Recent advances of inorganic fillers in mixed matrix membrane for gas separation, Goh, P., A. Ismail, S. Sanip, B. Ng and M. Aziz, Elsevier]' from (Goh et al. 2011))

membranes, the interface can be tailored to promote the compatibility of filler particles and polymer matrix (Shu 2007).

Several methods have been suggested and utilized by researchers to avoid an interfacial defect formation during membrane preparation. Using a low T_g (glass transition temperature) polymer as matrix of the membrane, casting at temperature above T_g of polymer, reducing T_g by incorporation of a plasticizer into the polymer solution, surface modification of filler particles using coupling agents, using melt processing for preparation of membranes and using hydrophobic material for both organic and inorganic phase are the methods that can be applied to combat defect problems and fabricate the ideal mixed matrix membranes (Aroon et al. 2010).

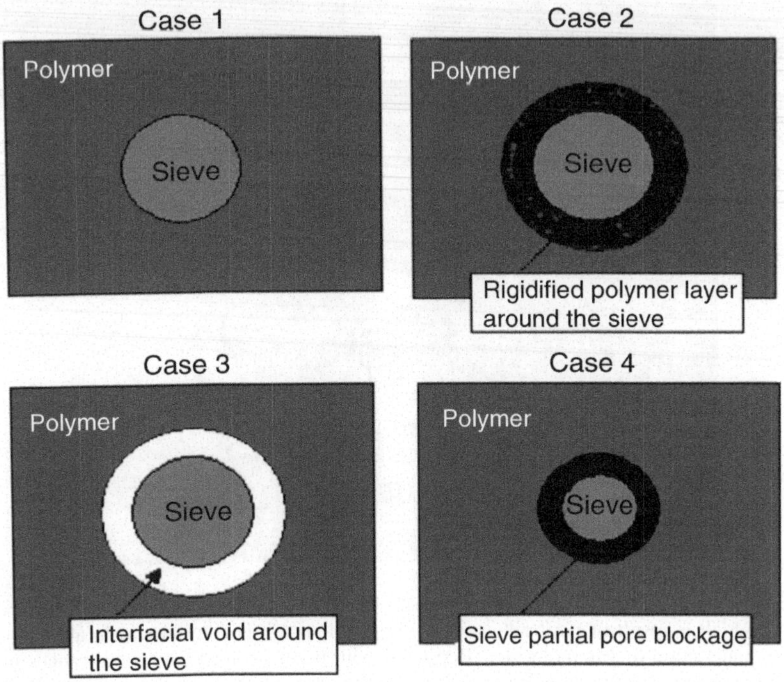

Fig. 4.14 Schematic view of the interfacial defects between filler particles and polymer, Case 1: Ideal mixed matrix membrane, Case 2, 3 and 4: nonideal mixed matrix membrane. (Reprinted with permission of [Polymeric mixed matrix membranes containing zeolites as a filler for gas separation applications: A review, Bastani, D., N. Esmaeili and M. Asadollahi, Elsevier]' from (Bastani et al. 2013))

4.5 Summary and Outlook

An increased CO_2 emission suggests that the need to find a novel technology which can reduce the CO_2 content. Membrane process is known as the novel capture method to separate certain component from a gas stream. In this chapter, after brief introduction on the membrane technology, mixed matrix membranes have been discussed. Different types of filler particles which are used to separate CO_2 have been covered. The application of zeolites, silica, carbon nanotubes, carbon molecular sieve, metal organic frameworks and graphene for capturing CO_2 has also been reviewed. Despite intensive research on developing the mixed matrix membranes, they still cannot satisfy the industrial expectations. The performance of the mixed matrix membranes can be affected by several issues such as agglomeration, non-favorable filler dispersion and poor adhesion between polymer-particle interfaces which make the membrane performance to decrease. To avoid these problems, different strategies can be

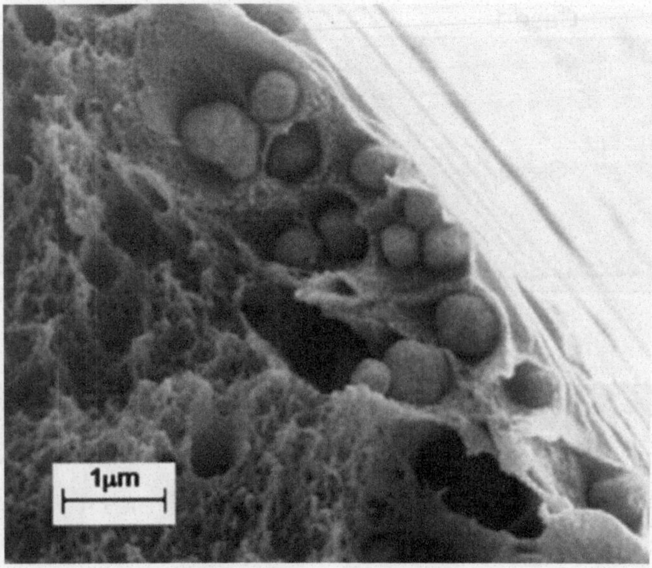

Fig. 4.15 Scanning electron microscope (SEM) of sieves-in-a-cage morphology of the zeolite particles which is result of poor adhesion at the zeolite-polymer interface and causes the non-selective penetration of gas molecule. (Reprinted with permission of [Recent advances of inorganic fillers in mixed matrix membrane for gas separation, Goh, P., A. Ismail, S. Sanip, B. Ng and M. Aziz, Elsevier]' from (Goh et al. 2011))

used like functionalization of the filler surface with CO_2-philic materials. Fabrication of thin film composite mixed matrix membranes, development of mathematical models to describe transport through membrane, using fillers with large aspect ratio and/or in nanoparticle form, testing the membranes under realistic practical conditions and study on new alternate fillers seem to be good solutions to achieve improved mixed matrix membranes separation performance.

References

Adams R, Carson C, Ward J, Tannenbaum R, Koros W (2010) Metal organic framework mixed matrix membranes for gas separations. Microporous Mesoporous Mater 131(1–3):13–20. https://doi.org/10.1016/j.micromeso.2009.11.035

Ahmad N, Leo C, Mohammad A, Ahmad A (2017) Modification of gas selective SAPO zeolites using imidazolium ionic liquid to develop polysulfone mixed matrix membrane for CO2 gas separation. Microporous Mesoporous Mater 244:21–30. https://doi.org/10.1016/j.micromeso.2016.10.001

Ahmadi M, Janakiram S, Dai Z, Ansaloni L, Deng L (2018) Performance of mixed matrix membranes containing porous two-dimensional (2D) and three-dimensional (3D) fillers for CO2 separation: a review. Membranes 8(3):50. https://doi.org/10.3390/membranes8030050

Anjum MW, de Clippel F, Didden J, Khan AL, Couck S, Baron GV, Denayer JF, Sels B, Vankelecom I (2015) Polyimide mixed matrix membranes for CO2 separations using carbon–silica nanocomposite fillers. J Membr Sci 495:121–129. https://doi.org/10.1016/j.memsci.2015.08.006

Arjmandi M, Pakizeh M (2014) Mixed matrix membranes incorporated with cubic-MOF 5 for improved polyetherimide gas separation membranes: theory and experiment. J Ind Eng Chem 20(5):3857–3868. https://doi.org/10.1016/j.jiec.2013.12.091

Aroon M, Ismail A, Matsuura T, Montazer-Rahmati M (2010) Performance studies of mixed matrix membranes for gas separation: a review. Sep Purif Technol 75(3):229–242. https://doi.org/10.1016/j.seppur.2010.08.023

Bae T-H, Long JR (2013) CO 2/N 2 separations with mixed-matrix membranes containing Mg 2 (dobdc) nanocrystals. Energy Environ Sci 6(12):3565–3569. https://doi.org/10.1039/C3EE42394H

Bae TH, Lee JS, Qiu W, Koros WJ, Jones CW, Nair S (2010) A high-performance gas-separation membrane containing submicrometer-sized metal–organic framework crystals. Angew Chem 122(51):10059–10062. https://doi.org/10.1002/ange.201006141

Bastani D, Esmaeili N, Asadollahi M (2013) Polymeric mixed matrix membranes containing zeolites as a filler for gas separation applications: a review. J Ind Eng Chem 19(2):375–393. https://doi.org/10.1016/j.jiec.2012.09.019

Bhaskar A, Banerjee R, Kharul U (2014) ZIF-8@ PBI-BuI composite membranes: elegant effects of PBI structural variations on gas permeation performance. J Mater Chem A 2(32):12962–12967. https://doi.org/10.1039/C4TA00611A

Bryan N, Lasseuguette E, van Dalen M, Permogorov N, Amieiro A, Brandani S, Ferrari M-C (2014) Development of mixed matrix membranes containing zeolites for post-combustion carbon capture. Energy Procedia 63:160–166. https://doi.org/10.1016/j.egypro.2014.11.016

Buck DW II, Alam M, Kim JY (2009) Injectable fillers for facial rejuvenation: a review. J Plast Reconstr Aesthet Surg 62(1):11–18. https://doi.org/10.1016/j.bjps.2008.06.036

Bushell AF, Attfield MP, Mason CR, Budd PM, Yampolskii Y, Starannikova L, Rebrov A, Bazzarelli F, Bernardo P, Jansen JC (2013) Gas permeation parameters of mixed matrix membranes based on the polymer of intrinsic microporosity PIM-1 and the zeolitic imidazolate framework ZIF-8. J Membr Sci 427:48–62. https://doi.org/10.1016/j.memsci.2012.09.035

Cao L, Tao K, Huang A, Kong C, Chen L (2013) A highly permeable mixed matrix membrane containing CAU-1-NH 2 for H 2 and CO 2 separation. Chem Commun 49(76):8513–8515. https://doi.org/10.1039/C3CC44530E

Car A, Stropnik C, Peinemann K-V (2006) Hybrid membrane materials with different metal–organic frameworks (MOFs) for gas separation. Desalination 200(1–3):424–426. https://doi.org/10.1016/j.desal.2006.03.390

Carter D, Tezel F, Kruczek B, Kalipcilar H (2017) Investigation and comparison of mixed matrix membranes composed of polyimide matrimid with ZIF–8, silicalite, and SAPO–34. J Membr Sci 544:35–46. https://doi.org/10.1016/j.memsci.2017.08.068

Cho S, Yu H-R, Choi TH, Jung M-J, Lee Y-S (2018) Surface functionalization and CO2 uptake on carbon molecular sieves: experimental observation and theoretical study. Appl Surf Sci 447:8–14. https://doi.org/10.1016/j.apsusc.2018.03.153

Clarizia G, Algieri C, Regina A, Drioli E (2008) Zeolite-based composite PEEK-WC membranes: gas transport and surface properties. Microporous Mesoporous Mater 115(1–2):67–74. https://doi.org/10.1016/j.micromeso.2008.01.048

Dai Y, Johnson J, Karvan O, Sholl DS, Koros W (2012) Ultem®/ZIF-8 mixed matrix hollow fiber membranes for CO2/N2 separations. J Membr Sci 401:76–82. https://doi.org/10.1016/j.memsci.2012.01.044

Dalane K, Dai Z, Mogseth G, Hillestad M, Deng L (2017) Potential applications of membrane separation for subsea natural gas processing: a review. J Nat Gas Sci Eng 39:101–117. https://doi.org/10.1016/j.jngse.2017.01.023

D'Alessandro DM, Smit B, Long JR (2010) Carbon dioxide capture: prospects for new materials. Angew Chem Int Ed 49(35):6058–6082. https://doi.org/10.1002/anie.201000431

Ding R, Li W (2018) A brief introduction to corrosion protective films and coatings based on graphene and graphene oxide. J Alloys Compd 764:1039–1055. https://doi.org/10.1016/j.jallcom.2018.06.133

Dong G, Hou J, Wang J, Zhang Y, Chen V, Liu J (2016) Enhanced CO2/N2 separation by porous reduced graphene oxide/Pebax mixed matrix membranes. J Membr Sci 520:860–868. https://doi.org/10.1016/j.memsci.2016.08.059

Dorosti F, Omidkhah M, Abedini R (2014) Fabrication and characterization of Matrimid/MIL-53 mixed matrix membrane for CO2/CH4 separation. Chem Eng Res Des 92(11):2439–2448. https://doi.org/10.1016/j.cherd.2014.02.018

Elma M, Yacou C, Wang DK, Smart S, Diniz da Costa JC (2012) Microporous silica based membranes for desalination. Water 4(3):629–649. https://doi.org/10.3390/w4030629

Friess K, Hynek V, Šípek M, Kujawski WM, Vopička O, Zgažar M, Kujawski MW (2011) Permeation and sorption properties of poly (ether-block-amide) membranes filled by two types of zeolites. Sep Purif Technol 80(3):418–427. https://doi.org/10.1016/j.seppur.2011.04.012

Ge L, Zhou W, Rudolph V, Zhu Z (2013) Mixed matrix membranes incorporated with size-reduced cu-BTC for improved gas separation. J Mater Chem A 1(21):6350–6358. https://doi.org/10.1039/C3TA11131H

Ghadimi A, Mohammadi T, Kasiri N (2014) A novel chemical surface modification for the fabrication of PEBA/SiO2 nanocomposite membranes to separate CO2 from syngas and natural gas streams. Ind Eng Chem Res 53(44):17476–17486. https://doi.org/10.1021/ie503216p

Goh P, Ismail A, Sanip S, Ng B, Aziz M (2011) Recent advances of inorganic fillers in mixed matrix membrane for gas separation. Sep Purif Technol 81(3):243–264. https://doi.org/10.1016/j.seppur.2011.07.042

Golemme G, Bruno A, Manes R, Muoio D (2006) Preparation and properties of superglassy polymers—zeolite mixed matrix membranes. Desalination 200(1–3):440–442. https://doi.org/10.1016/j.desal.2006.03.396

Guiver MD, Robertson GP, Dai Y, Bilodeau F, Kang YS, Lee KJ, Jho JY, Won J (2002) Structural characterization and gas-transport properties of brominated matrimid polyimide. J Polym Sci A Polym Chem 40(23):4193–4204. https://doi.org/10.1002/pola.10516

Hu J, Cai H, Ren H, Wei Y, Xu Z, Liu H, Hu Y (2010) Mixed-matrix membrane hollow fibers of Cu3 (BTC) 2 MOF and polyimide for gas separation and adsorption. Ind Eng Chem Res 49(24):12605–12612. https://doi.org/10.1021/ie1014958

Huang Z, Li Y, Wen R, May Teoh M, Kulprathipanja S (2006) Enhanced gas separation properties by using nanostructured PES-zeolite 4A mixed matrix membranes. J Appl Polym Sci 101(6):3800–3805. https://doi.org/10.1002/app.24041

Huang G, Isfahani AP, Muchtar A, Sakurai K, Shrestha BB, Qin D, Yamaguchi D, Sivaniah E, Ghalei B (2018) Pebax/ionic liquid modified graphene oxide mixed matrix membranes for enhanced CO2 capture. J Membr Sci 565:370–379. https://doi.org/10.1016/j.memsci.2018.08.026

Ismail A, Goh P, Sanip S, Aziz M (2009) Transport and separation properties of carbon nanotube-mixed matrix membrane. Sep Purif Technol 70(1):12–26. https://doi.org/10.1016/j.seppur.2009.09.002

Japip S, Wang H, Xiao Y, Chung TS (2014) Highly permeable zeolitic imidazolate framework (ZIF)-71 nano-particles enhanced polyimide membranes for gas separation. J Membr Sci 467:162–174. https://doi.org/10.1016/j.memsci.2014.05.025

Jha P, Way JD (2008) Carbon dioxide selective mixed-matrix membranes formulation and characterization using rubbery substituted polyphosphazene. J Membr Sci 324(1–2):151–161. https://doi.org/10.1016/j.memsci.2008.07.005

Jia M, Feng Y, Qiu J, Zhang X, Yao J (2018) Amine-functionalized MOFs@ GO as filler in mixed matrix membrane for selective CO2 separation. Sep Purif Technol 213:63–69. https://doi.org/10.1016/j.seppur.2018.12.029

Joglekar M, Itta AK, Kumar R, Wenz GB, Mayne J, Williams PJ, Koros WJ (2019) Carbon molecular sieve membranes for CO2/N2 separations: evaluating subambient temperature performance. J Membr Sci 569:1–6. https://doi.org/10.1016/j.memsci.2018.10.003

Jusoh N, Yeong YF, Lau KK, Shariff AM (2017) Enhanced gas separation performance using mixed matrix membranes containing zeolite T and 6FDA-durene polyimide. J Membr Sci 525:175–186. https://doi.org/10.1016/j.memsci.2016.10.044

Karatay E, Kalıpçılar H, Yılmaz L (2010) Preparation and performance assessment of binary and ternary PES-SAPO 34-HMA based gas separation membranes. J Membr Sci 364(1–2):75–81. https://doi.org/10.1016/j.memsci.2010.08.004

Khdhayyer MR, Esposito E, Fuoco A, Monteleone M, Giorno L, Jansen JC, Attfield MP, Budd PM (2017) Mixed matrix membranes based on UiO-66 MOFs in the polymer of intrinsic microporosity PIM-1. Sep Purif Technol 173:304–313. https://doi.org/10.1016/j.seppur.2016.09.036

Konios D, Stylianakis MM, Stratakis E, Kymakis E (2014) Dispersion behaviour of graphene oxide and reduced graphene oxide. J Colloid Interface Sci 430:108–112. https://doi.org/10.1016/j.jcis.2014.05.033

Kulprathipanja S, Neuzil RW, Li NN (1988) Separation of fluids by means of mixed matrix membranes, Google Patents

Li Y, Chung T-S, Cao C, Kulprathipanja S (2005) The effects of polymer chain rigidification, zeolite pore size and pore blockage on polyethersulfone (PES)-zeolite a mixed matrix membranes. J Membr Sci 260(1–2):45–55. https://doi.org/10.1016/j.memsci.2005.03.019

Li T, Pan Y, Peinemann K-V, Lai Z (2013) Carbon dioxide selective mixed matrix composite membrane containing ZIF-7 nano-fillers. J Membr Sci 425:235–242. https://doi.org/10.1016/j.memsci.2012.09.006

Li X, Cheng Y, Zhang H, Wang S, Jiang Z, Guo R, Wu H (2015a) Efficient CO2 capture by functionalized graphene oxide nanosheets as fillers to fabricate multi-permselective mixed matrix membranes. ACS Appl Mater Interfaces 7(9):5528–5537. https://doi.org/10.1021/acsami.5b00106

Li X, Ma L, Zhang H, Wang S, Jiang Z, Guo R, Wu H, Cao X, Yang J, Wang B (2015b) Synergistic effect of combining carbon nanotubes and graphene oxide in mixed matrix membranes for efficient CO2 separation. J Membr Sci 479:1–10. https://doi.org/10.1016/j.memsci.2015.01.014

Lin Y, Xu Y, Loh CH, Wang R (2018) Development of robust fluorinated TiO 2/PVDF composite hollow fiber membrane for CO 2 capture in gas-liquid membrane contactor. Appl Surf Sci 436:670–681. https://doi.org/10.1016/j.apsusc.2017.11.263

Liu C, McCulloch B, Wilson ST, Benin AI, Schott ME (2009) Metal organic framework—polymer mixed matrix membranes, Google Patents

Liu XL, Li YS, Zhu GQ, Ban YJ, Xu LY, Yang WS (2011) An organophilic pervaporation membrane derived from metal–organic framework nanoparticles for efficient recovery of bio-alcohols. Angew Chem 123(45):10824–10827. https://doi.org/10.1002/ange.201104383

Luebke D, Myers C, Pennline H (2006) Hybrid membranes for selective carbon dioxide separation from fuel gas. Energy Fuel 20(5):1906–1913. https://doi.org/10.1021/ef060060b

Mahajan R, Koros WJ (2000) Factors controlling successful formation of mixed-matrix gas separation materials. Ind Eng Chem Res 39(8):2692–2696. https://doi.org/10.1021/ie990799r

Mahajan R, Koros WJ (2002) Mixed matrix membrane materials with glassy polymers. Part 2. Polym Eng Sci 42(7):1432–1441. https://doi.org/10.1002/pen.11042

Moghaddam ZS, Kaykhaii M, Khajeh M, Oveisi AR (2018) Synthesis of UiO-66-OH zirconium metal-organic framework and its application for selective extraction and trace determination of thorium in water samples by spectrophotometry. Spectrochim Acta A Mol Biomol Spectrosc 194:76–82. https://doi.org/10.1016/j.saa.2018.01.010

Moore TT, Koros WJ (2005) Non-ideal effects in organic–inorganic materials for gas separation membranes. J Mol Struct 739(1–3):87–98. https://doi.org/10.1016/j.molstruc.2004.05.043

Nabais AR, Ribeiro RP, Mota JP, Alves VD, Esteves IA, Neves LA (2018) CO2/N2 gas separation using Fe (BTC)-based mixed matrix membranes: a view on the adsorptive and filler properties of metal-organic frameworks. Sep Purif Technol 202:174–184. https://doi.org/10.1016/j.seppur.2018.03.028

Nafisi V, Hägg M-B (2014) Development of dual layer of ZIF-8/PEBAX-2533 mixed matrix membrane for CO2 capture. J Membr Sci 459:244–255. https://doi.org/10.1016/j.memsci.2014.02.002

Noble RD, Gin DL (2011) Perspective on ionic liquids and ionic liquid membranes. J Membr Sci 369(1–2):1–4. https://doi.org/10.1016/j.memsci.2010.11.075

Ordoñez MJC, Balkus KJ Jr, Ferraris JP, Musselman IH (2010) Molecular sieving realized with ZIF-8/Matrimid® mixed-matrix membranes. J Membr Sci 361(1–2):28–37. https://doi.org/10.1016/j.memsci.2010.06.017

Peng F, Lu L, Sun H, Wang Y, Liu J, Jiang Z (2005) Hybrid organic– inorganic membrane: solving the tradeoff between permeability and selectivity. Chem Mater 17(26):6790–6796. https://doi.org/10.1021/cm051890q

Perez EV, Balkus KJ Jr, Ferraris JP, Musselman IH (2009) Mixed-matrix membranes containing MOF-5 for gas separations. J Membr Sci 328(1–2):165–173. https://doi.org/10.1016/j.memsci.2008.12.006

Perez EV, Balkus KJ Jr, Ferraris JP, Musselman IH (2014) Metal-organic polyhedra 18 mixed-matrix membranes for gas separation. J Membr Sci 463:82–93. https://doi.org/10.1016/j.memsci.2014.03.045

Perez EV, Kalaw GJ, Ferraris JP, Balkus KJ, Musselman IH (2017) Amine-functionalized (Al) MIL-53/VTEC™ mixed-matrix membranes for H 2/CO 2 mixture separations at high pressure and high temperature. J Membr Sci 530:201–212. https://doi.org/10.1016/j.memsci.2017.02.003

Rada ZH, Abid HR, Sun H, Shang J, Li J, He Y, Liu S, Wang S (2018) Effects of-NO2 and-NH2 functional groups in mixed-linker Zr-based MOFs on gas adsorption of CO2 and CH4. Prog Nat Sci Mater Int 28(2):160–167. https://doi.org/10.1016/j.pnsc.2018.01.016

Robeson LM (1991) Correlation of separation factor versus permeability for polymeric membranes. J Membr Sci 62(2):165–185. https://doi.org/10.1016/0376-7388(91)80060-J

Sahoo NG, Rana S, Cho JW, Li L, Chan SH (2010) Polymer nanocomposites based on functionalized carbon nanotubes. Prog Polym Sci 35(7):837–867. https://doi.org/10.1016/j.progpolymsci.2010.03.002

Sanaeepur H, Kargari A, Nasernejad B (2014) Aminosilane-functionalization of a nanoporous Y-type zeolite for application in a cellulose acetate based mixed matrix membrane for CO 2 separation. RSC Adv 4(109):63966–63976. https://doi.org/10.1039/C4RA08783F

Sanip S, Ismail A, Goh P, Soga T, Tanemura M, Yasuhiko H (2011) Gas separation properties of functionalized carbon nanotubes mixed matrix membranes. Sep Purif Technol 78(2):208–213. https://doi.org/10.1016/j.seppur.2011.02.003

Seoane B, Coronas J, Gascon I, Benavides ME, Karvan O, Caro J, Kapteijn F, Gascon J (2015) Metal–organic framework based mixed matrix membranes: a solution for highly efficient CO 2 capture? Chem Soc Rev 44(8):2421–2454. https://doi.org/10.1039/C4CS00437J

Shen J, Liu G, Huang K, Li Q, Guan K, Li Y, Jin W (2016) UiO-66-polyether block amide mixed matrix membranes for CO2 separation. J Membr Sci 513:155–165. https://doi.org/10.1016/j.memsci.2016.04.045

Shin JH, Yu HJ, An H, Lee AS, Hwang SS, Lee SY, Lee JS (2019) Rigid double-stranded siloxane-induced high-flux carbon molecular sieve hollow fiber membranes for CO2/CH4 separation. J Membr Sci 570:504–512. https://doi.org/10.1016/j.memsci.2018.10.076

Shu S (2007) Engineering the performance of mixed matrix membranes for gas separations. Georgia Institute of Technology, Atlanta

Song Q, Nataraj S, Roussenova MV, Tan JC, Hughes DJ, Li W, Bourgoin P, Alam MA, Cheetham AK, Al-Muhtaseb SA (2012) Zeolitic imidazolate framework (ZIF-8) based polymer nanocomposite membranes for gas separation. Energy Environ Sci 5(8):8359–8369. https://doi.org/10.1039/C2EE21996D

Strathmann H (2005) Membrane technology and applications (Richard W. Baker, ed). Wiley (2004), Elsevier

Sun H, Wang T, Xu Y, Gao W, Li P, Niu QJ (2017) Fabrication of polyimide and functionalized multi-walled carbon nanotubes mixed matrix membranes by in-situ polymerization for CO2 separation. Sep Purif Technol 177:327–336. https://doi.org/10.1016/j.seppur.2017.01.015

Tantekin-Ersolmaz ŞB, Atalay-Oral Ç, Tatlıer M, Erdem-Şenatalar A, Schoeman B, Sterte J (2000) Effect of zeolite particle size on the performance of polymer–zeolite mixed matrix membranes. J Membr Sci 175(2):285–288. https://doi.org/10.1016/S0376-7388(00)00423-3

Venna SR, Lartey M, Li T, Spore A, Kumar S, Nulwala HB, Luebke DR, Rosi NL, Albenze E (2015) Fabrication of MMMs with improved gas separation properties using externally-functionalized MOF particles. J Mater Chem A 3(9):5014–5022. https://doi.org/10.1039/C4TA05225K

Vinoba M, Bhagiyalakshmi M, Alqaheem Y, Alomair AA, Pérez A, Rana MS (2017) Recent progress of fillers in mixed matrix membranes for CO2 separation: a review. Sep Purif Technol 188:431–450. https://doi.org/10.1016/j.seppur.2017.07.051

Vu DQ, Koros WJ, Miller SJ (2003) Mixed matrix membranes using carbon molecular sieves: I. Preparation and experimental results. J Membr Sci 211(2):311–334. https://doi.org/10.1016/S0376-7388(02)00429-5

Wang H, Holmberg BA, Yan Y (2002) Homogeneous polymer–zeolite nanocomposite membranes by incorporating dispersible template-removed zeolite nanocrystals. J Mater Chem 12(12):3640–3643. https://doi.org/10.1039/B207394C

Wang S, Liu Y, Huang S, Wu H, Li Y, Tian Z, Jiang Z (2014) Pebax–PEG–MWCNT hybrid membranes with enhanced CO2 capture properties. J Membr Sci 460:62–70. https://doi.org/10.1016/j.memsci.2014.02.036

Wang M, Wang Z, Zhao S, Wang J, Wang S (2017) Recent advances on mixed matrix membranes for CO2 separation. Chin J Chem Eng 25(11):1581–1597. https://doi.org/10.1016/j.cjche.2017.07.006

Weng T-H, Tseng H-H, Wey M-Y (2010) Fabrication and characterization of poly (phenylene oxide)/SBA-15/carbon molecule sieve multilayer mixed matrix membrane for gas separation. Int J Hydrog Energy 35(13):6971–6983. https://doi.org/10.1016/j.ijhydene.2010.04.024

Wijenayake SN, Panapitiya NP, Versteeg SH, Nguyen CN, Goel S, Balkus KJ Jr, Musselman IH, Ferraris JP (2013) Surface cross-linking of ZIF-8/polyimide mixed matrix membranes (MMMs) for gas separation. Ind Eng Chem Res 52(21):6991–7001. https://doi.org/10.1021/ie400149e

Wong-Foy AG, Matzger AJ, Yaghi OM (2006) Exceptional H2 saturation uptake in microporous metal– organic frameworks. J Am Chem Soc 128(11):3494–3495. https://doi.org/10.1021/ja058213h

Xin Q, Li Z, Li C, Wang S, Jiang Z, Wu H, Zhang Y, Yang J, Cao X (2015) Enhancing the CO 2 separation performance of composite membranes by the incorporation of amino acid-functionalized graphene oxide. J Mater Chem A 3(12):6629–6641. https://doi.org/10.1039/C5TA00506J

Yang T, Chung T-S (2013a) High performance ZIF-8/PBI nano-composite membranes for high temperature hydrogen separation consisting of carbon monoxide and water vapor. Int J Hydrog Energy 38(1):229–239. https://doi.org/10.1016/j.ijhydene.2012.10.045

Yang T, Chung T-S (2013b) Room-temperature synthesis of ZIF-90 nanocrystals and the derived nano-composite membranes for hydrogen separation. J Mater Chem A 1(19):6081–6090. https://doi.org/10.1039/C3TA10928C

Yang T, Xiao Y, Chung T-S (2011) Poly–/metal-benzimidazole nano-composite membranes for hydrogen purification. Energy Environ Sci 4(10):4171–4180. https://doi.org/10.1039/C1EE01324F

Yong HH, Park HC, Kang YS, Won J, Kim WN (2001) Zeolite-filled polyimide membrane containing 2, 4, 6-triaminopyrimidine. J Membr Sci 188(2):151–163. https://doi.org/10.1016/S0376-7388(00)00659-1

Yu J, Li L, Liu N, Lee R (2013) An approach to prepare defect-free PES/MFI-type zeolite mixed matrix membranes for CO2/N2 separation. J Mater Sci 48(10):3782–3788. https://doi.org/10.1007/s10853-013-7178-z

Zarshenas K, Raisi A, Aroujalian A (2016) Mixed matrix membrane of nano-zeolite NaX/poly (ether-block-amide) for gas separation applications. J Membr Sci 510:270–283. https://doi.org/10.1016/j.memsci.2016.02.059

Zhang Y, Balkus KJ Jr, Musselman IH, Ferraris JP (2008a) Mixed-matrix membranes composed of Matrimid® and mesoporous ZSM-5 nanoparticles. J Membr Sci 325(1):28–39. https://doi.org/10.1016/j.memsci.2008.04.063

Zhang Y, Musselman IH, Ferraris JP, Balkus KJ Jr (2008b) Gas permeability properties of Matrimid® membranes containing the metal-organic framework cu–BPY–HFS. J Membr Sci 313(1–2):170–181. https://doi.org/10.1016/j.memsci.2008.01.005

Zhang C, Dai Y, Johnson JR, Karvan O, Koros WJ (2012) High performance ZIF-8/6FDA-DAM mixed matrix membrane for propylene/propane separations. J Membr Sci 389:34–42. https://doi.org/10.1016/j.memsci.2011.10.003

Zhao Y, Jung BT, Ansaloni L, Ho WW (2014) Multiwalled carbon nanotube mixed matrix membranes containing amines for high pressure CO2/H2 separation. J Membr Sci 459:233–243. https://doi.org/10.1016/j.memsci.2014.02.022

Zhao D, Ren J, Wang Y, Qiu Y, Li H, Hua K, Li X, Ji J, Deng M (2017) High CO2 separation performance of Pebax®/CNTs/GTA mixed matrix membranes. J Membr Sci 521:104–113. https://doi.org/10.1016/j.memsci.2016.08.061

Zhu H, Jie X, Wang L, Liu D, Cao Y (2017) Polydimethylsiloxane/postmodified MIL-53 composite layer coated on asymmetric hollow fiber membrane for improving gas separation performance. J Appl Polym Sci 134(26). https://doi.org/10.1002/app.44999

Zhuang G-L, Tseng H-H, Wey M-Y (2014) Preparation of PPO-silica mixed matrix membranes by in-situ sol–gel method for H2/CO2 separation. Int J Hydrog Energy 39(30):17178–17190. https://doi.org/10.1016/j.ijhydene.2014.08.050

Zornoza B, Esekhile O, Koros WJ, Téllez C, Coronas J (2011a) Hollow silicalite-1 sphere-polymer mixed matrix membranes for gas separation. Sep Purif Technol 77(1):137–145. https://doi.org/10.1016/j.seppur.2010.11.033

Zornoza B, Seoane B, Zamaro JM, Téllez C, Coronas J (2011b) Combination of MOFs and zeolites for mixed-matrix membranes. ChemPhysChem 12(15):2781–2785. https://doi.org/10.1002/cphc.201100583

Chapter 5
Ionic Liquids for Carbon Dioxide Capture

Mohammad Mesbah, Shabnam Pouresmaeil, Sanaz Abouali Galledari, Masumeh Momeni, Shohreh Shahsavari, and Ebrahim Soroush

Abstract The worldwide increasing emission of carbon dioxide and undeniable effects of global warming have raised the necessity to enhance existing CO_2 capture processes. In recent decade, ionic liquids (ILs), which are a group of salts that are liquid at temperatures below 100 °C, has been the subject of many CO_2 capturing research studies by various scholars across the word. These exciting salts have the potential to improve selectivity and sorption capacity of conventional CO_2 scrubbing solvents, CO_2 adsorbents and membranes used for CO_2 separation. In this chapter, we briefly review the advantages and disadvantages of using different ionic liquids and their various modifications for sorption of CO_2. First, their physical characteristics including viscosity, thermal stability and biodegradability in their pure state and after absorption of carbon dioxide will be discussed. Then absorption capacity of pure conventional ionic liquids and their effective parameters such as cationic structure are introduced and compared with Functionalized and promoted ionic liquids. In this section, different functionalization and polymerization methods are reviewed and the challenges in each scheme are briefly discussed. Then the CO_2 adsorption studies, which involved ionic liquids either as adsorbent medium or as enhancing agent, are reviewed. Solid ionic liquids as potential alternative for conventional adsorbents are briefly introduced and afterwards the chapter will be mostly dedicated to the different ionic liquid-supported solid porous materials in

M. Mesbah (✉)
Young Researchers and Elites Club, Science and Research Branch, Islamic Azad University, Tehran, Iran
e-mail: m_mesbah@alum.sharif.edu

S. Pouresmaeil
Biotechnology Group, Faculty of Chemical Engineering, Tarbiat Modares University, Tehran, Iran

S. A. Galledari · M. Momeni
Department of Chemical Engineering, Sahand University of Technology, Tabriz, Iran

S. Shahsavari
Young Researchers and Elite Club, Shiraz Branch, Islamic Azad University, Shiraz, Iran

E. Soroush
Young Researchers and Elites Club, Ahvaz Branch, Islamic Azad University, Ahvaz, Iran

Inamuddin et al. (eds.), *Sustainable Agriculture Reviews 38*, Sustainable Agriculture Reviews 38, https://doi.org/10.1007/978-3-030-29337-6_5

particular metal organic frameworks (MOFs). In this section, the methods of doping ionic liquids on the adsorbent surface, type of porous medium for ionic liquid interaction and the selectivity of resultant ionic liquid- supported material are considered.

Keywords Ionic liquid · CO_2 · Carbon capture · Room temperature ionic liquid · Task specific ionic liquid · Supported ionic liquid membrane · Supported ionic liquid materials · Functionalized adsorbents with ionic liquids

Abbreviations

AC	Active carbon
CIL	Conventional ionic liquids
FIL	Functionalized ionic liquid
FTIR	Fourier-transform infrared spectroscopy
IL	Ionic liquid
MOF	Metal-organic framework
PIL	Polymerized ionic liquid
PMMA	Porous poly(methylmethacrylate)
PTFE	Polytetrafluoroethylene
PVFD	Polyvinylidene Fluoride
SILM	Supported ionic liquid membrane
SMH	Silica microhoneycomb
TSIL	Task-specific ionic liquid

5.1 Introduction

Anthropogenic CO_2 emission and global climate change have been considered as vital issues which can affect on our daily life. To reduce this global emission, researchers have focused on finding new materials which can potentially separate CO_2. The use of innovative absorbents to replace traditional solvents have gained much attention. Todays, ionic liquids (ILs) are known as alternative green solvents for amine solutions. Ionic liquids are ionic compounds which comprise from various cations and anions. These ionic salts are non-volatile, non-flamable and do not crystallize at room temperature (Itoh 2017). Different cations and anions can be tailored to synthesize different type of ionic liquids with desired physico-chemical properties. This flexibility has led researchers to design potential and energy-efficient absorbents for CO_2 capture that have better absorption ability. In this chapter, efforts have been done on introduction of different types of ionic liquids, their properties and also their potential for CO_2 capture.

5.2 Thermophysical Properties of Ionic Liquids

5.2.1 *Viscosity*

It has been widely accepted that the viscosity of ionic liquids would sharply increase after CO_2 absorption. Several solutions are suggested to tackle this problem such as using hybrid solvents like ionic liquid- amine or ionic liquid-water. The demerits of this method are increase in vapour pressure of the solvent, high risk of corrosion and difficult recyclability of ionic liquid, which make this solution speculative. In other words, the hybrid solvents do not enjoy the superiorities of pure ionic liquid anymore. Another solution which has been proposed is using ionic liquids which absorb carbon dioxide physically. This seems to be a practical solution to the aforementioned problem; however, the low capacity of this absorption leads to a high solvent circulation rate. As a result, the feasibility of the industrial uses of ionic liquid will be put the question. Ionic liquids loss, the unendurable viscosity and their high cost, restrict the industrial application of ionic liquids to some extent. It has been reported that ionic liquids absorb acid gases very slowly (Tang et al. 2005c). In fact, the high viscosity leads to limitation in gas diffusion into ionic liquids whereby the rate of absorption could be affected. Consider the case of NH_2-functionalized ionic liquids, where reaching equilibrium takes about 48 h at 303 K, due to the low rate of absorption (Karadas et al. 2010).

The high viscosity of ionic liquids is due to the molecular weight as well as intermolecular interaction (Götz et al. 2016). The viscosity of ionic liquids increases with the size of anion and also through increasing the alkyl group in the cation for ionic liquids with the same size of anions. Increasing alkyl groups causes cations to interact more with each other through van der Waals forces and hence requiring additional energy for molecular motion (Singh et al. 2012) As a result of their high viscosity, ionic liquids have some certain operating problems. There are apparently a few practical approaches for overcoming this issues. First, the structure of ionic liquids can be simulated with the aid of computer programs for design and evaluation of ionic liquids for carbon dioxide absorption; which is a cost-effective way of finding new ionic liquids. Second, The cheaper materials can be used for synthesizing some target ionic liquids. Moreover, adding co-solvents to ionic liquids and increasing operational temperatures are other possible solutions. Each method has its principles and limitations.

Karadas et al. (2010) mixed ionic liquid with suitable organic compounds to reduce the viscosity of ionic liquids. It is asserted that such hybrid solvents lower the amount of required ionic liquid and the solvent cost as a consequence. However, the resulted hybrid solvents have some problems such as volatility, which is a common issue with organic solvents. It is necessary to do further investigations for avoiding such drawbacks. Not all co-solvents are able to be mixed with the ionic liquids; therefore, the selection of solvents faces some limitations.

There are a few researches who study on the gas-liquid mass transfer and hydrodynamics of ionic liquids. As mentioned before, the high viscosity of ionic liquids affects their behaviour. The temperature increase leads to better efficiency of the ionic liquid process and different hydrodynamic behavior at relatively high temperatures (higher than 373 K); due to the lower viscosities (Kaji et al. 2009; Zhang et al. 2013, 2014). However, the range of operating temperature for ionic liquids in the open literature is usually between 288 K and 308 K.

Götz et al. (2016) may be the pioneers in investigating the viscosity of ionic liquids at high temperatures (ranging from 473 K to 573 K). They found that all tested ionic liquids (i.e. [BMIM][Tf_2N], [N_{111}][Tf_2N]) show moderate viscosities above 473 K. According to this study, although the increase in temperature could extremely affect the viscosity, it could reduce the surface tension very slowly. By way of illustration, consider the viscosity of [BMMIM][Tf_2N], which decreased 99% through temperature increase from 293 K to 473 K whereas the surface tension dropped only 23%.

The VFA equation is used to describe the viscosity of ionic liquids in response to temperature:

$$\eta = \eta_0 \exp\left(\frac{\alpha}{T - T_0}\right) \tag{5.1}$$

Where, η is viscosity (c_P), T represents the absolute temperature, and T_O, $\eta_{o,}$ and α are the empirical constants. According to Xu and Zhang (2017) the temperature dependency of ILs thermal conductivity also obeys the VFT equation. Moreover the existence of acetate ion could reduce the viscosity; in contrast, chloride ions increase the viscosity of ionic liquids (Xu and Zhang 2017).

5.2.2 *Thermal Stability*

Researchers have widely investigated the low-temperature ionic liquids regarding their thermal stability (Ngo et al. 2000; Huddleston et al. 2001; Kosmulski et al. 2004; Van Valkenburg et al. 2005). According to these studies, different chemical compounds, which either are added to ionic liquids or are existent in the form of impurities, take ambient moisture as an example, make ionic liquids well suited for industrial applications. Common impurities such as water and chlorides in ionic liquids have trivial impact on their thermal decomposition, but the impact of other additives is not completely clarified. It should be noted that the purification process of low-temperature ionic liquids is difficult; moreover, the least common controlled parameter of thermal stability in literature is the impurity level.

In many studies, both research and review papers, the temperature levels of thermal stability have been repeatedly mentioned, while there is not any adequate experimental details and this is ambiguous for most of the readers who seek lasting stability data. A thermal stability endurance of low-temperature ionic liquids has been studied by Kosmulski et al. (2004) under the existence of air. In some salts, no considerable difference between the atmosphere of oxygen and nitrogen was shown by fast scans in the beginning of the degradation. There are some other salts, the thermal decomposition of which is expedited in the existence of oxygen; however, all aforementioned outcomes are not definitely related to the enduring stability higher than 400 °C.

5.2.3 Biodegradability

The biodegradability properties of ionic liquids are some of the vital aspects of their environmental impacts. Easily biodegradable chemicals experience fast as well as complete mineralization according to organization for economic co-operation and development's guidelines. Primary and ultimate biodegradability potential are seen as week and extensive molecular cleavage, respectively (Petkovic et al. 2011).

The chemical groups including linear alkyl chains, carboxylic acid groups, aldehyde, hydroxyl, amides and esters considerably boost biodegradability. Boethling et al. reviewed this issue which is applicable for the purpose of designing biodegradable ionic liquids (Boethling et al. 2007). Take either pyridinium (Harjani et al. 2009b) or imidazolium (Gathergood et al. 2004) ionic liquids as an example; on the side chain of which, the added ester functional group make cations more biodegradable.

It has been reported that ionic liquids having alkyl chains with four carbon atoms are poorly biodegradable (Docherty et al. 2007; Stolte et al. 2008), even when a sulfonate group or unsaturated groups such as vinyl or alkyl are included (Harjani et al. 2009a). The hydroxylation of the side chain for degradation of [C_4mbpy]Br was drawn up; however, the intermediate chemical product (3-methylpyridine) could not be conclusively identified (Pham et al. 2008). Relative to the imidazolium-based ionic liquids, the corresponding pyridinium ones, have higher potential of biodegradability (Grabińska-Sota and Kalka 2003). Another degradation pathway of [$C_n$mbpy]Br (n = 4, 6 or 8) has recently been proposed by Docherty et al. (2010), whereby alkyl side chain become unsaturated and the aromatic ring is hydroxylated; consequently, the procedure established by Pham et al. is being opposed (2008).

According to researches, lots of degradation pathways are practical and their efficiency is a function of both microbial metabolic capacities and their community. Therefore, it is necessary to consider "regional" impact to come to a conclusion. It is reported that some phosphorium ionic liquids are tolerable for microbial attack. In this regard, take trihexylphosphine- and Tricyclohexylphosphine-derived cations as the typical examples. These ionic liquids have different ester side chains; therefore, they combine with various anions (Atefi et al. 2009). Nevertheless, it has been shown that the cholinium alkanoates and cholinium cation are highly and readily biodegradable compounds, respectively (Petkovic et al. 2010).

The knowledge about the area of ionic liquid applications and disposal methods, their contamination ranges and bioaccumulation factors in the environment and their fate is limited and this makes the designing of more reliable bioassays difficult and complex, which are needed to demonstrate ionic liquids' biodegradability potential.

Behaviour of ionic liquids in soil has not been discussed sufficiently; therefore, the studies focusing on this issue are in increasing demand, particularly the high persistent imidazolium family and ecotoxicological dangers (Studzińska et al. 2009). Studies have revealed that the desorption and sorption of ionic liquids in soil is affected by the ionic interactions (Stepnowski et al. 2007), and their lipophilicity, the length of the side chains, and Coulombic interactions are established by their mobility (Mrozik et al. 2009). Nevertheless, filamentous fungi apparently have a pivotal role in biodegradation, mostly owing to the fact that there are various fungal strains living usually in soil and they can bear high concentrations of ionic liquids (Petkovic et al. 2010). Most of the studies come to the same conclusion that the cations and anions experience totally unique fate mechanism and they also pursue their certain degradation pathways; therefore, in the design of ionic liquids, both cation and anion have a fundamental role (Petkovic et al. 2011).

5.3 Pure Ionic Liquids for CO_2 Capture

5.3.1 Conventional Ionic Liquids

Room temperature ionic liquids has been used for CO_2 capture since 1999. It has been reported that the solubility of CO_2 in 1-butyl-3-methyl- imidazolium hexafluorophosphate ([bmim][PF_6]) ionic liquid is high, to the extent that CO_2 solubility of 0.72 mole fraction was achieved at 93 bar and 40 °C. According to the results, when high amount of CO_2 was dissolved in the liquid phase (i.e. ionic liquid), there was only a negligible amount of ionic liquid in the gas phase (Blanchard et al. 1999, 2001). This fact displays high potential of ionic liquids in CO_2 capturing processes. Similar results were achieved for the CO_2 solubility of some common imidazolium-based ionic liquids (Chen et al. 2006); in addition to determination of CO_2 solubility in eight hydroxyl ammonium based ionic liquids (Yuan et al. 2007). It has been discovered that the cations of ionic liquids holding longer alkyl side chain on their imidazolium ring, increase the solubility of CO_2 (Baltus et al. 2004; Zhang et al. 2005; Chen et al. 2006); however, alkyl chain cations typically have trivial impact on CO_2 solubility (Supasitmongkol and Styring 2010). In contrast, the impact of anions of ionic liquids is much higher than their cations on CO_2 solubility (Hasib-ur-Rahman et al. 2010). According to observational evidence, the affinity of CO_2 molecules for anions is higher than their affinity for cations (Anthony et al. 2005). As an instant, the solubility of CO_2 in ionic liquids whose has [bmim]$^+$ cation and different anions is in the order of: [methide]$^-$ > [Tf_2N]$^-$ > [CF_3SO_3]$^-$ > [PF_6]$^-$ > [BF_4]$^-$ > [D CA]$^-$ > [NO_3]$^-$ (Cadena et al. 2004). An increase in the amount of fluorine in anions makes CO_2 more soluble because of high coulombic interactions, which are in charge of liquid organization (Zhang et al. 2008).

Figures 5.1 and 5.2 depict some of the common ionic liquid anions and cations . Table 5.1 shows gas absorption properties of different ionic liquids for gas separation.

Both selectivity and solubility of various gases have paramount importance in the process of gas separation through typical ionic liquids. Flue gas, syngas, and natural gas contain gases such as CH_4, C_2H_6, C_2H_4, NO, CO, N_2, H_2 and O_2. except for CO_2. The selectivity of CO_2 in conventional ionic liquids (CILs) is higher than the other gases (Anthony et al. 2002; Anderson et al. 2007; Bara et al. 2009a; Zhao et al. 2011). The selectivity of ionic liquids toward CO_2 is higher than that of to

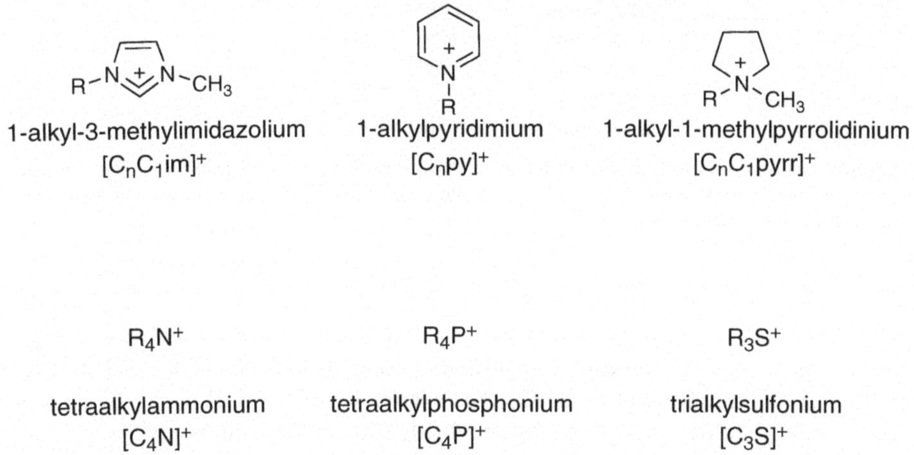

Fig. 5.1 Some cations commonly used for ionic liquids that provide sufficiently strong hydrogen bond donor and causes increasing in the solubility of CO_2, Reprinted with permission of [Ionic liquids: a brief history, Welton, T., Springer]' from (Welton 2018)

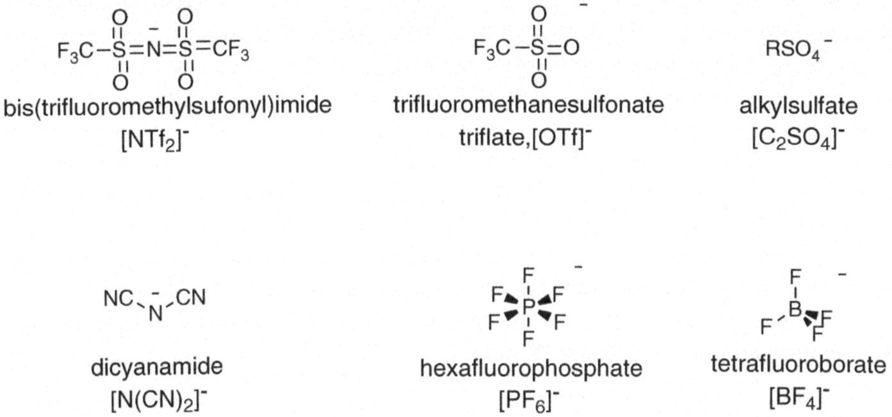

Fig. 5.2 Some anions which can be used for ionic liquids that provide a suitably strong hydrogen bond acceptor. (Reprinted with permission of [Ionic liquids: a brief history, Welton, T., Springer]' from (Welton 2018))

Table 5.1 Gas absorption properties of different ionic liquids

		Gas absorption capacity (mol/L)			
Ionic liquid	Absorption type	CO_2	O_2	N_2	H_2
[C_4mim][Tf_2N]	Physisorption	0.0780	–	–	0.0020
[C_4mim][PF_6]	Physisorption	0.0756	0.0039	–	0.0024
[C_4mim][BF_4]	Physisorption	0.0642	0.0032	0.0027	0.0021
[C_8mim][Tf_2N]	Physisorption	0.1069	–	–	–
[C_2mim][eFAP]	Physisorption	0.0864	–	–	–
[N_{1114}][Tf_2N]	Chemisorption	0.0815	–	–	0.0028
[C_4mim][AC]	Chemisorption	1.579	–	–	–
[P_{66614}][pro]	Chemisorption	0.7613	–	–	–
[P_{66614}][ala]	Chemisorption	0.9859	–	–	–
[P_{66614}][Gly]	Chemisorption	0.7156	–	–	–
[MTBDH][TFE]	Chemisorption	2.326	–	–	–
[MTBDH][Im]	Chemisorption	2.273	–	–	–

Reprinted with permission of [Combination of ionic liquids with membrane technology: A new approach for CO_2 separation, Dai, Z., R. D. Noble, D. L. Gin, X. Zhang and L. Deng, Elsevier]' from (Dai et al. 2016)

other light hydrocarbons as an example molecular simulation shows high interaction energies and high molecular quadrupole moments between CO_2 and [MDEA][Cl] ionic liquid (Zhang et al. 2012). The selectivity of H_2S/CH_4 and CO_2/CH_4 in four ionic liquids including Nmethyl-2-hydroxyethylammonium propionate (2mHEAPr), [C_4mim][CH_3SO_3], [C_4mim][NTf_2] and [P_{66614}] [NTf_2] have recently been assessed. It was stated that the selectivity of sour gas/CH_4 is associated with ionic liquids' polarity and this was explained by the parameter of Kamlet–Taft b which is applicable for choosing a highly selective ionic liquid (Carvalho and Coutinho 2011). Aforementioned aspects of ionic liquids show their potential in the process of gas separation. Moreover, the simple process of solvent regeneration through stripping with inert gas, and pressure swing or thermal treatment is the other merit of ionic liquids as energy consumption in this processes is considerably low. However, conventional ionic liquids cannot vie with state-of-the-art physical solvents; for example, polyethylene glycol of dimethylethers. The reason for this is that CO_2 capacity of conventional ionic liquids is low even when techniques such as selecting proper anions and fluorination of cautions are applied to enhance the CO_2 solubility (Heintz et al. 2009).In particular, CO_2 absorption capacity is lower in the lower pressures. This explains why in natural gas or coal based power plants, CO_2 capturing from the flue gas through typical ionic liquids is not favourable. In order to deal with these kinds of disadvantages, chemisorption or both chemisorption and physisorption of task-specific ionic liquids could be cultivated.

5.3.2 *Task-Specific Ionic Liquids*

Task-specific ionic liquids (FTIR) initiate proper moieties such as amine into ordinary ionic liquids and as a consequence, the absorption capacity of CO_2 could be enhanced. [NH_2p-bim][BF_4] ionic liquid which consists of an imidazolium ion was synthesized by Bates (Bates et al. 2002). In this case the cation is covalently bounded to a primary amine moiety. The reaction mechanism between the task-specific ionic liquids and CO_2 is also proposed and that was parallel to the reaction of CO_2 and organic amins. The capacity of CO_2 in [NH_2pbim][BF_4] approximately reaches the maximum theoretical value of 0.5 mol CO_2/mol ionic liquid, which is so-called '1:2 mechanism' and this is like alkanolamine. The low volatility of task-specific ionic liquids makes it possible to regenerate these ionic liquids easily by heating at the temperature of 80–100 °C under vacuum at several ocassions. This work has made it possible to design and synthesize ionic liquids with new approaches. For CO_2 capturing, a new kind of anion-tethered task-specific ionic liquids, which is called tetrabutylphosphonium amino acid [$P(C_4)_4$][AA], was made. Within synthesis process, tetrabutylphosphonium hydroxide [$P(C_4)_4$][OH] reacts with amino acids such as L-lysine, L-serine, L-b-alanine, L-alanine and glycine and the results show that CO_2 capacities are similar to those of achieved by [NH_2p-bim] [BF_4] (Zhang et al. 2006). Afterwards, prolinate [P_{66614}][Pro] and methioninate [P_{66614}][Met] _ two rihexy(tetradecyl) phosphonium ionic liquids_ were made and their CO_2 absorption were in the stoichiometry of 1:1, which was more than the absorption previously achieved by either aqueous amine absorbents or cation-functionalized ionic liquid. The 1:1 stoichiometry is confirmed by ab initio calculations and Fourier-transform infrared spectroscopy (FTIR) spectrum (Fig. 5.3) (Gurkan et al. 2010).

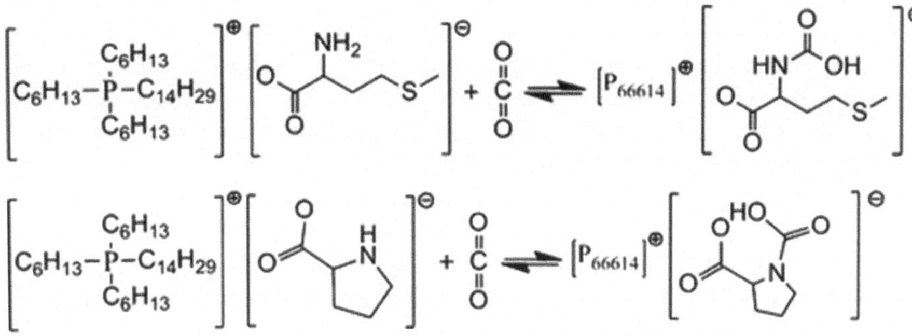

Fig. 5.3 Schematics of CO_2 reaction with amino acid based Ionic liquids, [P666614][Met] (top) and [P66614][Pro] (bottom). ionic liquids are synthesized from reaction of [P66614][OH] with the corresponding amino acids. [P66614][Met] and [P66614][Pro] react with CO_2 in 1:1 stoichiometry which is a more favourable stoichiometry than one CO_2 for two amines (1:2) that accurse in reaction between CO_2 and amines. (Reprinted with permission of [Equimolar CO_2 absorption by anion-functionalized ionic liquids, Gurkan, B. E., J. C. de la Fuente, E. M. Mindrup, L. E. Ficke, B. F. Goodrich, E. A. Price, W. F. Schneider and J. F. Brennecke, American Chemical Society]' from (Gurkan et al. 2010) Society)

Later, some amino-functionalized anion-tethered ionic liquids were made, including isoleucinate [P_{66614}][Ile], leucinate [P_{66614}][Leu], valinate [P_{66614}][Val], sarcosinate [P_{66614}][Sar], alanate [P_{66614}][Ala], and trihexyl(tetradecyl) phosphonium glycinate [P_{66614}][Gly] and almost all of them pursue the mechanism of 1:1 (Goodrich et al. 2010). Although the viscosity of ionic liquids drastically increases through the CO_2 chemical absorption, this issue may be alleviated by lowering the amount of available hydrogen atoms on the anions. The other anion tethered ionic liquids which had been functionalized with amins, including threoninate [P_{66614}][Thr], taurinate [P_{66614}][Tau], prolinate [P_{66614}][Pro], methioninate [P_{66614}][Met], lysinate [P_{66614}][Lys], glutaminate [P_{6661}4][Gln], and trihexyl(tetradecyl) phosphonium asparaginate [P_{66614}][Asn], were also synthesized for CO_2 separation (Goodrich et al. 2011).

Unlike other amino acid based ionic liquids which have a considerable increase in viscosity, typically two order of magnitudes upon CO_2 absorption, [P_{66614}][Pro] after CO_2 absorption shows a relatively small rise, only by a factor of 2 (Soutullo et al. 2007; Gutowski and Maginn 2008). The ring structure of [P_{66614}][Pro] is responsible for this specific behavior because of the fact that the ring structure restricts both availability and quantity of hydrogen atoms engaging in the network of hydrogen bonds. According to the studies about CO_2 kinetics in aminofunctionalized ionic liquids, the absorption of CO_2 is extremely improved comparing with the physical absorption in [bmim][BF_4] (Sánchez et al. 2011).

Both cation and anion with alkaline groups were functionalized through 20 natural amino acids whereby dual aminofunctionalized phosphonium ionic liquids ([aP_{4443}][AA]) were made to boost the CO_2 absorption (Zhang et al. 2009). According to the results of CO_2 chemical absorption through [aP_{4443}][Val], [a_{P4443}][Ala], [aP_{4443}][leu], and [aP_{4443}[Gly], the ratio of which during 80 min is approximately 1 mol CO_2/mol ionic liquid, and this is in concurrent with the suggested absorption mechanism (Bates et al. 2002; Zhang et al. 2006). [aemmim][Tau] which is another dual amino ionic liquid with taurine anion and imidazolium cation that have been functionalized with amino acid was also synthesized. The results showed that, at 303.15 K and about 1 bar, the capacity of absorbing CO_2 for this ionic liquid is 0.9 mol CO_2/mol ionic liquid. As dissolved CO_2 can be readily released at high temperature and vacuum condition, [aemmim][Tau] can be reused to the extent that there was not considerable loss after six recycles (Xue et al. 2011); however, these dual aminofunctionalized phosphonium ionic liquids also possess the disadvantage of high viscosity (The viscosity of [aP_{4443}][AA] is 758–1985 mPa.s at 298 K, which is similar to other task-specific ionic liquids. This is partly due to the H-O bonds formed between either anion and cations or anions themselves and this is confirmed by molecular simulation (Liu et al. 2009).

The CO_2 capacity increases considerably through task-specific ionic liquids holding amine groups; however, in comparison with organic solvents, it has obvious downsides apart from the high viscosity which is lower CO_2 gravimetric capacity to the extent that it is confined to almost 10% even when the partial pressure of CO_2 is high. This issue could be explained by the ionic liquids' characterizations of high molecular weight and low mole absorption capacities. Anion-functionalized protic ionic liquids were described as being suitable for both the reversible and fast CO_2

capturing; moreover, they show great ability in CO_2 capturing at atmospheric pressure. For instance, each mole of $[MTBDH^+]_2[HFPD^{2-}]$ ionic liquid can absorb at least 2 mole of CO_2 and this is equal to the gravimetric capacity of superior to 16% (Wang et al. 2010).

TEGO IL K5 is a task-specific ionic liquid that physically absorb CO_2 and has about 15 ethylene oxide units. At temperatures near ambient, the ether functional group has been described as CO_2 selective absorber. The multi-ether functional groups in TEGO IL K5 motivated to use it for CO_2 and H_2S removal from dry fuel gas. At the CO_2 partial pressure of 30 bar and a temperature range of 300–500 K, the solubility of carbon dioxide in this ionic liquid ranges from 0.27 to 0.62 mole fraction (Heintz et al. 2009). [(ETO)$_2$IM[Tf$_2$N] is another ionic liquid functionalized with ether, which has been compared with conventional ionic liquids, [dmim][MP], [bmim][SCN], and [bmim][BF$_4$], at pressure $\leq$ 300 bar and temperature $\leq$ 373 K. These ionic liquids have a lower CO_2 solubility of 67–123 g CO_2/kg comparing with 198 g CO_2/kg poly(ethylene glycol) dimethyl ether at 313 K and 40 bar (Revelli et al. 2010). However, poly(ethylene glycol) dimethyl ether dissolves heavy hydrocarbons which is considered an important deficiency. This is while ionic liquids are generally immiscible with hydrocarbons (Revelli et al. 2010) and can endure higher temperatures (Heintz et al. 2009).

It has been reported that (p-vinylbenzyl) trimethylammonium hexafluorophosphate monomer, [VBTMA][PF$_6$], has a high CO_2 solubility with 0.47 g CO_2/g ionic liquid at temperature of almost 298 K and pressure of atmosphere. This is a higher capacity, comparing with the pyridinium- or imidazolium-based ionic liquids. For instance, under similar conditions, the [C$_2$mim][ES] has a capacity of merely 0.017 g CO_2/g ionic liquid (Supasitmongkol and Styring 2010). Therefore, in order to increase the potential of ionic liquids for competing with conventional solvents, more research is required for enhancing their gravimetric capacities.

5.3.3 *Polymerized Ionic Liquids in CO_2 Capture*

Leaching the liquid across membrane pores at the time when the pressure drops within the matrix medium exceeds the liquid stabilizing forces is one of the most important deficiencies of supported ionic liquid membranes (SILMs). For overcoming this problem and reduce the absorption and desorption Tang et al. (2005a) made a polymeric form of ionic liquid which increased the sorption of CO_2 compared to conventional ionic liquids. The results revealed that the sorption capacity of CO_2 for tetraalkylammonium-based ILs such as P[VBTMA][BF$_4$], is 7.6 times greater than those of conventional ionic liquid like [C$_4$mim][BF$_4$]. These polymers absorb and desorb carbon dioxide much faster than conventional ionic liquids. In addition, they can selectively absorb CO_2 in N_2/CO_2 mixed gas and at 78.97 kPa and 22 °C do not absorb O_2 or N_2 (Tang et al. 2005a, d). The results also showed that by grafting polyethylene glycol (PEG) on to ionic polymers like P[VBTMA][BF$_4$], CO_2-selective membranes with high mechanical, chemical and thermal stability can be prepared (Hu et al. 2006). Bara et al. (2007) showed high permeability, solubility, and diffusion of CO_2, CH_4, and N_2 in two types of poly-conventional ionic liquids. The results indicated completely reversible sorption/desorption and proved these poly-ionic liquids are very potential to be used as membrane materials and sorbents for capturing CO_2. Tang et al. (Tang et al. 2005a, b, c; Blasig et al. 2007; Tang et al. 2009) investigated the solubility of CO_2 in an ammonium-based ionic liquid and other polymer types, probing the structural properties on the sorption of CO_2. The results indicated that the polymerized ionic liquids with different cations have CO_2 sorption capacities of which increases in the following sequence: imidazolium< phosphnium< pyridinium< ammonium; and the polymerized ionic liquids with different anions increase in the following order: $Tf_2N < PF_6 < BF_4^-$. P[MATMA][BF_4] with a polymethylmethacrylate basis have lower CO_2 absorption capacity than P[VBTMA][BF_4] with a polystyrene basis. The capacity for CO_2 sorption is decreased by Long alkyl-based substituents on the cross-linking and cation. In addition, the results indicated that the process of CO_2 sorption involves dissolving in the polymeric medium and Langmuir sorption in micro-voids.

Bara et al. (2007, 2008a) bounded conventional ionic liquid monomers to the substituents of nitrile-terminated alkyl or oligo(ethylene glycol) and these polymeric membranes enhanced separation factors by 50% for CO_2/N_2 and CO_2/CH_4 separation in comparison with *n*-alkyl substituents that have similar CO_2 permeability but comparable length. Much higher permeability was observed for OEG-functionalized poly- conventional ionic liquids comparing with those that have C_nCN functional groups. They integrated poly-conventional ionic liquid and conventional ionic liquid as a composite membrane for gas separation and absorption of CO_2. The results showed that comparing with the similar poly-conventional ionic liquid membrane which lacks any free ion pairs by incorporating a sufficient quantity of conventional ionic liquid and introduction of free ion pairs in the polymeric matrix, the CO_2/N_2 selectivity and CO_2 permeability increases up 25%–33% and to 300%–600%, respectively (Bara et al. 2008b).

At ambient temperature the polymeric ionic liquids are very viscous and have the capability to be easily employed as membrane materials; therefore, higher CO_2 sorption capacity will be provided in comparison with the small molecular ionic liquids, if the methods of combing supported ionic liquid membranes and the polymerized ionic liquids are applied. The application of polymerized-ionic liquids as membrane materials for absorption of CO_2 have been summarized by Bara et al. (2009a, b). The authors concluded that such membranes are an appropriate option for absorption of carbon dioxide.

Supported ionic liquid membranes have been known as potential materials for the selective separation of different species such as CO_2 in gas mixtures. In such materials, ionic liquid is immobilized inside the pores of the porous polymeric or inorganic membrane. The high viscosity of ionic liquids can effectively prevent the possible to move through the pores, when the membrane is exposured to high pressure. The properties like high ion conductivity, high solvent power, their negligible vapor pressure and their high affinity with CO_2 make ionic liquids as a potential liquid to be used in supported liquid membranes (Lozano et al. 2011). Hanioka et al.

(2008) prepared supported liquid membrane made of polytetrafluoroethylene (PTFE) membrane and ionic liquid which have an amine moity. Three types of ionic liquid including [C_3NH_2mim] [Tf_2N], [C_4mim][Tf_2N], and [C_3NH_2mim][CF_3SO_3] were utilized to fabricate supported ionic liquid membranes. They reported that amine moity facilitates the transport of CO_2 through the membrane and the prepared supported ionic liquid membranes showed high selevtivity and high stability for CO_2 separation. Myers et al. (2008) investigated CO_2 separation from H_2 using facilitated transport supported ionic liquid membrane. They synthesized 1-(3-aminopropyl)-3-methylimidazolium bis(trifluoromethylsulfonyl)imide to used for fabrication of the supported liquid membrane. They measured the separation performance of the prepared membrane with a mixture of 20 mol% CO_2 and 20 mol% H_2 balanced with argon. The results indicated that the CO_2/H_2 selectivity was higher than 15. The CO_2 permeability was high about 1000 Barrer (1 Barrer = 10^{-6} cm^3(STP).cm/cm^2.s.cmHg) at temperature of 250 °C. Liu et al. (2016) fabricated the liquid supported membrane using ionic liquids [BMIM][NTf_2], [BMIM][TfO] [BMIM][AC], and [BMIM][DCA]. These ionic liquids were fixed into Polyvinylidene Fluoride (PVDF) membrane and the performance of these membranes were investigated using CO_2, H_2 and N_2 gas permeation test. The best separation performance for CO_2/H_2 separation achieved when [BMIM][TfO] was fixed in polyvinylidene fluoride. CO_2 permeability of this supported ionic liquid membrane was enough high about 1966 Barrer and CO_2/H_2 selectivity was 16.2 at 40 °C. In case of CO_2/N_2 separation, [BMIM][AC] supported ionic liquid membranes indicated better performance than others. The CO_2 permeability of [BMIM][AC] supported ionic liquid membrane was about 520 Barrer and its CO_2/N_2 selectivity was 21.8 at temperature of 30 °C. They also concluded that gas separation performance increases with the increase in temperature. Several novel works have been done to improve the gas separation performance of supported ionic liquid membraness. Abdelrahim et al. (2017) have recently prepared biomimetic supported ionic liquid membranes using the thermophilic Sulfurihydrogenibium yellowstonense carbonic anhydrase SspCA enzyme to enhance the CO_2 transport at elevated temperature. The membrane showed CO_2 permeability of 733.73 Barrer and CO_2/N_2 selectivity of 35.6. In addition to supported ionic liquid membranes, polymerized ionic liquid (PIL) membrane, polymer/ionic liquid gel membrane and membrane contactor with ionic liquid solvent are the other types of combining ionic liquids with membrane technology to capture CO_2 (Dai et al. 2016).

5.4 CO_2 Capture with Ionic Liquids Functionalized Solvents

The solubility of CO_2 in conventional ionic liquids is insufficient; therefore, through functionalizing with an appropriate moiety such as amine, the capacity of CO_2 absorption can be enhanced (Carvalho et al. 2009). There are two general approaches for creating functionalized ionic liquids (FILs), attaching an ion to the functional group and functionalizing the anion through a -NH_2 alkaline group like [NH_2p-bim][BF_4].

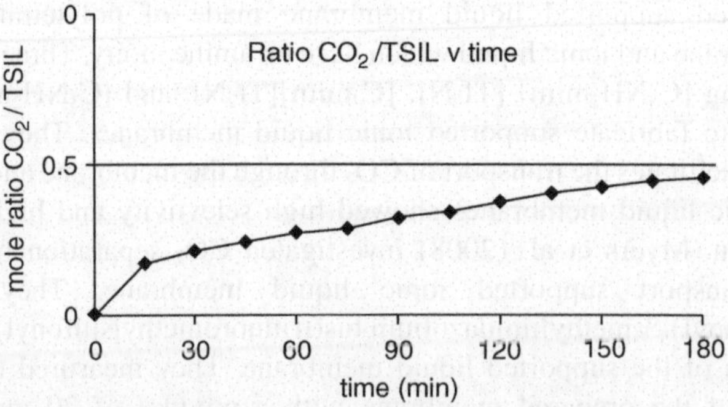

Fig. 5.4 The molar ratio of CO_2/TSIL vs. time which approaches the theoretical maximum of 0.5 mol CO_2/mol solvent. (Reprinted with permission of [CO_2 capture by a task-specific ionic liquid, Bates, E. D., R. D. Mayton, I. Ntai and J. H. Davis, American Chemical Society]' from (Bates et al. 2002))

A solvent developed by Bates et al. (2002), [NH_2p-bim][BF_4], consists of an imidazolium ion which is covalently bonded to a primary amine moiety. The reaction mechanism, which has been proposed for interaction of functionalized ionic liquid and CO_2 was analogous with the reaction between CO_2 and conventional organic amine. Figure 5.4 shows that, similar to the capacity of conventional alkanolamines, the capacity of [NH_2p-bim][BF_4] approaches the theoretical maximum of 0.5 mol CO_2/mol solvent . The absorption process of CO_2 is reversible; in this regard, CO_2 desorbs from ionic liquid through heating under vacuum to temperatures between 80 and 100 °C for a couple of hours. With no considerable efficiency loss, the regenerated ionic liquid can be used for five cycles of CO_2 absorption.

5.4.1 Functionalizing the Anion of Ionic Liquid with Alkaline Group-NH₂

Zhang et al. (2006) used alkaline group-NH_2 to functionalize the anion, and made tetrabutylphosphonium amino acid ([$P(C_4)_4$][AA]) through the reaction of amino acids including l-lysine, l-serine, l-b-alanine, l-alanine, and glycine with tetrabutylphosphonium hydroxide [$P(C_4)_4$][OH]. This ionic liquid have a carbon dioxide absorption capacity of 0.5 mol CO_2/mol ionic liquid, similar to the capacity of [NH_2p-bim][BF_4]. The absorption of CO_2 into amino-functionalized ionic liquid; for example, 1-(3-aminopropyl)-3-methylimidazolium tetrafluoroborate [APmim][BF_4], was investigated kinetically by Sánchez et al. (2011) at the temperature of 303 and 333 K, and concentrations of 45–253 mol.m^{-3}. Depending on the concentration of [APmim][BF_4], the experimental enhancement factors based on the CO_2

absorbed fluxes ranges between 1.2 and 2.3. Han et al. (Li et al. 2008) studied the basicity of ionic liquid urea/choline chloride and 1,1,3,3-tetramethylguanidinium perchlorate [TMG][ClO_4] and 1-aminoethyl-3-methyl imidazolium tetrafluoroborate [AEMIM][BF_4]. They concluded that the basicity can be significantly reduced by adding CO_2 to ionic liquid at ambient pressure and after eliminating the CO_2 via bubbling N_2 across the solvents, the basicity of ionic liquids was recovered. This can aid to comprehend the mechanism of reaction between CO_2 and ionic liquid.

(3-aminopropyl) tributylphosphonium hydroxide ([aP_{4443}]-[OH]) was neutralized by 20 natural amino acids and consequently, a dual amino-functional ionic liquid; i.e. (3-aminopropyl)tributylphosphonium amino acid ([aP_{4443}][AA]), was achieved. (3-aminopropyl) tributyl phosphonium bromide ([aP_{4443}]Br) along with anion-exchange resin were used to generate [aP_{4443}][OH]. The CO_2 absorption of [aP_{4443}][AA] was studied after it was supported on porous SiO2. It was obtained that, in 80 min, CO_2 chemical absorption ratio through [aP_{4443}][Val], [aP_{4443}][Ala], [aP_{4443}][leu], and [aP_{4443}[Gly], [aP_{4443}][leu] was approximately 1 molCO_2/mol ionic liquid and this result was concurrent with suggested absorption mechanism by Ref. (Bates et al. 2002; Zhang et al. 2006). The capacity of CO_2 absorption and consequently the ionic liquid rate go down, but it is possible to recover ionic liquid for many times while there is no significant waste after 5 cycles.

The synthesis of a novel dual amino ionic liquid [aemmim][Tau] through amino-functionalized imidazolium cation and taurine anion were carried out by Mu et al. (Xue et al. 2011). This ionic liquid achieves the CO2 absorption capacity of 0.9 mol CO_2/mol ionic liquid at the pressure of 0.1 MPa and the temperature of 303.15 K. The results showed that under vacuum condition or high temperatures, CO_2 was easily desorbed from ionic liquid; therefore, [aemmim][Tau] can be decreased when no substantial waste was detected in absorption capacity after 6 recycles.

5.4.2 *Attach the Anion of Ionic Liquid with Functional Group*

An intermolecular proton transfer was found by Gurkan et al. (2010). They hindered the process of intermolecular proton transfer through one larger phosphonium cations and eventually, [P_{66614}][Met] and [P_{66614}][Pro] were made. Different reaction mechanisms between CO_2 and these ionic liquids are suggested. In these ionic liquids, CO_2 reaches the approximate maximum capacity of 1.0 mol CO_2/mol ionic liquid, which has been described by Fig. 5.5. Subsequently, six of amine-functionalized anion-bonded ionic liquids were made, including isoleucinate ([P_{66614}][Ile]), leucinate ([P_{66614}][Leu]), valinate ([P_{66614}][Val]), sarcosinate ([P_{66614}] [Sar]), alanate ([P_{66614}][Ala]), and trihexyl(tetradecyl)phosphonium glycinate ([P_{66614}] [Gly]); then, it was displayed that the capacity of four of them was superior to 0.5 mol CO_2/mol ionic liquid at $P_{CO2} < 0.1$ MPa. This explains why mechanism of 1:1 whereby CO_2 reacts with one mole ionic liquid to produce the carbamic acid

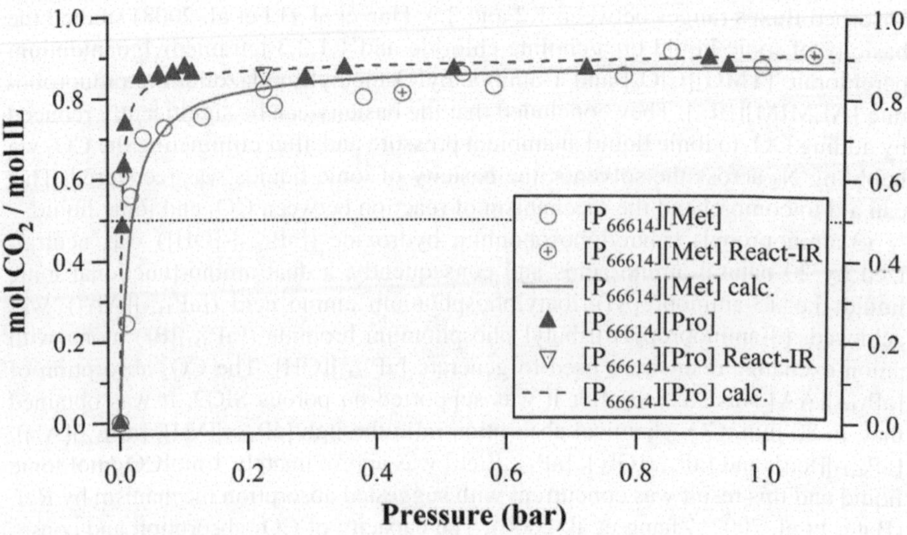

Fig. 5.5 Absorption of CO_2 by [P66614][Pro] and [P66614][Met] ionic liquids at 22 °C in which absorption of CO_2 reached the approximate maximum capacity of 1.0 mol CO_2/mol IL by increasing in pressure. The lines represent the calculated value and are based on Langmuir model. (Reprinted with permission of [Equimolar CO_2 absorption by anion-functionalized ionic liquids, Gurkan, B. E., J. C. de la Fuente, E. M. Mindrup, L. E. Ficke, B. F. Goodrich, E. A. Price, W. F. Schneider and J. F. Brennecke, American Chemical Society]' from (Gurkan et al. 2010))

is more probable than further reaction with another ionic liquid (the 1:2 mechanism) to make a carbamate. Although chemically absorbed CO_2 leads to the drastic increase in the viscosity of ionic liquid, this issue could be alleviated moderately through lowering the number of available hydrogen atoms on the anion (Goodrich et al. 2010). Moreover, based on thermal isotherms, predicted value of absorption heat for the reaction of [P_{66614}][Pro] with CO_2 followed 1:1 mechanism, was −63 kJ. mol^{-1} CO_2. [P_{66614}][Pro] has a ring structure which makes a restriction for hydrogen atoms to create hydrogen tethering network; this may explain why its viscosity was not increased after absorption.

To put it differently, absorption capacity of CO_2 could be improved through rising pressure, while the maximum capacity of aqueous amine solutions would be achievable at 0.2 MPa (Shen and Li 1992; Sánchez et al. 2007). The reason which makes these ionic liquids face serious limitations in industry is their high viscosity after the absorption of CO_2 (Soutullo et al. 2007; Gutowski and Maginn 2008). The molecular orbital simulation was used by Yu and Zhang (2007) in order to the thorough study of both the inter-ionic interaction and structure of ionic liquid; as a result, they discovered that anion-caution interaction increases (The reason of viscosity increase) through NH_2 based intermolecular hydrogen bonds. However, it is worthwhile investigating the in-deep reaction mechanism.

5.5 CO_2 Adsorption

5.5.1 *Conventional Adsorbents*

Another method for separation of carbon dioxide is adsorption technology. Conventional dry solid adsorbents which used in this method are divided into non-carbonaceous such as zeolites, metal oxide carbonates, silica, metal-organic frameworks, alkali metal and porous polymers and carbonaceous materials. Carbonaceous materials, such as activated carbons, grapheme, ordered porous carbons and activated carbon fibers, due to their lightweight, large pore volumes and specific surface areas specifications, have particularly good abilities for gas adsorption (Han et al. 2012; Seema et al. 2014). Also, because of having reasonable material cost, low-temperature operation of adsorption and desorption process (below 373 K) and low energy consumption they are useful for carbon dioxide capture. Moreover, Carbonaceous materials are not moisture sensitive and they can be used at atmospheric pressure (Lee et al. 2015).

5.5.2 *Solid Ionic Liquids*

Todays, some researches have been conducted on using solid ionic liquids in adsorption process. This is an extremely novel area. Dowson et al. (2016) claimed that solid ionic liquids with their exceptionally rapid uptake and desorb speeds (three or four orders of magnitude faster than room-temperature ionic liquids) can be used as an adsorbent in the pressure-swing adsorption process. On this basis, they construct an experimental pressure-swing adsorption setup using solid ionic liquids as adsorbents. One of the most important results from this study was about the quantity of ionic liquids, which they used. In fact, they realized that compared to corresponding liquid base process less solid ionic liquids would be needed. As we know, thermal stability is an important factor in each process. The results indicated that at elevated temperatures, solid ionic liquids show less degradation compared to room-temperature ionic liquids. Also, high energy consumption and long-time diffusion in liquid bulk are other disadvantages of absorption and desorption by the use of room-temperature ionic liquids. To overcome these concerns of absorption, solid ionic liquids are considered as alternatives. Solid ionic liquids are ionic liquids with melting points significantly above room temperature. Solid ionic liquids provide benefits of allowing accessing to large ionic surface areas and reducing diffusion difficulty and accelerating desorbing speed (Dowson et al. 2016; Wang et al. 2017).

Also, structurally simpler and cheaper synthesize of solid ionic liquids rather than room-temperature ionic liquids, lack of sorbent pumping problems and considerably easier drying can be mentioned as other benefits of solid ionic liquids(Dowson et al. 2016; Reed et al. 2017).

5.5.3 Supported Ionic Liquids onto a Solid Porous Material

Solid sorbents can be utilized over a wide ranges of temperatures and the less energy is needed for their regeneration (Yong et al. 2002; Wang et al. 2011). Due to this, most potent solid materials have been developed for sorption of CO_2. Sorption capacity, selectivity, the cost and regenerability of sorbent should be investigated as the important factors for CO_2 sorption on solid media (Samanta et al. 2011). Activated carbonaceous materials, zeolites, carbonates, and metal-organic frameworks (MOFs) are known as the promising materials for post combustion CO_2 capture (Siriwardane et al. 2005; Arstad et al. 2008; Kim et al. 2013). During recent years, by introducing of ionic liquids, these novel materials have been used to immobilizing on solid porous substrates (Kolding et al. 2012; Wang et al. 2013; Wan et al. 2014).

Supported ionic liquids concept is a relatively new area of research and has been developed for a more efficient use of ionic liquids, where a thin layers of ionic liquid are immobilized onto a solid porous material (support particle) which in full surface coverage and adhesion are two of the most important properties of the support particle. In fact, in the supported ionic liquids technique, the advantages of ionic liquids (such advantages as high solvent capacity, non-volatility, high thermal and chemical stabilities, extremely low vapour pressure and non-flammability) are combined with those of support materials(Lemus et al. 2011; Reed et al. 2017; Ding and Jiang 2018).

To increase the CO_2 absorption efficiency, reduce the consumtion and even make recycling easier, ionic liquids can be immobilized on various substrates. It can provide the high contact area between ionic liquids and gas species, hence increase the gas transport rate. Immobilizing ionic liquids onto porous supports leads to take the advantages of ionic liquids and overcome their shortcomings simultaneously (Hasib-ur-Rahman et al. 2010; Lemus et al. 2011).

Depending on the particular application required, the choice of support type and immobilization method are essential. Immobilization of ionic liquids can be carried out in different approaches, either covalent or no covalent bonding with the supports, such as simple physisorption, self-assembly, surface grafting, and polymerization whereby the leaching of ionic liquids from supports occurs within the process. In general, polymerization and surface gratifying are both complex and expensive. A self-assembly approach could be considered as an efficient method owing to the fact that the process is simple and economical. Besides, the ionic liquids usage is highly efficient (Valkenberg et al. 2000; Zhao et al. 2002; Sasaki et al. 2005; Virtanen et al. 2009; Zhang et al. 2009; Wang et al. 2014).

Supports, based on their pore diameter can be grouped into three classes: micro porous, $d < 2.0$ nm; mesoporous, $2.0 < d < 50$ nm; and macro porous, $d > 50$ nm (Zhang et al. 2009). It was noticed that ionic liquid fills pores of the support based on their availability; therefore, ionic liquid fills micro, meso, and macro pores sequentially.

In supported ionic liquids, ionic liquids can be spread on both internal and external of the supports surfaces and provide large specific surface area and mechanical properties of the support; consequently, both the problem of mass transfer restrictions and high demand for considerable amount of ionic liquids would be tackled (Lemus et al. 2011; Zhou et al. 2016).

Assuming the low cost of support material, this technique considerably reduces sorbent inventory costs and improves the ionic liquid efficiency. In fact, this is a consequence of decrising in ionic liquid quantity required by increasing ionic surface area resulted from coating (Reed et al. 2017). Besides, the porosity and surface area and ionic liquid activation play a significant role in the adsorption(Cota and Martinez 2017). Activated carbon, zeolites, and metal-organic frameworks among others, are such supports that could be considered as viable options (Reed et al. 2017).

Result of adsorption study which is carried out with Yusuf, N., et al., in the actual adsorber unit, presented that the impregnated activated carbon with ionic liquids (means that activated carbon used as support) reacts faster and adsorbs larger amounts of carbon dioxide rather than fresh activated carbon in a short period of time at a high flow rate. In fact, narrow particle size distribution is assumed to be a reason of improving the adsorption capacity(Yusuf et al. 2019).

Also, metal-organic frameworks, because of their high specific surface porosity and affinity for carbon dioxide adsorption, represent a very remarkable candidate for ionic liquids supporting (Fig. 5.6). Good physicochemical properties of both metal-organic frameworks and ionic liquids and their infinite number of possible combinations make them a very good combination for the best carbon dioxide /gas adsorption performance(Cota and Martinez 2017).

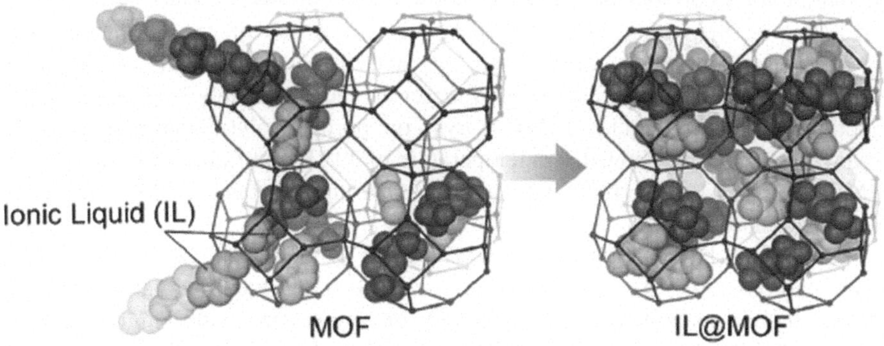

Fig. 5.6 The illustration of metal-organic frameworks which modified with ionic liquid. Metal-organic frameworks because of their high specific surface area and porosity, are very good applicant for supporting ionic liquids. (Reprinted with permission of [Recent advances in the synthesis and applications of metal organic frameworks doped with ionic liquids for CO_2 adsorption, Cota, I. and F. F. J. C. C. R. Martinez, American Chemical Society]' from (Cota and Martinez 2017)) *MOF* Metal-organic framework

One of the potential sorbents for CO_2 is magnesium oxide which possess good CO_2 sorption capacity and can be used for wide ranges of temperatures (Lee et al. 2008; Li et al. 2010; Vu et al. 2014). To prepare mesoporous $MgO–Al_2O_3$ based materials, surfactants such as P123 and F127 can be used (Wang et al. 2008; Cai et al. 2011; Ge et al. 2013). Because of self-aggregation properties of ionic liquids, they act as templates to achieve microporous materials. Mixture of different templates can also be utilised to promote the physicochemical properties of the material (Chen et al. 1997; Dai et al. 2001; Kuang et al. 2004; Brezesinski et al. 2005; Sel et al. 2006). Tanaka et al. (2011) increased the CO_2 capture capacity of nanoporous silica microspheres by dispersion of 1-hexadecyl-3-methylimidazolium chloride ionic liquid into their structures. They observed that CO_2 capture performance enhanced in comparison to pure ionic liquid. This behaviour is due to the occupation of ordered molecular domains by CO_2. Wan et al. (2014) induced ionic liquids into mesoporous MCM-41 and alumina substrates. In their study, identical quantity of tetrabutylphosphonium-based ionic liquid with 3-aminopyrazole anions or imidazole was loaded onto MCM-41 and alumina substrates (up to 33 wt.% ionic liquid). Despite the fact that MCM-41 had higher surface area than that of alumina, the supported alumina showed higher CO_2 capacity. The positive ξ-potential of alumina, produce a more suitable ionic liquid dispersion with the active anion which surrounded by the cationic alkyl chains and prevent the purge of adsorbed CO_2 by the N_2 carrier gas. Arellano et al. (2015) developed the study of Zinc-functionalized ionic liquid sorbents which supported on mesoporous silica. The capture capacity of CO_2 was high about 148.72 mg CO_2/g-sorbent at temperature of 313 K and atmospheric pressure. The effects of various supported sorbents have been studied on the performance of CO_2 sorption and desorption by Ren et al. (2012). They firstly synthesized six amino acid salts ([apaeP444][AA])-type ionic liquids and then, immobilized them on porous silica. Their performance was tested at temperature 298 K and 1 atm pressure with a pure CO_2 gas and the gas flow rate of 50 mL/min. Of these supported sorbents, SiO_2-[apaeP444][Gly] and SiO_2-[apaeP444][Lys] indicated high CO_2 uptakes. In another study, Wang et al. (2013) immobilized amino acid-functionalized ionic liquids [EMIM][Lys] on porous poly(methylmethacrylate) (PMMA). The results show that at temperature of 298 K and 1 atm, CO_2 capacity of [EMIM][Lys]-PMMA sorbent was about 73.48 mg/g-sorbnet or 0.87 mol/mol-AAIL within 40 min.

Imidazolium-based ionic liquids was utilized by Zhu et al. (2011) to modify active carbons (ACs) for CO_2 capture. The ionic liquid-modified active carbons showed higher performance to adsorb CO_2 in comparision with amino-modified polymers. They also observed that the smaller particle size (38–150 μm) of ionic liquid-modified active carbons indicated the highest adsorption capacity of 187.26 mg/g-sorbent at 298.15 K temperature and 4.663 atm pressure in pure CO_2. In the study done by Erto et al. (2015) two commercial active carbons impregnated with [Hmim][BF4] and [Emim][Gly] was investigated for separation of CO_2 from synthetic flue gas. They observed that [Emim][Gly] enhanced the capture capacity, while [Hmim][BF4] reduced the capacity. Other researchers have also studied the performance of different sorbents to separate CO_2. 3-aminopropyltriethoxysilane-modified graphite oxide was synthesized by Hong (Hong et al. 2013) to capture CO_2. Tamilarasan (Tamilarasan and Ramaprabhu 2015; Tamilarasan and Ramaprabhu 2016) also studied amine-rich ionic liquid ([DAMT][BF4]) and ionic liquid ([BMIM][BF4]) grafted on graphene for CO_2 adsorption. Excellent CO_2 adsorption-desorption performance was achieved, when Hiremath et al. applied amino acid functionalized imidazolium based ionic liquid onto mesoporous silica (Hiremath et al. 2016). The metal-organic framework (MOF) material MIL-101 was incorporated with imidazoliumbased poly(ionic liquid)s by Ding and Jiang. The resultant material indicated good CO_2 capture capability (Ding and Jiang 2018).

Nkinahamira et al. (2017) grafted two quaternary ammonium ionic liquids, tetra-n-heptyl ammonium bromide ([thepAm][Br]) and (vinyl benzyl) trimethylammonium chloride ([VBTMA][Cl]), on mesoporous MCM-41 to promote the capacity of CO_2 absorption and ionic liquids selectivity. This method led to reducing in mesoporosity, which caused to reduction in CO_2 adsorption capacity, while increase in the selectivity of CO_2/CO. The results also indicated that the adsorbents which grafted with [VBTMA][Cl] in comparision with those grafted with [thepAm][Br], had better performance with regard to CO_2/CO selectivity of 95 and excess CO_2 loading of 3.90 mmol/g. Trihexyl(tetradecyl)-phosphonium alanate ([P66614][Ala]) ionic liquid was used by Yoshida et al. to impregnating Silica microhoneycombs (SMHs). CO_2 capacity of ~0.26 mmol-$CO_2 \cdot g^{-1}$ was observed in a CO_2 33% − N_2 67% gas mixture at 303 K and atmospheric pressure (Yoshida et al. 2017). Zhu et al. (2018) modified porous silica gel using different types of ionic liquids to separate CO_2. They grafted and impregnated ionic liquids onto support surface and the obtained results showed that grafting of ionic liquids onto porous silica gel can slightly reduce CO_2 adsorption capacity in comparison with impregnated silica. The impregnation quantity effects, diffusion coefficient and the effect of pore size distribution of adsorbents on the CO_2/N_2 selectivity when it has been used several kinds of ionic liquids were also investigated.

References

Abdelrahim MYM, Martins CF, Neves LA, Capasso C, Supuran CT, Coelhoso IM, Crespo JG, Barboiu M (2017) Supported ionic liquid membranes immobilized with carbonic anhydrases for CO2 transport at high temperatures. J Membr Sci 528:225–230. https://doi.org/10.1016/j.memsci.2017.01.033

Anderson JL, Dixon JK, Brennecke JF (2007) Solubility of CO2, CH4, C2H6, C2H4, O2, and N2 in 1-Hexyl-3-methylpyridinium Bis (trifluoromethylsulfonyl) imide: comparison to other ionic liquids. Acc Chem Res 40(11):1208–1216. https://doi.org/10.1021/ar7001649

Anthony JL, Maginn EJ, Brennecke JF (2002) Solubilities and thermodynamic properties of gases in the ionic liquid 1-n-butyl-3-methylimidazolium hexafluorophosphate. J Phys Chem B 106(29):7315–7320. https://doi.org/10.1021/jp020631a

Anthony JL, Anderson JL, Maginn EJ, Brennecke JF (2005) Anion effects on gas solubility in ionic liquids. J Phys Chem B 109(13):6366–6374. https://doi.org/10.1021/jp046404l

Arellano IH, Huang J, Pendleton P (2015) High capacity CO 2 sorbents based on zinc-functionalized ionic liquid confined in morphologically diverse porous matrices. RSC Adv 5(80):65074–65083. https://doi.org/10.1039/C5RA12738F

Arstad B, Fjellvåg H, Kongshaug KO, Swang O, Blom R (2008) Amine functionalised metal organic frameworks (MOFs) as adsorbents for carbon dioxide. Adsorption 14(6):755–762. https://doi.org/10.1007/s10450-008-9137-6

Atefi F, Garcia MT, Singer RD, Scammells PJ (2009) Phosphonium ionic liquids: design, synthesis and evaluation of biodegradability. Green Chem 11(10):1595–1604. https://doi.org/10.1039/B913057H

Baltus RE, Culbertson BH, Dai S, Luo H, DePaoli DW (2004) Low-pressure solubility of carbon dioxide in room-temperature ionic liquids measured with a quartz crystal microbalance. J Phys Chem B 108(2):721–727. https://doi.org/10.1021/jp036051a

Bara JE, Lessmann S, Gabriel CJ, Hatakeyama ES, Noble RD, Gin DL (2007) Synthesis and performance of polymerizable room-temperature ionic liquids as gas separation membranes. Ind Eng Chem Res 46(16):5397–5404. https://doi.org/10.1021/ie0704492

Bara JE, Gin DL, Noble RD (2008) Effect of anion on gas separation performance of polymer–room-temperature ionic liquid composite membranes. Ind Eng Chem Res 47(24):9919–9924. https://doi.org/10.1021/ie801019x

Bara JE, Gabriel CJ, Hatakeyama ES, Carlisle TK, Lessmann S, Noble RD, Gin DL (2008a) Improving CO2 selectivity in polymerized room-temperature ionic liquid gas separation membranes through incorporation of polar substituents. J Membr Sci 321(1):3–7. https://doi.org/10.1016/j.memsci.2007.12.033

Bara JE, Hatakeyama ES, Gin DL, Noble RD (2008b) Improving CO2 permeability in polymerized room-temperature ionic liquid gas separation membranes through the formation of a solid composite with a room-temperature ionic liquid. Polym Adv Technol 19(10):1415–1420. https://doi.org/10.1002/pat.1209

Bara JE, Camper DE, Gin DL, Noble RD (2009a) Room-temperature ionic liquids and composite materials: platform technologies for CO2 capture. Acc Chem Res 43(1):152–159. https://doi.org/10.1021/ar9001747

Bara JE, Carlisle TK, Gabriel CJ, Camper D, Finotello A, Gin DL, Noble RD (2009b) Guide to CO2 separations in imidazolium-based room-temperature ionic liquids. Ind Eng Chem Res 48(6):2739–2751. https://doi.org/10.1021/ie8016237

Bates ED, Mayton RD, Ntai I, Davis JH (2002) CO2 capture by a task-specific ionic liquid. J Am Chem Soc 124(6):926–927. https://doi.org/10.1021/ja017593d

Blanchard LA, Hancu D, Beckman EJ, Brennecke JF (1999) Green processing using ionic liquids and CO 2. Nature 399(6731):28. https://doi.org/10.1038/19887

Blanchard LA, Gu Z, Brennecke JF (2001) High-pressure phase behavior of ionic liquid/CO2 systems. J Phys Chem B 105(12):2437–2444. https://doi.org/10.1021/jp003309d

Blasig A, Tang J, Hu X, Shen Y, Radosz M (2007) Magnetic suspension balance study of carbon dioxide solubility in ammonium-based polymerized ionic liquids: poly (p-vinylbenzyltrimethyl ammonium tetrafluoroborate) and poly ([2-(methacryloyloxy) ethyl] trimethyl ammonium tetrafluoroborate). Fluid Phase Equilib 256(1–2):75–80. https://doi.org/10.1016/j.fluid.2007.03.007

Boethling R, Sommer E, DiFiore D (2007) Designing small molecules for biodegradability. Chem Rev 107(6):2207–2227. https://doi.org/10.1021/cr050952t

Brezesinski T, Erpen C, Iimura K-i, Smarsly B (2005) Mesostructured crystalline ceria with a bimodal pore system using block copolymers and ionic liquids as rational templates. Chem Mater 17(7):1683–1690. https://doi.org/10.1021/cm0479180

Cadena C, Anthony JL, Shah JK, Morrow TI, Brennecke JF, Maginn EJ (2004) Why is CO2 so soluble in imidazolium-based ionic liquids? J Am Chem Soc 126(16):5300–5308. https://doi.org/10.1021/ja039615x

Cai W, Yu J, Anand C, Vinu A, Jaroniec M (2011) Facile synthesis of ordered mesoporous alumina and alumina-supported metal oxides with tailored adsorption and framework properties. Chem Mater 23(5):1147–1157. https://doi.org/10.1021/cm102512v

Carvalho PJ, Coutinho JA (2011) The polarity effect upon the methane solubility in ionic liquids: a contribution for the design of ionic liquids for enhanced CO 2/CH 4 and H 2 S/CH 4 selectivities. Energy Environ Sci 4(11):4614–4619. https://doi.org/10.1039/C1EE01599K

Carvalho PJ, Álvarez VH, Marrucho IM, Aznar M, Coutinho JA (2009) High pressure phase behavior of carbon dioxide in 1-butyl-3-methylimidazolium bis (trifluoromethylsulfonyl) imide and 1-butyl-3-methylimidazolium dicyanamide ionic liquids. J Supercrit Fluids 50(2):105–111. https://doi.org/10.1016/j.supflu.2009.05.008

Chen F, Huang L, Li Q (1997) Synthesis of MCM-48 using mixed cationic– anionic surfactants as templates. Chem Mater 9(12):2685–2686. https://doi.org/10.1021/cm9703942

Chen Y, Zhang S, Yuan X, Zhang Y, Zhang X, Dai W, Mori R (2006) Solubility of CO2 in imidazolium-based tetrafluoroborate ionic liquids. Thermochim Acta 441(1):42–44. https://doi.org/10.1016/j.tca.2005.11.023

Cota I, Martinez FF (2017) Recent advances in the synthesis and applications of metal organic frameworks doped with ionic liquids for CO2 adsorption. 351:189–204. https://doi.org/10.1016/j.ccr.2017.04.008

Dai L-R, Wang T-W, Bu L-T, Chen G (2001) Mixed surfactant templating route for mesoporous silica. Colloids Surf A Physicochem Eng Asp 181(1–3):151–157. https://doi.org/10.1016/S0927-7757(00)00791-3

Dai Z, Noble RD, Gin DL, Zhang X, Deng L (2016) Combination of ionic liquids with membrane technology: A new approach for CO2 separation. J Membr Sci 497:1–20. https://doi.org/10.1016/j.memsci.2015.08.060

Ding M, Jiang H-L (2018) Incorporation of imidazolium-based poly (ionic liquid) s into a metal–organic framework for CO2 capture and conversion. 8(4):3194–3201. https://doi.org/10.1021/acscatal.7b03404

Docherty KM, Dixon JK, Kulpa CF Jr (2007) Biodegradability of imidazolium and pyridinium ionic liquids by an activated sludge microbial community. Biodegradation 18(4):481–493. https://doi.org/10.1007/s10532-006-9081-7

Docherty KM, Joyce MV, Kulacki KJ, Kulpa CF (2010) Microbial biodegradation and metabolite toxicity of three pyridinium-based cation ionic liquids. Green Chem 12(4):701–712. https://doi.org/10.1039/B919154B

Dowson GRM, Reed DG, Bellas JM, Charalambous C, Styring P (2016) Fast and selective separation of carbon dioxide from dilute streams by pressure swing adsorption using solid ionic liquids. 192:511–527. https://doi.org/10.1039/C6FD00035E

Erto A, Silvestre-Albero A, Silvestre-Albero J, Rodríguez-Reinoso F, Balsamo M, Lancia A, Montagnaro F (2015) Carbon-supported ionic liquids as innovative adsorbents for CO2 separation from synthetic flue-gas. J Colloid Interface Sci 448:41–50. https://doi.org/10.1016/j.jcis.2015.01.089

Gathergood N, Garcia MT, Scammells PJ (2004) Biodegradable ionic liquids: part I. concept, preliminary targets and evaluation. Green Chem 6(3):166–175. https://doi.org/10.1039/B315270G

Ge J, Deng K, Cai W, Yu J, Liu X, Zhou J (2013) Effect of structure-directing agents on facile hydrothermal preparation of hierarchical γ-Al2O3 and their adsorption performance toward Cr (VI) and CO2. J Colloid Interface Sci 401:34–39. https://doi.org/10.1016/j.jcis.2013.03.028

Goodrich BF, De La Fuente JC, Gurkan BE, Zadigian DJ, Price EA, Huang Y, Brennecke JF (2010) Experimental measurements of amine-functionalized anion-tethered ionic liquids with carbon dioxide. Ind Eng Chem Res 50(1):111–118. https://doi.org/10.1021/ie101688a

Goodrich BF, de la Fuente JC, Gurkan BE, Lopez ZK, Price EA, Huang Y, Brennecke JF (2011) Effect of water and temperature on absorption of CO2 by amine-functionalized anion-tethered ionic liquids. J Phys Chem B 115(29):9140–9150. https://doi.org/10.1021/jp2015534

Götz M et al (2016) Hydrodynamics of organic and ionic liquids in a slurry bubble column reactor operated at elevated temperatures. Chem Eng J 286:348–360. https://doi.org/10.1016/j.cej.2015.10.044

Grabińska-Sota E, Kalka J (2003) An assessment of the toxicity of pyridinium chlorides and their biodegradation intermediates. Environ Int 28(8):687–690. https://doi.org/10.1016/S0160-4120(02)00112-5

Gurkan BE, de la Fuente JC, Mindrup EM, Ficke LE, Goodrich BF, Price EA, Schneider WF, Brennecke JF (2010) Equimolar CO2 absorption by anion-functionalized ionic liquids. J Am Chem Soc 132(7):2116–2117. https://doi.org/10.1021/ja909305t

Gutowski KE, Maginn EJ (2008) Amine-functionalized task-specific ionic liquids: a mechanistic explanation for the dramatic increase in viscosity upon complexation with CO2 from molecular simulation. J Am Chem Soc 130(44):14690–14704. https://doi.org/10.1021/ja804654b

Han KK, Zhou Y, Chun Y, Zhu JH (2012) Efficient MgO-based mesoporous CO2 trapper and its performance at high temperature. 203:341–347. https://doi.org/10.1016/j.jhazmat.2011.12.036

Hanioka S, Maruyama T, Sotani T, Teramoto M, Matsuyama H, Nakashima K, Hanaki M, Kubota F, Goto M (2008) CO2 separation facilitated by task-specific ionic liquids using a supported liquid membrane. J Membr Sci 314(1–2):1–4. https://doi.org/10.1016/j.memsci.2008.01.029

Harjani JR, Farrell J, Garcia MT, Singer RD, Scammells PJ (2009a) Further investigation of the biodegradability of imidazolium ionic liquids. Green Chem 11(6):821–829. https://doi.org/10.1039/B900787C

Harjani JR, Singer RD, Garcia MT, Scammells PJ (2009b) Biodegradable pyridinium ionic liquids: design, synthesis and evaluation. Green Chem 11(1):83–90. https://doi.org/10.1039/B811814K

Hasib-ur-Rahman M, Siaj M, Larachi F (2010) Ionic liquids for CO2 capture—development and progress. Chem Eng Process Process Intensif 49(4):313–322. https://doi.org/10.1016/j.cep.2010.03.008

Heintz YJ, Sehabiague L, Morsi BI, Jones KL, Luebke DR, Pennline HW (2009) Hydrogen sulfide and carbon dioxide removal from dry fuel gas streams using an ionic liquid as a physical solvent. Energy Fuel 23(10):4822–4830. https://doi.org/10.1021/ef900281v

Hiremath V, Jadhav AH, Lee H, Kwon S, Seo JG (2016) Highly reversible CO2 capture using amino acid functionalized ionic liquids immobilized on mesoporous silica. Chem Eng J 287:602–617. https://doi.org/10.1016/j.cej.2015.11.075

Hong S-M, Kim SH, Lee KB (2013) Adsorption of carbon dioxide on 3-aminopropyl-triethoxysilane modified graphite oxide. Energy Fuel 27(6):3358–3363. https://doi.org/10.1021/ef400467w

Hu X, Tang J, Blasig A, Shen Y, Radosz M (2006) CO2 permeability, diffusivity and solubility in polyethylene glycol-grafted polyionic membranes and their CO2 selectivity relative to methane and nitrogen. J Membr Sci 281(1–2):130–138. https://doi.org/10.1016/j.memsci.2006.03.030

Huddleston JG, Visser AE, Reichert WM, Willauer HD, Broker GA, Rogers RD (2001) Characterization and comparison of hydrophilic and hydrophobic room temperature ionic liquids incorporating the imidazolium cation. Green Chem 3(4):156–164. https://doi.org/10.1039/B103275P

Itoh T (2017) Biotransformation in ionic liquid. In: Future directions in biocatalysis. Elsevier, pp 27–67. https://doi.org/10.1016/B978-0-444-63743-7.00002-0

Kaji R, Zhao D, Licence P, Azzopardi B (2009) Studies of the interaction of ionic liquid and gas in a small-diameter bubble column. Ind Eng Chem Res 48(17):7938–7944. https://doi.org/10.1021/ie8014439

Karadas F, Atilhan M, Aparicio S (2010) Review on the use of ionic liquids (ILs) as alternative fluids for CO2 capture and natural gas sweetening. Energy Fuel 24(11):5817–5828. https://doi.org/10.1021/ef1011337

Kim TK, Lee KJ, Cheon JY, Lee JH, Joo SH, Moon HR (2013) Nanoporous metal oxides with tunable and nanocrystalline frameworks via conversion of metal–organic frameworks. J Am Chem Soc 135(24):8940–8946. https://doi.org/10.1021/ja401869h

Kolding H, Fehrmann R, Riisager A (2012) CO 2 capture technologies: current status and new directions using supported ionic liquid phase (SILP) absorbers. SCIENCE CHINA Chem 55(8):1648–1656. https://doi.org/10.1007/s11426-012-4683-x

Kosmulski M, Gustafsson J, Rosenholm JB (2004) Thermal stability of low temperature ionic liquids revisited. Thermochim Acta 412(1–2):47–53. https://doi.org/10.1016/j.tca.2003.08.022

Kuang D, Brezesinski T, Smarsly B (2004) Hierarchical porous silica materials with a trimodal pore system using surfactant templates. J Am Chem Soc 126(34):10534–10535. https://doi.org/10.1021/ja0470618

Lee SC, Chae HJ, Lee SJ, Choi BY, Yi CK, Lee JB, Ryu CK, Kim JC (2008) Development of regenerable MgO-based sorbent promoted with K2CO3 for CO2 capture at low temperatures. Environ Sci Technol 42(8):2736–2741. https://doi.org/10.1021/es702693c

Lee S-Y, Park S-J, Chemistry E (2015) A review on solid adsorbents for carbon dioxide capture. 23:1–11. https://doi.org/10.1016/j.jiec.2014.09.001

Lemus J, Palomar J, Gilarranz MA, Rodriguez JJ (2011) Characterization of supported ionic liquid phase (SILP) materials prepared from different supports. 17(3):561–571. https://doi.org/10.1007/s10450-011-9327-5

Li W, Zhang Z, Han B, Hu S, Song J, Xie Y, Zhou X (2008) Switching the basicity of ionic liquids by CO 2. Green Chem 10(11):1142–1145. https://doi.org/10.1039/B811624E

Li L, Wen X, Fu X, Wang F, Zhao N, Xiao F, Wei W, Sun Y (2010) MgO/Al2O3 sorbent for CO2 capture. Energy Fuel 24(10):5773–5780. https://doi.org/10.1021/ef100817f

Liu X, Zhou G, Zhang S, Yao X (2009) Molecular dynamics simulation of dual amino-functionalized imidazolium-based ionic liquids. Fluid Phase Equilib 284(1):44–49. https://doi.org/10.1016/j.fluid.2009.06.004

Liu Z, Liu C, Li L, Qin W, Xu A (2016) CO2 separation by supported ionic liquid membranes and prediction of separation performance. Int J Greenhouse Gas Control 53:79–84. https://doi.org/10.1016/j.ijggc.2016.07.041

Lozano L, Godínez C, De Los Rios A, Hernández-Fernández F, Sánchez-Segado S, Alguacil FJ (2011) Recent advances in supported ionic liquid membrane technology. J Membr Sci 376(1–2):1–14. https://doi.org/10.1016/j.memsci.2011.03.036

Mrozik W, Jungnickel C, Ciborowski T, Pitner WR, Kumirska J, Kaczyński Z, Stepnowski P (2009) Predicting mobility of alkylimidazolium ionic liquids in soils. J Soils Sediments 9(3):237–245. https://doi.org/10.1007/s11368-009-0057-1

Myers C, Pennline H, Luebke D, Ilconich J, Dixon JK, Maginn EJ, Brennecke JF (2008) High temperature separation of carbon dioxide/hydrogen mixtures using facilitated supported ionic liquid membranes. J Membr Sci 322(1):28–31. https://doi.org/10.1016/j.memsci.2008.04.062

Ngo HL, LeCompte K, Hargens L, McEwen AB (2000) Thermal properties of imidazolium ionic liquids. Thermochim Acta 357:97–102. https://doi.org/10.1016/S0040-6031(00)00373-7

Nkinahamira F, Su T, Xie Y, Ma G, Wang H, Li J (2017) High pressure adsorption of CO2 on MCM-41 grafted with quaternary ammonium ionic liquids. Chem Eng J 326:831–838. https://doi.org/10.1016/j.cej.2017.05.173

Petkovic M, Ferguson JL, Gunaratne HN, Ferreira R, Leitao MC, Seddon KR, Rebelo LPN, Pereira CS (2010) Novel biocompatible cholinium-based ionic liquids—toxicity and biodegradability. Green Chem 12(4):643–649. https://doi.org/10.1039/B922247B

Petkovic M, Seddon KR, Rebelo LPN, Pereira CS (2011) Ionic liquids: a pathway to environmental acceptability. Chem Soc Rev 40(3):1383–1403. https://doi.org/10.1039/C004968A

Pham TPT, Cho C-W, Jeon C-O, Chung Y-J, Lee M-W, Yun Y-S (2008) Identification of metabolites involved in the biodegradation of the ionic liquid 1-butyl-3-methylpyridinium bromide by activated sludge microorganisms. Environ Sci Technol 43(2):516–521. https://doi.org/10.1021/es703004h

Reed DG, Dowson GR, Styring P (2017) Cellulose-supported ionic liquids for low-cost pressure swing CO2 capture. 5:13. https://doi.org/10.3389/fenrg.2017.00013

Ren J, Wu L, Li B-G (2012) Preparation and CO2 sorption/desorption of N-(3-aminopropyl) aminoethyl tributylphosphonium amino acid salt ionic liquids supported into porous silica particles. Ind Eng Chem Res 51(23):7901–7909. https://doi.org/10.1021/ie2028415

Revelli A-L, Mutelet F, Jaubert J-N (2010) High carbon dioxide solubilities in imidazolium-based ionic liquids and in poly (ethylene glycol) dimethyl ether. J Phys Chem B 114(40):12908–12913. https://doi.org/10.1021/jp1057989

Samanta A, Zhao A, Shimizu GK, Sarkar P, Gupta R (2011) Post-combustion CO2 capture using solid sorbents: a review. Ind Eng Chem Res 51(4):1438–1463. https://doi.org/10.1021/ie200686q

Sánchez LG, Meindersma G, De Haan A (2007) Solvent properties of functionalized ionic liquids for CO2 absorption. Chem Eng Res Des 85(1):31–39. https://doi.org/10.1205/cherd06124

Sánchez LG, Meindersma G, De Haan A (2011) Kinetics of absorption of CO2 in amino-functionalized ionic liquids. Chem Eng J 166(3):1104–1115. https://doi.org/10.1016/j.cej.2010.12.016

Sasaki T, Zhong C, Tada M, Iwasawa Y (2005) Immobilized metal ion-containing ionic liquids: preparation, structure and catalytic performance in Kharasch addition reaction. (19):2506–2508. https://doi.org/10.1039/B500349K

Seema H, Kemp KC, Le NH, Park S-W, Chandra V, Lee JW, Kim KS (2014) Highly selective CO2 capture by S-doped microporous carbon materials. 66:320–326. https://doi.org/10.1016/j.carbon.2013.09.006

Sel O, Kuang D, Thommes M, Smarsly B (2006) Principles of hierarchical meso-and macropore architectures by liquid crystalline and polymer colloid templating. Langmuir 22(5):2311–2322. https://doi.org/10.1021/la0520841

Shen KP, Li MH (1992) Solubility of carbon dioxide in aqueous mixtures of monoethanolamine with methyldiethanolamine. J Chem Eng Data 37(1):96–100. https://doi.org/10.1021/je00005a025

Singh VV, Nigam AK, Batra A, Boopathi M, Singh B, Vijayaraghavan R (2012) Applications of ionic liquids in electrochemical sensors and biosensors. Int J Electrochem 2012:1–19. https://doi.org/10.1155/2012/165683

Siriwardane RV, Shen M-S, Fisher EP, Losch J (2005) Adsorption of CO2 on zeolites at moderate temperatures. Energy Fuel 19(3):1153–1159. https://doi.org/10.1021/ef040059h

Soutullo MD, Odom CI, Wicker BF, Henderson CN, Stenson AC, Davis JH (2007) Reversible CO2 capture by unexpected plastic-, resin-, and gel-like ionic soft materials discovered during the combi-click generation of a TSIL library. Chem Mater 19(15):3581–3583. https://doi.org/10.1021/cm0705690

Stepnowski P, Mrozik W, Nichthauser J (2007) Adsorption of alkylimidazolium and alkylpyridinium ionic liquids onto natural soils. Environ Sci Technol 41(2):511–516. https://doi.org/10.1021/es062014w

Stolte S, Abdulkarim S, Arning J, Blomeyer-Nienstedt A-K, Bottin-Weber U, Matzke M, Ranke J, Jastorff B, Thöming J (2008) Primary biodegradation of ionic liquid cations, identification of degradation products of 1-methyl-3-octylimidazolium chloride and electrochemical wastewater treatment of poorly biodegradable compounds. Green Chem 10(2):214–224. https://doi.org/10.1039/B713095C

Studzińska S, Kowalkowski T, Buszewski B (2009) Study of ionic liquid cations transport in soil. J Hazard Mater 168(2–3):1542–1547. https://doi.org/10.1016/j.jhazmat.2009.03.029

Supasitmongkol S, Styring P (2010) High CO_2 solubility in ionic liquids and a tetraalkylammonium-based poly (ionic liquid). Energy Environ Sci 3(12):1961–1972. https://doi.org/10.1039/C0EE00293C

Tamilarasan P, Ramaprabhu S (2015) Integration of polymerized ionic liquid with graphene for enhanced CO 2 adsorption. J Mater Chem A 3(1):101–108. https://doi.org/10.1039/C4TA04808C

Tamilarasan P, Ramaprabhu S (2016) Amine-rich ionic liquid grafted graphene for sub-ambient carbon dioxide adsorption. RSC Adv 6(4):3032–3040. https://doi.org/10.1039/C5RA22029G

Tanaka S, Kida K, Fujimoto H, Makino T, Miyake Y (2011) Self-assembling imidazolium-based ionic liquid in rigid nanopores induces anomalous CO2 adsorption at low pressure. Langmuir 27(13):7991–7995. https://doi.org/10.1021/la201431z

Tang J, Tang H, Sun W, Plancher H, Radosz M, Shen Y (2005a) Poly (ionic liquid) s: a new material with enhanced and fast CO_2 absorption. Chem Commun (26):3325–3327. https://doi.org/10.1039/B501940K

Tang J, Tang H, Sun W, Radosz M, Shen Y (2005b) Low-pressure CO2 sorption in ammonium-based poly (ionic liquid) s. Polymer 46(26):12460–12467. https://doi.org/10.1016/j.polymer.2005.10.082

Tang J, Tang H, Sun W, Radosz M, Shen Y (2005c) Poly (ionic liquid) s as new materials for CO_2 absorption. J Polym Sci A Polym Chem 43(22):5477–5489. https://doi.org/10.1002/pola.21031

Tang, J., Sun, W., Tang, H., Radosz, M., & Shen, Y. (2005d). Enhanced CO2 absorption of poly (ionic liquid) s. Macromolecules, 38(6), 2037–2039. https://doi.org/10.1021/ma047574z

Tang J, Shen Y, Radosz M, Sun W (2009) Isothermal carbon dioxide sorption in poly (ionic liquid) s. Ind Eng Chem Res 48(20):9113–9118. https://doi.org/10.1021/ie900292p

Valkenberg MH, de Castro C, Holderich WF (2000) Immobilisation of chloroaluminate ionic liquids on silica materials. 14(1–4):139–144. https://doi.org/10.1023/A:1009023520210

Van Valkenburg ME, Vaughn RL, Williams M, Wilkes JS (2005) Thermochemistry of ionic liquid heat-transfer fluids. Thermochim Acta 425(1–2): 181–188. https://doi.org/10.1016/j.tca.2004.11.013

Virtanen P, Mikkola J-P, Toukoniitty E, Karhu H, Kordas K, Eränen K, Wärnå J, Salmi TJCT (2009) Supported ionic liquid catalysts—from batch to continuous operation in preparation of fine chemicals. 147:S144–S148. https://doi.org/10.1016/j.cattod.2009.07.020

Vu A-T, Park Y, Jeon PR, Lee C-H (2014) Mesoporous MgO sorbent promoted with KNO3 for CO2 capture at intermediate temperatures. Chem Eng J 258:254–264. https://doi.org/10.1016/j.cej.2014.07.088

Wan MM, Zhu HY, Li YY, Ma J, Liu S, Zhu JH (2014) Novel CO2-capture derived from the basic ionic liquids orientated on mesoporous materials. ACS Appl Mater Interfaces 6(15):12947–12955. https://doi.org/10.1021/am5028814

Wang G, Zhang L, Dai H, Deng J, Liu C, He H, Au CT (2008) P123-assisted hydrothermal synthesis and characterization of rectangular parallelepiped and hexagonal prism single-crystalline MgO with three-dimensional wormholelike mesopores. Inorg Chem 47(10):4015–4022. https://doi.org/10.1021/ic7015462

Wang C, Luo H, Jiang D e, Li H, Dai S (2010) Carbon dioxide capture by superbase-derived protic ionic liquids. Angew Chem Int Ed 49(34):5978–5981. https://doi.org/10.1002/anie.201002641

Wang Q, Luo J, Zhong Z, Borgna A (2011) CO2 capture by solid adsorbents and their applications: current status and new trends. Energy Environ Sci 4(1):42–55. https://doi.org/10.1039/C0EE00064G

Wang X, Akhmedov NG, Duan Y, Luebke D, Hopkinson D, Li B (2013) Amino acid-functionalized ionic liquid solid sorbents for post-combustion carbon capture. ACS Appl Mater Interfaces 5(17):8670–8677. https://doi.org/10.1021/am402306s

Wang J, Ji F, Xing J, Cui S, Bao Y, Hao W (2014) Lipase immobilization onto the surface of PGMA-b-PDMAEMA-grafted magnetic nanoparticles prepared via atom transfer radical polymerization. 22(11–12): 1333–1339.https://doi.org/10.1016/j.cjche.2014.09.029

Wang L, Xu Y, Wang S, Wei Y (2017) CO2 and CH4 sorption by solid-state [P4 4 4 4][NTf2] ionic liquid based on quartz crystal microbalance experiments under different pressures. 31(4): 4179–4185.https://doi.org/10.1021/acs.energyfuels.6b03104

Welton T (2018) Ionic liquids: a brief history. Biophys Rev 10:691–706. https://doi.org/10.1007/s12551-018-0419-2

Xu H, Zhang D (2017) Viscosities and conductivities of [BMIM] Zn (Ac) xCly (;) ionic liquids at different temperatures. J Chem 2017:1–5. https://doi.org/10.1155/2017/7428097

Xue Z, Zhang Z, Han J, Chen Y, Mu T (2011) Carbon dioxide capture by a dual amino ionic liquid with amino-functionalized imidazolium cation and taurine anion. Int J Greenhouse Gas Control 5(4):628–633. https://doi.org/10.1016/j.ijggc.2011.05.014

Yong Z, Mata V, Rodrigues A r E (2002) Adsorption of carbon dioxide at high temperature—a review. Sep Purif Technol 26(2–3):195–205. https://doi.org/10.1016/S1383-5866(01)00165-4

Yoshida S, Takahashi K, Kudo S, Iwamura S, Ogino I, Mukai SR (2017) CO2 separation in a flow system by silica microhoneycombs loaded with an ionic liquid prepared by the ice-templating method. Ind Eng Chem Res 56(10):2834–2839. https://doi.org/10.1021/acs.iecr.6b05001

Yu G, Zhang S (2007) Insight into the cation–anion interaction in 1, 1, 3, 3-tetramethylguanidinium lactate ionic liquid. Fluid Phase Equilib 255(1):86–92. https://doi.org/10.1016/j.fluid.2007.03.033

Yuan X, Zhang S, Liu J, Lu X (2007) Solubilities of CO2 in hydroxyl ammonium ionic liquids at elevated pressures. Fluid Phase Equilib 257(2):195–200. https://doi.org/10.1016/j.fluid.2007.01.031

Yusuf NYM, Masdar MS, Isahak WNRW, Nordin D, Husaini T, Majlan EH, Wu SY, Rejab SAM, Lye CC (2019) Impregnated carbon–ionic liquid as innovative adsorbent for H2/CO2 separation from biohydrogen. 44(6):3414–3424. https://doi.org/10.1016/j.ijhydene.2018.06.155

Zhang S, Yuan X, Chen Y, Zhang X (2005) Solubilities of CO2 in 1-butyl-3-methylimidazolium hexafluorophosphate and 1, 1, 3, 3-tetramethylguanidium lactate at elevated pressures. J Chem Eng Data 50(5):1582–1585. https://doi.org/10.1021/je050046d

Zhang J, Zhang S, Dong K, Zhang Y, Shen Y, Lv X (2006) Supported absorption of CO2 by tetrabutylphosphonium amino acid ionic liquids. Chem Eur J 12(15):4021–4026. https://doi.org/10.1002/chem.200501015

Zhang X, Liu Z, Wang W (2008) Screening of ionic liquids to capture CO2 by COSMO-RS and experiments. AICHE J 54(10):2717–2728. https://doi.org/10.1002/aic.11573

Zhang Y, Zhang S, Lu X, Zhou Q, Fan W, Zhang X (2009) Dual amino-functionalised phosphonium ionic liquids for CO2 capture. Chem Eur J 15(12):3003–3011. https://doi.org/10.1002/chem.200801184

Zhang X, Zhang X, Dong H, Zhao Z, Zhang S, & Huang Y (2012) Carbon capture with ionic liquids: overview and progress. Ener Environ Sci 5(5):6668–6681. https://doi.org/10.1039/C2EE21152A

Zhang X, Dong H, Bao D, Huang Y, Zhang X, Zhang S (2013) Effect of small amount of water on CO2 bubble behavior in ionic liquid systems. Ind Eng Chem Res 53(1):428–439. https://doi.org/10.1021/ie4020827

Zhang Y, Doroodchi E, Moghtaderi B (2014) Utilization of ventilation air methane as an oxidizing agent in chemical looping combustion. Energy Convers Manag 85:839–847. https://doi.org/10.1016/j.enconman.2014.01.005

Zhao D, Wu M, Kou Y, Min E (2002) Ionic liquids: applications in catalysis. 74(1–2):157–189. https://doi.org/10.1016/S0920-5861(01)00541-7

Zhao Y, Zhang X, Dong H, Zhen Y, Li G, Zeng S, Zhang S (2011) Solubilities of gases in novel alcamines ionic liquid 2-[2-hydroxyethyl (methyl) amino] ethanol chloride. Fluid Phase Equilib 302(1–2):60–64. https://doi.org/10.1016/j.fluid.2010.08.017

Zhou Y, Liu J, Xiao M, Meng Y, Sun L (2016) Designing supported ionic liquids (ILs) within inorganic nanosheets for CO2 capture applications. 8(8):5547–5555. https://doi.org/10.1021/acsami.5b11249

Zhu T, Bi W, Row KH (2011) Adsorption of carbon dioxide on ionic liquids-modified active carbons and amino-modified polymer. Korean J Chem Eng 28(3):914–916. https://doi.org/10.1007/s11814-010-0343-z

Zhu J, He B, Huang J, Li C, Ren T (2018) Effect of immobilization methods and the pore structure on CO2 separation performance in silica-supported ionic liquids. Microporous Mesoporous Mater 260:190–200. https://doi.org/10.1016/j.micromeso.2017.10.035

Chapter 6
Carbon Sequestration in Alkaline Soils

Muhammad Rashid, Qaiser Hussain, Khalid Saifullah Khan, Mohammad I. Alwabel, Munir Ahmad, Sarosh Alvi, Muhammad Riaz, Song Xiongyun, Abdul Manaf, Muhammad Azeem, and Saqib Bashir

Abstract Healthy soils ensure food security through sustainable agricultural production and also support in mitigating the climate change hazardous like global warming and greenhouse gases emission. However, the anthropogenic distresses of global carbon cycle cause serious risks of global warming due to the rise in the atmospheric concentration of carbon dioxide and other greenhouse gases. Alkaline soils occupy around one-third of the land but contain only a small quantity of soil organic carbon which is crucial in sustaining soil health and productivity. The mismanagement and misuse of alkaline soils result in soil erosion, poor soil productivity, lower water retention, weak soil biodiversity, and desertification which ultimately cause a substantial loss of soil organic carbon. These soils have pronounced capacity

M. Rashid · Q. Hussain (✉) · K. S. Khan
Institute of Soil Science, Pir Mehr Ali Shah Arid Agriculture University, Rawalpindi, Punjab, Pakistan
e-mail: qaiser.hussain@uaar.edu.pk

M. I. Alwabel · M. Ahmad
Soil Sciences Department, College of Food and Agricultural Sciences, King Saud University, Riyadh, Saudi Arabia

S. Alvi
Soil and Water Testing Laboratory, Rawalpindi, Pakistan

M. Riaz
Department of Environmental Sciences & Engineering, Government College University, Faisalabad, Pakistan

S. Xiongyun
College of Resources and Environment, Qingdao Agricultural University, Qingdao, People's Republic of China

A. Manaf
Department of Agronomy, Pir Mehr Ali Shah Arid Agriculture University, Rawalpindi, Punjab, Pakistan

M. Azeem
Department of Agriculture, Hazara University, Mansehra, Pakistan

S. Bashir
Department of Soil & Environmental Science, Ghazi University, DG Khan, Pakistan

Inamuddin et al. (eds.), *Sustainable Agriculture Reviews 38*, Sustainable Agriculture Reviews 38, https://doi.org/10.1007/978-3-030-29337-6_6

of carbon sequestration which requires proper management and land use practices. The adoption of conservation tillage, crop rotation, cover crops, and organic amendments can minimize soil erosion and enhance soil quality. Additionally, these practices compared with conventional operations can enhance soil quality and water retention and reduce the emission of greenhouse gases. The fallowing of lands must be discouraged because appropriate vegetative cover promotes carbon storage in alkaline soils.

Keywords C sequestration · Alkaline soils · Global C cycle · Soil management · Land degradation · Conservation tillage · Crop cover and rotation · Soil erosion · Water conservation · Soil fertility management

6.1 Introduction

Soil profiles vary in their properties and could be either acidic or alkaline; nevertheless, the near surface layers generally tend to be more acidic than sub-surface layers. Alkaline soils are characterized by a pH level above 7 and predominantly contain a great quantity of Ca, Mg and, Na. The availability of nutrients is often limited which results in poor crop growth and yield. Alkaline soils are mostly situated in arid areas of the world occupying approximately 700 million hectares (FAO 2002). The soils with deposition of $CaCO_3$ containing loess materials, or developed from calcareous parent materials are also alkaline in nature. However, at a local scale, alkaline soils may develop in valley floors with carbonates and other salts accumulate due to alkaline groundwater. The acidity and alkalinity of soils are of critical importance which could contribute in multiple ways including the supply of plant nutrients and cation exchange capacity of the soil. The high alkalinity may lead to various environmental problems, difficulties with growing plants, or deteriorating soil physical properties. Generally, the soil composition controls the buffering capacity of the soil; however, the geographic, geological, biological, climatological factors and anthropogenic activities determine the nature of soils. Therefore, the pH of the soil varies with each climatic zone around the globe (Fig. 6.1). Although most of the plants can grow in slightly alkaline soils, only a few species will survive in highly alkaline soils because plant nutrients turn into insoluble forms and become unavailable to plants.

The carbon dioxide absorption capacity of alkaline soils was estimated, 62–622 g C/m^2/year, which may contribute significantly in the global carbon cycle (Grace 2004), and ignoring this potential option in soil carbon dynamics (Dawson and Smith 2007) would be surely lop-sided. But it is wondering why such an important process has been ignored for soil carbon sequestration. The dryland soils, contain 241 Pg soil organic carbon which is about 40 times more than anthropogenic activities (6.3 Pg) (IPCC 2001), however, these soils contain a higher quantity of soil inorganic carbon (Batjes 1998) therefore the management of both carbon pools is necessary to reducing carbon dioxide emission under alkaline soils (Lal 2002). Since alkaline soils cover a significant land surface, therefore, storage of soil organic

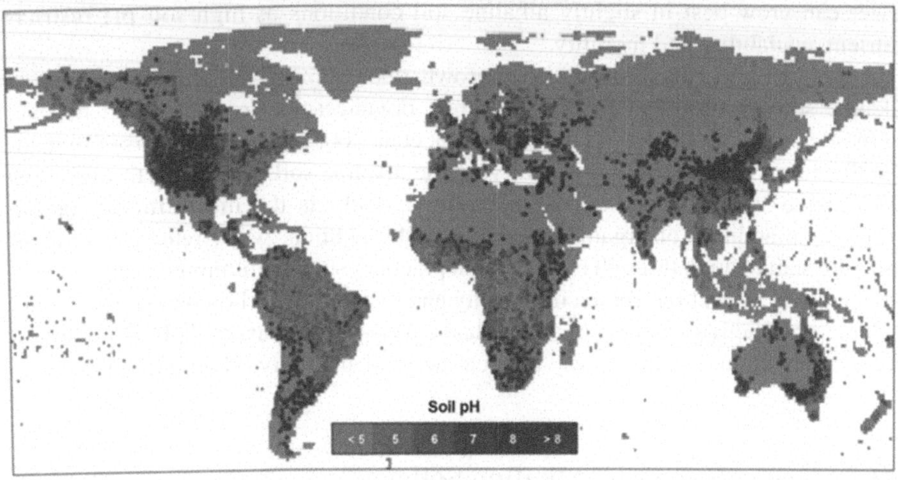

Fig. 6.1 Global map of soil pH (Slessarev 2016)

carbon in alkaline soils may contribute significantly to mitigate the current increase in atmospheric carbon dioxide level. Soil organic carbon can enhance soil quality attributes like the stability of soil aggregates, soil fertility and, nutrient cycling. The improvement of these parameters will reduce soil susceptibility to degradation processes e.g., erosion, maintenance of agricultural productivity and ecosystem quality (Maestre et al. 2012). The effect of exchangeable cations on soil organic carbon (Baldock and Nelson 2000) and soil organic matter stability has been reported by several researchers (Kaiser 2012; O'Brien et al. 2015), they found that polyvalent cations produce electrostatic bridging between soil mineral surfaces and SOC, which is an important mechanism of soil organic carbon stabilization in alkaline conditions (Stewart et al. 2008). The stabilization of organic carbon due to Mg in neutral pH soils is well known (Kaiser 2012) but, Ca provides better soil organic carbon stability because Ca is chief predictor of soil organic carbon retention in soil (O'Brien et al. 2015) which was also confirmed by Oades (1988) who reported better and longer soil organic carbon retention in alkaline calcareous soils which could be attributed to decreased solubility of organic carbon due to the presence of Ca (Muneer and Oades 1989).

6.2 Management of Alkaline Soils

Maintaining the soil pH within the appropriate range is considered as the least understood aspect for sustainable soil health. In general, the optimum pH for mineral soils for higher crop production is between 5.8 and 6.5, while the soils with high organic content (higher than 5%), the optimum soil pH is between 5.0 and 5.5. However, some of the crops can grow under moderate acidic soils (pH 5.5–6.0) and fewer can grow best in slightly alkaline soil conditions as high soil pH restricts nutrient availability and mobility.

Although high pH restricts the plant growth due to limited nutrients availability in alkaline soils, some plant roots can secrete organic acids to reduce soil pH and increase nutrient availability (Janicka-Russak et al. 2008). It is usually easier to manage the nutrient deficiencies by acidifying the alkaline soils; however, to overcome the alkaline soil pH, building organic carbon stocks is the most efficient option through the addition of organic amendments. In addition, water retention and soil structure stabilization through management practices such as minimum tillage to provide physical protection, enhancing quality and quantity of carbon inputs, improving soil microbial activity and establishing plant cover will decrease soil pH The addition of elemental sulfur is a direct way to neutralize soil alkalinity (Goulding 2016).

6.3 Carbon Stocks in Alkaline Soils

Soil carbon consists of organic and inorganic carbon. Soil organic carbon is dominant in humid areas having acidic conditions while soil inorganic carbon is prevalent under dry lands having alkaline conditions. Soil inorganic carbon comprised of: (i) primary carbonates; created through weathering of parent material and; (ii) secondary carbonates originated by the dissolution of carbon dioxide forming carbonic acid which precipitates again with, Ca^{+2} and, Mg^{+2} (Lal 2006). Alkaline soils possess lower organic carbon, however, a considerable quantity of inorganic carbon is present as soil carbonates (Denef et al. 2008) which sequester inorganic carbon in alkaline soils (Entry et al. 2004). Few researchers considered precipitation of secondary carbonates is the main mechanism in carbon sequestration while others consider dissolution of carbonates as a dominant mechanism (Sanderman 2012). Application of Ca-enriched water for irrigation resulted in the formation of $CaCO_3$ and carbon dioxide evolution in the atmosphere (Schlesinger 2000). If appropriate land management and use policies are adopted (Marks et al. 2009), the carbon sequestration in alkaline soil may alleviate not only carbon dioxide emission into the atmosphere but also can enhance soil quality indicators e.g., aggregate stability, fertility, and nutrient cycling.

The formation of $CaCO_3$ is governed by soil carbon dioxide, pH, Ca^{2+} content and soil moisture. Alkaline soil pH would promote the production of bicarbonate which will lead to the enhanced inorganic carbon, while an increase in soil carbon dioxide/decrease in soil pH would drive dissolution of bicarbonate which will decrease soil inorganic carbon stocks (Wang 2015), that is why a sufficient quantity of inorganic carbon is often found in arid and semiarid alkaline soils (Yang and Li 2011). Soil inorganic carbon concentration in soil mostly depends upon the type of land use. In alkaline soils, excessive salts accumulation negatively affect soil biological and enzyme activity (Rao and Pathak 1996) which can be improved by proper soil cover (Rao and Ghai 1985).

The addition of organic carbon through the decomposition of plant residues depends on rainfall and temperature in alkaline soils (Viscarra-Rossel et al. 2014). However, some other soil physical and chemical properties also affect organic carbon retention. Spain et al. (1983) studied that organic carbon retention is negatively correlated with soil alkalinity and the soil pH is the second most important parameter after rainfall that determines variations in organic carbon stocks in temperate and tropical areas since alkaline soils are mostly found in the low rainfall regions, therefore, both high pH and, low rainfall may limit retention of organic carbon in these soils. Soil pH and the Ca concentration are two important factors influencing dissolved organic carbon in alkaline soils (Tavakkoli et al. 2015). High soil pH may result in an increased level of dissolved organic carbon as higher Ca concentration may stabilize organic carbon by the formation of Ca-organic carbon bridges. The soil texture may also influence organic carbon storage in soil significantly since coarse soil particles contain greater amounts of organic carbon in semiarid soils attributed to retention of carbon in macroaggregates which are stabilized by precipitated $CaCO_3$ (Cunliffe et al. 2016). Likewise, the soil microbial communities substantially affect the soil organic carbon since microbes digest up to 90% of the organic residues and respire them as carbon dioxide while, up to 30% of organic inputs are converted into humus, depending on soil type and climatic conditions. Alkaline soils are characterized by low rainfall, subsequently reducing the plant productivity which in turn affects the carbon sequestration in these soils.

Therefore, good soil management and rainwater harvesting practices in alkaline are important to maintain soil organic carbon pools. Management of alkaline soils requires a strong linkage of carbon with Ca^{+2} to protect the loss of carbon (Rengasamy and Olsson 1991). Addition of sulfur in alkaline is commonly used as an inorganic amendment which increases Ca^{+2} by dissolving native $CaCO_3$; capturing the atmospheric CO_2 into the soil not only reduces greenhouse gases but, also revitalizes soil for sustainable crop production.

6.4 Factor Affecting Soil Carbon Degradation and Decomposition

The conversion of vegetation and litter into organic carbon, dissolved carbon, and carbon dioxide depend on various soil properties including microbial communities, environmental conditions and, the chemistry of crop residues (Blanco-Canqui and Lal 2004). The effect of temperature on organic matter decomposition is variable due to its temperature sensitivity (Knorr et al. 2005) but some researchers revealed that both stable and labile organic matter pools have same temperature sensitivity (Fang et al. 2005). The species and age of the vegetation affect the decomposition of organic matter depending on its C:N ratio. The decomposition of stable part of organic matter depends on its lignin constituents (Martens 2000) which influence soil microbial activities. The physical and chemical protection of organic matter from microbial activities via soil aggregates may help to decrease the rate of organic matter decomposition. However, the physicochemical interactions between organic and its mineral components, and the stability of aggregate related carbon decomposition is not well documented (Blanco-Canqui and Lal 2004).

It has been well known that micro and macro-aggregates formation enhance organic carbon stability due to recalcitrant organic compounds and, precipitation of hydroxides; oxides; phosphates and inorganic carbon as carbonates increase soil aggregation, in alkaline soils (Bronick and Lal 2005). A significant quantity of organic compounds is captured in micro and macro-aggregates by carbon binding to mineral particles and hence protected from decomposition for a long period of time, unless land use disturbances (Post and Kwon 2000). The manipulation of soil for various agricultural practices increase the disturbance of soil macro-aggregates which consequently, affect the stability of carbon and accelerate carbon decomposition (Six et al. 2004), therefore, the variation in the rate of decomposition depends on the approach of soil organic carbon to soil microbes and, the potential of soil to retain carbon through soil aggregation. The mineralization of organic compounds promotes the loss of organic carbon as carbon dioxide and, CH_4 into the atmosphere and, as particulate organic carbon and dissolved organic carbon by soil and water erosion.

6.5 Carbon Sequestration

Soil carbon plays a significant role in sustainable soil health, higher crop production, and ecosystem safety. The transfer of the atmospheric carbon dioxide into the oceans, soils, vegetation, and geologic formations and long term protection by controlling its emission is called carbon sequestration. Sundermeier et al. (2005) stated that soil carbon sequestration, is the capture of atmospheric carbon dioxide in to soil, through the addition of crop and other organic biomasses; to combat climate change and global warming which may diminish greenhouse gases emission (EPA 2008) which ultimately, enhance soil quality and, ensure long term soil productivity. Metz et al. (2001) proposed seven climate change mitigation pathways including; carbon capture/storage, energy efficiency, reduction in carbon dioxide fuels, nuclear power, renewable energy, augmentation of biological sinks, and reducing the level of other than carbon dioxide greenhouse gases. Among these measures, augmentation of biological sinks and carbon capture/storage can effectively reduce carbon dioxide from the atmosphere while the other strategies could only, prevent further emissions. The carbon sequestration in soils could occur directly or indirectly (Soil Science Society of America 2001); direct carbon sequestration means chemical reaction to transform carbon dioxide into inorganic carbon-compounds like Ca and Mg carbonates and indirect soil carbon sequestration is a fixation of atmospheric carbon dioxide into plant biomass; via photosynthesis. However, some plant biomass is indirectly sequestered as organic carbon during decomposition. The change in organic carbon is significantly affected by agricultural activities and land use management practices, consequently disturbing the organic carbon density and distribution. Soil management measures including mulching, conservation tillage, crop

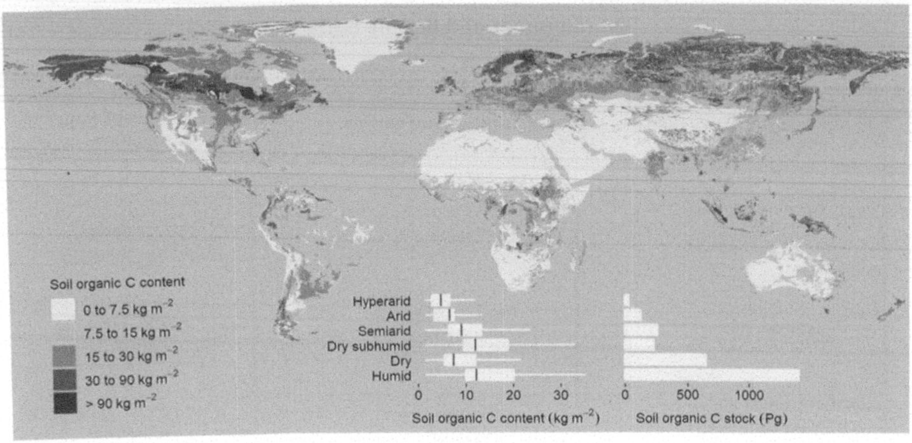

Fig. 6.2 Soil organic carbon content in dryland Soils. (Source: Plaza et al. 2018)

rotation, and cover crops can enhance organic carbon pool (Lal 2004). The sequestration of atmospheric carbon into SOM may (i) improve soil quality through enhanced fertility, (ii) improve soil structure and aggregate stability, (iii) increase water holding capacity, (iv) decrease nutrient loss, (v) reduce soil erosion, (vi) increase the ability of soil to remediate the toxic elements, and (vii) increase crop production (Fig. 6.2).

6.6 Strategies for Carbon Sequestration in Alkaline Soils

The alkaline soils around the world are often known for less vegetation and unproductive nature and considered unfavorable for carbon sink. However, a few strategies have been proposed by various researchers around the globe to enhance the carbon sequestration in alkaline soils (Fig. 6.3). A brief description of these strategies has been provided below:

6.6.1 Application of Amendments Containing Divalent Metals

Generally, alkaline soils can potentially sequester the atmospheric carbon dioxide through the chemical reaction of divalent metals as given below:

$$\mathrm{M}^{2+} + 2OH^{-} + \text{Carbon Dioxide} \rightarrow MCO_3 + \mathrm{H_2O}$$

Where M^{2+} denotes, a divalent metal react with CO_3^{2-} and form a stable metal carbonate. In alkaline soils, the Ca^{2+} and Mg^{2+} ions are presents in an excess amount,

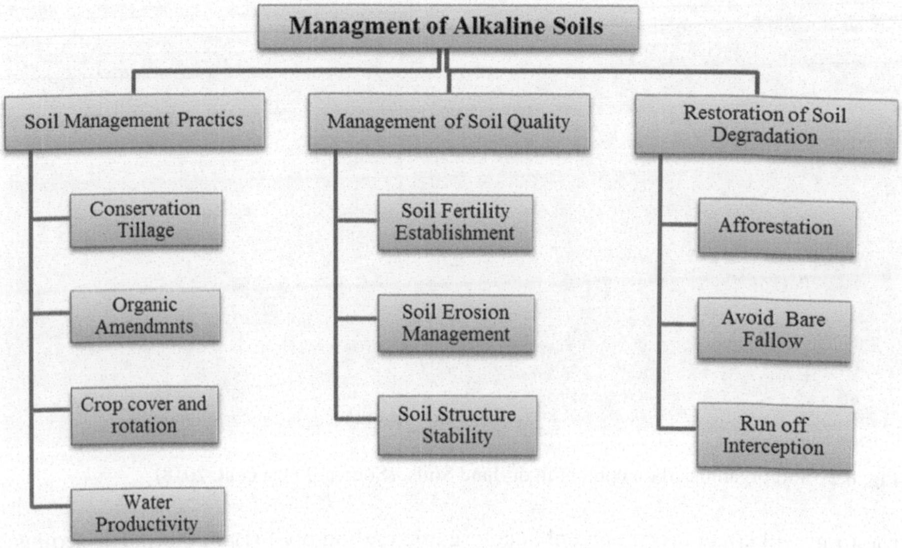

Fig. 6.3 Strategies for management of alkaline Soils

which can be detached and develop stable metal carbonates resulting an increase in soluble Na^+ in the soils which ultimately create sodicity (Pessarakli 1991; Singh et al. 2013). However, application of amendments containing more quantity of soluble divalent metals, in alkaline soils could enhance carbon sequestration. For instance, gypsum addition and other amendments containing a higher amount of Ca and Mg to alkaline soils can potentially fix the atmospheric carbon dioxide, in the form of carbonates and, organic matter (Han and Tokunaga 2014; Goel and Behl 1995). A considerable work has already been done to explore the response of gypsum on ameliorating the sodicity of soil which enhances crop production (Abdel-Fattah 2012; Hamza and Anderson 2003; Rasouli et al. 2013); since no work has reported the study of sequestration of atmospheric carbon dioxide through gypsum application, therefore, application of gypsum and other Ca and Mg containing amendments can serve as a useful strategy to mitigate climate change and ameliorate the sodicity in alkaline soils. Among other amendments, wood ash could potentially be exploited for carbon sequestration in alkaline soils since wood ash possesses higher amounts of alkaline metals such as Ca, Mg and K (Vance 1996). Similarly Zhao et al. (2017) revealed that application of wood ash to calcareous soil resulted in reduced carbon dioxide emission, due to carbonization of Ca and Mg present in wood ash; consequently increasing the organic carbon pool. Maljanen et al. (2006) stated that the application of wood ash acidified forest soils resulted in greater carbon dioxide fluxes. These contradictory findings could be attributed to differences in ash application rates, initial soil pH, soil types, and wood ash types. However, very little information is available about the effects of wood ash on carbon sequestration in alkaline soil. Besides gypsum and wood ash; the application of $CaCO_3$, CaO, and Ca Mg $(CO_3)_2$ as an amendment in alkaline soils can enhance the uptake of carbon dioxide by the soil, through carbonation reactions resulting in higher carbon sequestration.

6.6.2 Conservation Farming

Conservation farming is defined as minimized manipulation of soil by tillage and establishment of soil cover and, crop rotation (Hobbs 2007). This practice has great ability to sequester the carbon in soil through reducing soil erosion, enhancing water infiltration, soil surface aggregation, reducing compaction, increasing organic matter and carbon (FAO 2008). It can help in moderating soil temperatures and, suppressing weeds (Hobbs 2007) because conventional tillage disrupts the pores created by plant roots and soil microbial activity, therefore, the bare surfaces are exposed and soil aggregates exposed to breakdown from raindrops. This phenomenon gets even worst under alkaline soils where high temperature enhances organic matter decomposition ultimately destroying soil structure. The breakdown of soil aggregates clogs soil pores which reduce water infiltration and accelerate runoff resulting soil erosion reason why moldboard plow and disc harrow is considered a major cause of carbon loss in drylands because these processes destroy soil aggregates and accelerated decomposition. Hence, zero or reduced tillage and mulching is recommended under drylands to sequester carbon (Unger 1990) but positive effects these practices cannot reveal carbon sequestration in short period of time. Similarly, West and Post (2002) revealed that the shifting from conventional till to no-till has sequestered 0.57 ± 0.14 t C/ha/year and White and Rice (2009) revealed an improvement in insoluble carbon, enzyme activity and aggregate stability under no-till. Despite the fact that no-tillage is very useful for carbon sequestration there are few drawbacks of no-till such as a reduction in germination and soil temperature and higher chances of pest and disease attacks. Contrary, Baker et al. (2007) observed deeper penetration of roots under no-till system that is higher carbon sequestration due to no-till and further claimed that in some of the studies with sampling depths of more than 300 mm under no-till had exhibited no consistent deposition of organic carbon but a significant difference was recorded with a higher concentration of organic carbon in the upper soil layer while higher organic carbon concentration in the bottom layer with conventional tillage. Hence soil sampling must be done through entire soil profile instead top layer to avoid such variations because rhizodeposition describes all carbon loss from roots; including lysates, sloughed cells, insoluble mucilages, and mucigels and, water-soluble low molecular weight exudates, and secretions (Whipps and Lynch 1983). Therefore, it can be concluded that shifting from conventional tillage to no-tillage could sequester carbon up to substantial levels, in alkaline soils by decreasing the carbon dioxide emissions.

6.6.3 Cover Crops

Cover crops are defined as the crops cultivated with the objective to protect soil from erosion and, nutrient depletion by leaching and runoff (Reeves 1994). In alkaline soil, legumes and sorghum are generally grown as cover crops which enhance carbon sequestration through stabilization of soil structure and organic matter by increasing soil carbon and nitrogen concentration due to the decomposition of crop residues, root exudates and, symbiotic nitrogen fixation (Williams et al. 2016). Further, some legumes can be incorporated in soil on the basis of their ability to fix atmospheric nitrogen, leaf shedding and greater belowground biomass (Ganeshamurthy 2009). Venkateswarlu et al. (2007) indicated that alkaline arid lands can produce up to 3.75 t ha^{-1} biomass annually, without competing with the main crop and incorporation of this huge biomass into the soil, resulted in significant improvement of organic carbon, microbial biomass and, available nutrient. However, major constraint regarding the organic carbon stabilization in drylands is mono-cropping as it is a common practice in most of the drylands/alkaline soils resulting in fallowing of land for a significant period of time which may accelerate soil erosion and carbon loss (Venkateswarlu et al. 2007) (Tables 6.1 and 6.2).

Table 6.1 Carbon sequestration potential of various C sequestration strategies

Strategies	C sequestration (Ton/ha/year)
Cropland	0.10–0.20
Conservation farming	0.05–0.10
Mulching	0.10–0.20
Composting	0.05–0.10
Land fallowing	0.10–0.20
Nutrition management	0.10–0.20
Soil conservation	0.05–0.10
Reclamation of alkalinity and sodicity	0.10–0.20
Water management	0.05–0.10

Source: Lal et al. (1998)

Table 6.2 Soil management potential of carbon sequestration

Practices	C sequestration (kg ha^{-1})
Composting	1000–2000
Soil cover	800–1200
Zero tillage	100–500
Crop rotation	0–200
Manuring	0–200

Source: Lal (2004)

6.6.4 Manures and Composts

The manure and composts are most commonly used organic amendments which contain substantial amounts of carbon. The application of manure and composts to the alkaline soils not only improves soil physio-chemical properties but also improves soil structure and organic carbon pools (Gupta and Venkateswarlu 1994). Liu et al. (2013) reported an increased stable and labile organic carbon in 0–60 cm depth. The average concentration of particulate, dissolved and microbial biomass carbon in organic manure treated soil up to 0–60 cm depth was increased by 64.9–91.9, 42.5–56.9, and 74.7–99.4%, respectively. Carbon sequestration is dependent on soil management practices and initial soil carbon level. Among all management practices addition of manures and composted organic material has direct influence to improve the status of organic carbon for a relatively long period of time due to their stable carbon pool; protected from rapid mineralization due to temperature extremes in alkaline soils. Manure and composts effects on soil pH depend, on initial soil pH. Chang et al. (1991) reported that the addition of manure decreased soil pH and, composting may mitigate environmental issues due to the addition of raw manures/organic wastes (Paul 1996). The addition of composts can improve soil aggregation, structure, and fertility (Zink and Allen 1998) and reduce soil pH significantly due to release of $NH4^+$, carbon dioxide and, organic acids during the process of microbial metabolism (Walker et al. 2003); and increases organic carbon and soil structure resulting stability particularly that of the macroaggregates (Fortuna et al. 2003). Considering all the above-mentioned benefits it is concluded that the application of manures and composts are effective in alkaline soils for improving not only organic carbon but also help to decrease soil pH to enhance soil chemical and physical properties.

6.6.5 Crop Rotation/Selection

The type of crop grown, on alkaline soils contribute significantly in the organic carbon addition because land use influences the carbon inputs into the soil due varying quantities and nature of biomass production (Johnstone et al. 2010). Crop rotation means the arrangement of cropping pattern grown in periodic sequence on the same soil. This sequence of cropping affects soil quality, growth pattern and crop production in multiple ways such as organic carbon addition, soil structure and, soil particle aggregation, nutrients cycling and occurrence of pests and diseases, soil erosion, organic carbon concentration (Lal 2001). Therefore, crop selection and rotation determine the carbon addition (Ingram and Fernandez 2001). Rotation of the cropping sequence is proved more effective than monoculture for better retention of carbon and nitrogen (Biederbeck et al. 1984). Lal et al. (1998) proposed that increasing cropping intensity in semiarid alkaline soils is effective to reduce greenhouse gases and to alleviate climate change.

Further, crop rotation establishes better soil cover due to higher C:N ratio which will delay decomposition of organic biomass providing a longer retention time in the soil since C:N ratio lesser than 25:1 will lead to slower nitrogen mineralization (Sparks 2005). Therefore, having legumes in a crop rotation will have positive effects on carbon retention in alkaline soils. Smith (2014) observed that crop rotation with pasture system increased organic carbon significantly due to extensive rooting biomass which remained in the soil for an enhanced period of time. Moreover, continuous cropping leads to higher addition of organic matter compared with fallowing, common in arid regions (Álvaro-Fuentes et al. 2009b).

6.6.6 Biochar Intervention

Biochar is a resistant and recalcitrant black carbonaceous material; synthesized via thermochemical decomposition for carbon sequestration (Jeffery et al. 2011), however, it can potentially be used for several other purposes. The response of biochar application is more conducive in low fertile soil as an additional carbon supply may enhance plant productivity (El-Naggar et al. 2015). It is considered very stable in soil compared with original biomass due to its slow mineralization capacity (Spokas et al. 2011) which makes biochar a great choice for carbon sequestration (Clough and Condron 2010). This is the main reason biochar has fascinated researchers to use it as a tool for carbon sequestration and decreasing atmospheric carbon dioxide concentrations. The half-lives of biochar range from 102 to 107 years, providing more stable and, recalcitrant material comparing to other organic amendments, such as compost and, manure, etc. (Zimmerman et al. 2011). This enhanced stability could be due to its compressed aromatic structure (Purakayastha et al. 2015). Wang et al. (2016) calculated that, the mean residence time of its labile fraction, with a pool size of 3% was 108 days while the mean residence time of labile fraction with a pool size of 97%, was 556 days, therefore, it contributes to long-term C sequestration in soil and its addition also inhibit rapid native carbon mineralization (Zimmerman et al. 2011). Sheng et al. (2016) investigated that acid soils released higher carbon dioxide after biochar incorporation compared to neutral and alkaline soils and the decomposition of native and biochar supplemented organic carbon was enhanced in acidic soils whereas, biochar was a superior amendment for carbon sequestration in alkaline and neutral soils. Qayyum et al. (2015) revealed that the release of carbon dioxide after addition of biochar was not higher compared with control proving higher carbon sequestration potential of biochar since some previous studies have reported variable carbon mineralization potential of biochars (Qayyum et al. 2012) considering variable properties of biochar due to various feedstock, pyrolysis temperature, and resident time may affect carbon mineralization rates and carbon sequestration potential (Manya 2012; Sohi et al. 2010). Organic matter plays an important role in the formation of soil structure in alkaline conditions and if organic matter decomposes rapidly will deteriorate soil structure exposing soil to erosion and degradation. The cations increase pH and osmotic pressure and limit the uptake of adequate nutrients from the soil for plant growth. In recent times some research studies have been carried to study the effect of biochar on alkaline soils (Liu and Zhang 2012; Song et al. 2014), proved that lower rates did not increase pH of the alkaline soil rather increased cation exchange capacity and controlled salinization process. Biochar also increased soil moisture retention avoiding drought stress at critical stages in arid environments (Van Zwieten et al. 2010).

6.6.7 *Controlling Soil Erosion*

Alkaline soils are susceptible to soil erosion due to inadequate soil cover and torrential rainstorms. The intensity of rainstorms destroys the soil structure in alkaline soils resulting in the limited amount of organic carbon. Certain other factors including uneven land terrain aggravate the susceptibility to soil erosion and poor soil structure in drylands (García-Ruiz 2010). It is further alarming that erosion may increase up to 25–55%, during the twenty-first century (Delgado et al. 2013) which may expose carbonates to higher decomposition which may increase the carbon loss in the form of carbon dioxide due to climate change. Soil carbon losses can occur due to mineralization and soil erosion making this relationship very complicated. As a result of erosion, soil carbon may be eroded and, deposited to the nearby area which may have a positive effect on soil carbon sequestration. However, proper management of biomass left in soil may decrease the erosion in general and increase the organic carbon (Tiessen et al. 1994). Otherwise, in alkaline soils erosion can lead to exposure of the sub-soils, to erosive forces, rich in calciferous materials. Soil and crop management practices like plowing, application of fertilizers and, root exudates could lead to the dissolution of carbonates resulting in elevated emissions of carbon dioxide. In alkaline soils with higher amounts of carbonates and, bicarbonates and exposure of sub-surface could lead to the emissions of carbon dioxide. Lal (1990) suggested that increasing the resistance of soils against different erosive forces by establishing soil cover and providing a buffer against rainfall and runoff could prevent the soil losses and subsequently enhance carbon sequestration.

6.6.8 *Crop Residue Management*

Crop residues can either be completely removed from fields or they can be returned to the field directly or indirectly. The removal of crop residues would result in loss of carbon (Kätterer et al. 2012) therefore the utilization of crop residues will determine the extent of carbon sequestration due to the primary source of carbon constituting 40% of the total dry biomass therefore having the highest carbon sequestration potential (1.2–1.9 t C/ha/year) (Freibauer et al. 2004) hence, crop residue addition is related to its C input ability and soil retention time (Lugato et al. 2006). Kätterer et al. (2011) examined the stability of various soil amendments and, found that only 17% of crop residues retained in the soil after 50 years while farmyard manure (32%) and sewage sludge (54%) was present in the soil. The least retention of crop residues was could be due to its lesser recalcitrant portion than manure and sludge (von Lützow et al. 2006). In alkaline soils under dryland area, high temperature and poor soil structure promote the faster decomposition of crop residues incorporated in the soil, therefore the modification of crop residues to enhance its recalcitrance is important otherwise, surface application of crop residues without incorporating into soil can provide better results for both carbon sequestration as a mulch to preserve soil moisture because mulches are another efficient way to minimize soil erosion, surface runoff, and evaporation. Hence, crop residue management can reduce runoff, soil erosion, a temperature which may enhance water retention (Follett et al. 1987; Reicosky 1994).

6.7 Conclusion and Future Research Opportunities

The available data and analyses revealed that managing alkaline soil for carbon sequestration and combating climate. Better estimates of soil carbon stocks including inorganic and organic carbon may provide an opportunity for seeking a viable option to combat climate change in alkaline soils since existing estimates of organic carbon are unable to meet the challenges faced by carbon cycle. Alkaline soils cover a huge area with a diverse ecosystem and land use pattern which may contribute significantly in the global carbon cycle. Current misuse and, mismanagement of land has resulted in the destruction of soil structure and considerable loss of soil carbon. Appropriate soil and crop management can protect soils which can further be improved by systematic collection of more data regarding land use and its carbon sequestration potential in alkaline soils because the rising atmospheric concentration of carbon dioxide and global warming could also lead to heat stress that could offset the potential gain in carbon stocks. Rainwater harvesting, soil water retention and, crop water use efficiency needs to be maximized to ensure appropriate soil cover and reduce soil erosion susceptibility. The objective of sequestering a significant amount of carbon in alkaline soils is quite attainable which will result in social and environmental benefits.

References

Abdel-Fattah MK (2012) Role of gypsum and compost in reclaiming saline-sodic soils. IOSR J Agric Vet Sci 1(3):30–38. https://doi.org/10.9790/2380-0133038

Álvaro-Fuentes J, Lopez M, Arrue JL, Moret D, Paustin K (2009b) Tillage and cropping effects on soil organic carbon in mediterranean semiarid agroecosystems: testing the century model. Agric Ecosyst Environ 134:211–217. https://doi.org/10.1016/j.agee.2009.07.001

Baker CJ, Saxton KE, Ritchie WR, Chamen WCT, Reicosky DC, Ribeiro F, Justice SE, Hobbs PR (2007) No-tillage seeding in conservation agriculture. Published Jointly By CAB International and FAO

Baldock JA, Nelson PN (2000) Soil organic matter. In: Handbook of soil science. CRC Press, Boca Raton, pp 25–84
Batjes NH (1998) Mitigation of atmospheric carbon dioxide concentrations by increased carbon sequestration in the soil. Biol Fertil Soils 27:230–235. https://doi.org/10.1007/s003740050425
Biederbeck VO, Campbell CA, Zenter RP (1984) Effect of crop rotation and fertilization on some biological properties of a loam in Southwestern Saskatchewan. Can J Soil Sci 64:355–367. https://doi.org/10.4141/cjss84-037
Blanco-Canqui H, Lal R (2004) Mechanisms of carbon sequestration in soil aggregates. Crit Rev Plant Sci 23(6):481–504. https://doi.org/10.1080/07352680490886842
Bronick CJ, Lal R (2005) Soil structure and management: a review. Geoderma 124(1–2):3–22
Chang C, Sommerfeldt TG, Entz T (1991) Soil chemistry after eleven annual applications of cattle feedlot manure. J Environ Qual 20:475–480. https://doi.org/10.2134/jeq1991.0047242500200020022x
Clough TJ, Condron LM (2010) Biochar and the nitrogen cycle. J Environ Qual 39:1218–1223. https://doi.org/10.2134/jeq2010.0204
Cunliffe AM, Puttock AK, Turnbull L, Wainwright J, Brazier RE (2016) Dry land, calcareous soils store (and lose) significant quantities of near-surface organic carbon. J Geophys Res Earth Surf 121. https://doi.org/10.1002/2015JF003628
Dawson JJC, Smith P (2007) Carbon losses from soil and its consequences for land use management. Sci Total Environ 382:165–190. https://doi.org/10.1016/j.scitotenv.2007.03.023
Delgado JA, Nearing MA, Rice CW (2013) Conservation practices for climate change adaptation. Adv Agron 121:47–115. https://doi.org/10.1016/b978-0-12-407685-3.00002-5
Denef K, Stewart CE, Brenner J, Paustian K (2008) Does long-term center-pivot irrigation increase soil carbon stocks in semi-arid agro-ecosystems? Geoderma 145:121–129. https://doi.org/10.1016/j.geoderma.2008.03.002
El-Naggar AH, Usman AR, Al-Omran A, Ok YS, Ahmad M, Al-Wabel MI (2015) Carbon mineralization and nutrient availability in calcareous sandy soils amended with woody waste biochar. Chemosphere 138:67–73. https://doi.org/10.1016/j.chemosphere.2015.05.052
Entry JA, Sojka RE, Shewmaker GE (2004) Irrigation increases inorganic carbon in agricultural soils. Environ Manag 33:309–317. https://doi.org/10.1007/s00267-003-9140-3
EPA (2008) Carbon sequestration in agriculture and forestry. www.epa.gov/sequestration/index.html
Fang C, Smith P, Smith JU (2005) Is resistant soil organic matter more sensitive to temperature than the labile organic matter? Biogeosci Discuss 2:725–735. https://doi.org/10.5194/bgd-2-725-2005
FAO (2002) Harvesting carbon sequestration through land use change. A way out of rural poverty? The State of Food and Agriculture, Rome
FAO (2008) An international technical workshop investing in sustainable crop intensification the case for improving soil health. Integrated Crop Manage 6. 22–24 July Rome
Follett RH, Gupta SC, Hunt PG (1987) Conservation practices: relation to the management of plant nutrients for crop production. In: Follett RF et al (eds) Soil fertility and organic matter as critical components of agricultural production systems, SSSA special publication 19. ASA and SSSA, Madison, pp 19–52
Fortuna A, Harwood R, Kizilkaya K, Pauln EA (2003) Optimizing nutrient availability and potential carbon sequestration in an agroecosystem. Soil Biol Biochem 35:1005–1013. https://doi.org/10.1016/S0038-0717(03)00084-1
Freibauer A, Rounsevell MD, Smith P, Verhagen J (2004) Carbon sequestration in the agricultural soils of Europe. Geoderma 122:1–23. https://doi.org/10.1016/j.geoderma.2004.01.021
Ganeshamurthy AN (2009) Soil changes following long-term cultivation of pulses. J Agric Sci 147:699–706. https://doi.org/10.1017/S0021859609990104
García-Ruiz JM (2010) The effects of land uses on soil erosion in Spain: a review. Catena 81:1–11. https://doi.org/10.1016/j.catena.2010.01.001
Goel VL, Behl HM (1995) Fuel wood production potential of six prosopis species on an alkaline soil site. Biomass Bioenergy 8(1):17–20. https://doi.org/10.1016/0961-9534(95)00053-4

Goulding KWT (2016) Soil acidification and the importance of liming agricultural soils with particular reference to the United Kingdom. Soil Use Manag 32:390–399. https://doi.org/10.1111/sum.12270

Grace J (2004) Understanding and managing the global carbon cycle. J Ecol 92:189–202. https://doi.org/10.1111/j.0022-0477.2004.00874.x

Gupta JP, Venkateswarlu J (1994) Soil management and rain water conservation and use-V. Aridisols In: Soil management for sustainable agriculture in dry land areas. Bull Indian Soc Soil Sci 16:66–77

Hamza MA, Anderson WK (2003) Responses of soil properties and grain yields to deep ripping and gypsum application in a compacted loamy sand soil contrasted with a sandy clay loam soil in Western Australia. Aust J Agric Res 54:273–282. https://doi.org/10.1071/AR02102

Han YS, Tokunaga TK (2014) Calculating carbon mass balance from unsaturated soil columns treated with $CaSO_4$-minerals: test of soil carbon sequestration. Chemosphere 117:87–93. https://doi.org/10.1016/j.chemosphere.2014.05.084

Hobbs PR (2007) Conservation agriculture: what is it and why is it important for future sustainable food production? J Agric Sci 145:127–137. https://doi.org/10.1017/S0021859607006892

Ingram JSI, Fernandez ECM (2001) Managing carbon sequestration in soils: concepts and terminology. Agric Ecosyst Environ 87:111–117. https://doi.org/10.1016/S0167-8809(01)00145-1

IPCC (2001) Climate Change 2001: the scientific basis. Executive summary. Cambridge University Press, Cambridge

Janicka-Russak M, Kabala K, Burzynski M, Klobus G (2008) Response of plasma membrane H+-ATPase to heavy metal stress in Cucumis sativus root. J Exp Bot 59:3721–3728. https://doi.org/10.1093/jxb/ern219

Johnstone N, Haščič I, Popp D (2010) Renewable energy policies and technological innovation: evidence based on patent counts. Environ Resour Econ 45(1):133–155

Jeffery S, Verheijen FGA, van der Velde M, Bastos AC (2011) A quantitative review of the effects of biochar application to soils on crop productivity using meta-analysis. Agric Ecosyst Environ 144(1):175–187. https://doi.org/10.1016/j.agee.2011.08.015

Kaiser M (2012) The influence of mineral characteristics on organic matter content, composition, and stability of top soils under long-term arable and forest land use. J Geophys Res Biogeosci 117. https://doi.org/10.1029/2011JG001712

Kätterer T, Bolinder MA, Andrén O, Kirchmann H, Menichetti L (2011) Roots contribute more to refractory soil organic matter than above-ground crop residues, as revealed by a long-term field experiment. Agric Ecosyst Environ 141:184–192. https://doi.org/10.1016/j.agee.2011.02.029

Kätterer T, Bolinder MA, Berglund K, Kirchmann H (2012) Strategies for carbon sequestration in agricultural soils in Northern Europe. Acta Agric Scand Sect A Anim Sci 62:181–198. https://doi.org/10.1080/09064702.2013.779316

Knorr W, Prentice IC, House JI Hollan EA (2005) On the available evidence for the temperature dependence of soil organic carbon. Biogeosci Discuss 2:749–755. https://doi.org/10.5194/bgd-2-749-2005

Lal R (1990) In: Lal R, Stewart BA (eds) Soil erosion and land degradation: the global risks. Soil Degradation Springer-Verlag, New York, pp 129–172

Lal R (2001) World cropland soils as a source or sink for atmospheric C. Adv Agron 71:145–174. https://doi.org/10.1016/S0065-2113(01)71014-0

Lal R (2002) Soil carbon dynamics in cropland and rangeland. Environ Pollut 116:353–362. https://doi.org/10.1016/S0269-7491(01)00211-1

Lal R (2004) Soil carbon sequestration impacts on global climate change and food security. Science 304(5677):1623–1627. https://doi.org/10.1126/science.1097396

Lal R (2006) Managing soils for feeding a global population of 10 billion. J Sci Food Agric 86:2273–2284. https://doi.org/10.1002/jsfa.2626

Lal R, Kimble JM, Follet RF, Cole CV (1998) The potential of U.S. Cropland to sequester carbon and mitigate the greenhouse effect. Ann Arbor Press, Chelsea

Liu X. Zhang XC (2012) Effect of biochar on pH of alkaline soils in the loess plateau: results from incubation experiments. Int J Agric Biol 14:745–750. 12–005/SBC/2012/14 5-745-750

Liu E, Yan C, Mei X, Zhang Y, Fan T (2013) Long-term effect of manure and fertilizer on soil organic carbon pools in dryland farming in Northwest China. PLoS One 8(2):e56536. https://doi.org/10.1371/journal.pone.0056536

Lugato E, Berti A, Giardini L (2006) Soil organic carbon (SOC) dynamics with and without residue incorporation in relation to different nitrogen fertilisation rates. Geoderma 135:315–321

Maestre FT, Quero JL, Gotelli NJ, Escudero A, Ochoa V, Delga M (2012) Plant species richness and ecosystem multi-functionality in global dry lands. Science 335:214–218. https://doi.org/10.1126/science.1215442

Maljanen M, Nykänen H, Moilanen M, Martikainen PJ (2006) Greenhouse gas fluxes of coniferous forest floors as affected by Wood Ash addition. For Ecol Manag 237:143–149. https://doi.org/10.1016/j.foreco.2006.09.039

Manyà JJ (2012) Pyrolysis for biochar purposes: a review to establish current knowledge gaps and research needs. Environ Sci Technol 46(15):7939–7954

Marks E, Aflakpui GKS, Nkem J, Poch RM, Khouma M, Kokou K, Sagoe R, Sebastià MT (2009) Conservation of soil organic carbon, biodiversity and the provision of other ecosystem services along climatic gradients in West Africa. Biogeosciences 6:1825–1838. https://doi.org/10.5194/bg-6-1825-2009

Martens A (2000) management and crop residue influence soil aggregate stability. J Environ Qual 29:723–727. https://doi.org/10.2134/jeq2000.00472425002900030006x

Metz B, Davidson O, Swart R, Pan J (2001) Climate Change 2001: mitigation. Cambridge University Press

Muneer M, Oades JM (1989) The role of ca-organic interactions in soil aggregate stability.2. Field studies with C-14-labeled Straw, $CaCO_3$ and $CaSO_4{\cdot}H_2O$. Aust J Soil Res 27(2):401–409

O'Brien SL, Jastrow JD, Grimley DA, Gonzalez-Meler MA (2015) Edaphic controls on soil organic carbon stocks in restored grasslands. Geoderma 251:117–123. https://doi.org/10.1016/j.geoderma.2015.03.023

Oades JM (1988) The retention of organic-matter in soils. Biogeochemistry 5(1):35–70. https://doi.org/10.1007/BF02180317

Paul EA (1996) Soil microbiology, ecology, and biochemistry in perspective. In: Paul EA (ed) Soil microbiology and biochemistry, 3rd edn. Academic, San Diego, pp 13–17

Pessarakli M (1991) Formation of saline and sodic soils and their reclamation. J Environ Sci Health 26(7):1303–1320. https://doi.org/10.1080/10934529109375698

Plaza C, Zaccone C, Sawicka K, Méndez AM, Tarquis A, Gascó G, Gerard HBM, Schuur EAG, Maest FT (2018) Soil resources and element stocks in drylands to face global issues. Sci Rep 8:13788. https://doi.org/10.1038/s41598-018-32229-0

Post WM, Kwon KC (2000) Soil carbon sequestration and land-use change: processes and potential. Glob Chang Biol 6(3):317–327. https://doi.org/10.1046/j.1365-2486.2000.00308.x

Purakayastha TJ, Kumari S, Pathak H (2015) Characterization, stability and microbial effects of four biochars produced from crop residues. Geoderma 239:293–303. https://doi.org/10.1016/j.geoderma.2014.11.009

Qayyum MF, Steffens D, Reisenauer HP, Schubert S (2012) Kinetics of carbon mineralization of biochars compared with wheat straw in three soils. J Environ Qual 41:1210–1220. https://doi.org/10.2134/jeq2011.0058

Qayyum MF, Abid M, Danish S, Saeed MK (2015) Effects of various biochars on seed germination and carbon mineralization in an alkaline soil. Pak J Agric Sci 51(4):977–982

Rao DLN, Ghai SK (1985) Urease and dehydrogenase activity of alkali reclaimed soils. Aust J Soil Res 23:661–665. https://doi.org/10.1071/SR985066

Rao DLN, Pathak H (1996) Ameliorating influence of organic matter on biological activity of salt affected soils. Arid Soil Res Rehabil 10:311–319. https://doi.org/10.1080/15324989609381446

Rasouli F, Pouya AK, Karimian N (2013) Wheat yield and physico-chemical properties of a sodic soil from semi-arid area of Iran as affected by applied gypsum. Geoderma 193–194:246–255. https://doi.org/10.1016/j.geoderma.2012.10.001

Reeves DW (1994) Cover crops and rotations. In: Hatfield JL, Stewart BA (eds) Crops residue management, Advances in soil science. Lewis Publishers, Boca Raton, pp 125–172

Reicosky DW (1994) Crop residue management: Soil, crop, climate interactions. In: Hatfield JL, Steward BA (eds) Crops residue management. CRC Press. Inc, Boca Raton, pp 191–214

Rengasamy P, Olsson KA (1991) Sodicity and soil structure. Aust J Soil Res 29:935–954. https://doi.org/10.1071/SR9910935

Sanderman J (2012) Can management induced changes in the carbonate system drive soil carbon sequestration? A review with particular focus on Australia. Agric Ecosyst Environ 155:70–77. https://doi.org/10.1016/j.agee.2012.04.015

Schlesinger WH (2000) Carbon sequestration in soils: some cautions amidst optimism. Agric Ecosyst Environ 82:121–127. https://doi.org/10.1016/s0167-8809(00)00221-8

Sheng Y, Zhan Y, Lizhong Z (2016) Reduced carbon sequestration potential of biochar in acidic soil. Sci Total Environ 572:129–137. https://doi.org/10.1016/j.scitotenv.2016.07.140

Singh K, Singh B, Tuli R (2013) Sodic soil reclamation potential of Jatropha curcas: a long-term study. Ecol Eng 58:434–440. https://doi.org/10.1016/j.ecoleng.2013.07.006

Six J, Ogle SM, Breidt FJ, Conant RT, Mosier AR, Paustian K (2004) The potential to mitigate global warming with no-tillage management is only realized when practiced in the long term. Glob Chang Biol 10:155–160. https://doi.org/10.1111/j.1529-8817.2003.00730.x

Slessarev EW (2016) Water balance creates a threshold in soil pH at The Global Scale. Nature. https://doi.org/10.1038/nature20139

Smith JDV (2014) The effect of long-term no-tillage crop rotation practices on the soil organic matter functional pools. M.Sc. Agriculture dissertation. Stellenbosch University

Sohi S, Krull E, Lopez-Capel E, Bol R (2010) A review of biochar and its use and function in soil. Adv Agron 105:47–82. https://doi.org/10.1016/S0065-2113(10)05002-9

Soil Science Society of America (SSSA) (2001) Internet glossary of soil science. www.soils.org/sssagloss

Song YJ, Zhang XL, Ma B, Chang SX, Gong J (2014) Biochar addition affected the dynamics of ammonia oxidizers and nitrification in microcosms of a coastal alkaline soil. Biol Fertil Soils 50:321–332. https://doi.org/10.1007/s00374-013-0857-8

Spain AV, Isbell RF, Probert ME (1983) Soil organic matter in soils: an Australian viewpoint. CSIRO, Academic, Melbourne, pp 551–563

Sparks DL (2005) Metal and oxyanion sorption on naturally occurring oxide and clay mineral surfaces. In: Grassian VH (ed) Environmental catalysis. Taylor and Francis, Boca Raton, pp 3–36

Spokas KA, Novak JM, Stewart CE, Cantrell KB, Uchimiya M, DuSaire MG, Ro KS (2011) Qualitative analysis of volatile organic compounds on biochar. Chemosphere 85:869–882. https://doi.org/10.1016/j.chemosphere.2011.06.108

Stewart CE, Plante AF, Paustian K, Conant RT, Six J (2008) Soil carbon saturation: linking concept and measurable carbon pools. Soil Sci Soc Am J 72(2):379

Sundermeier A, Reeder R, Lal R (2005) Soil carbon sequestration fundamentals. Rep. No. OSU FactsheetAEX-510–05.OSUE, Columbus. http://ohioline.osu.edu/aex-fact/0510.html

Tavakkoli M, Eslami S, Motavasselian M (2015) Nutrient rich versus nutrient poor foods depressed patients based on Iranian traditional medicine resources. Avicenna J Phytomed 5(4):298–308. https://doi.org/10.22038/ajp.2015.4185

Tiessen H, Cuevas E, Chacon P (1994) The role of soil organic matter in sustaining soil fertility. Nature 371:783–785. https://doi.org/10.1038/371783a0

Unger PW (1990) Conservation tillage systems. In: Singh RP, Parr JF, Stewart BA (eds) Advances in soil science, Advances in soil science, vol 13, pp 27–68. https://doi.org/10.1007/978-1-4613-8982-8

Van Zwieten L, Kimber S, Morris S, Chan KY, Downie A, Rust J, Joseph S, Cowie A (2010) Effects of biochar from slow pyrolysis of papermill waste on agronomic performance and soil fertility. Plant Soil 327:235–246. https://doi.org/10.1007/s11104-009-0050-x

Vance ED (1996) Land application of wood-fired and combination boiler ashes: an overview. J Environ Qual 25:937–944

Venkateswarlu B, Srinivasarao C, Ramesh G, Venkateswarlu S, Katya JC (2007) Effects of long-term legume cover crop incorporation on soil organic carbon, microbial biomass, nutrient build-up and grain yields of sorghum/sunflower under rainfed conditions. Soil Use Manag 23:100–107. https://doi.org/10.1111/j.1475-2743.2006.00068.x

Viscarra-Rossel RA, Webster R, Bui E, Baldock JA (2014) Baseline map of organic carbon in Australian soil to support national carbon accounting and monitoring under Climate Change. Glob Chang Biol 20:2953–2970. https://doi.org/10.1111/gcb.12569

von Lützow M, Kögel-Knabner I, Ekschmitt K, Matzner E, Guggenberger G, Marschner B, Flessa H (2006) Stabilization of organic matter in temperate soils: mechanisms and their relevance under different soil conditions-a review. Eur J Soil Sci 57:426–445. https://doi.org/10.1111/j.1365-2389.2006.00809.x

Walker DJ, Clemente R, Roig A, Bernal MP (2003) The effects of soil amendments on heavy metal bioavailability in two contaminated mediterranean soils. Environ Pollut 122:303–312. https://doi.org/10.1016/S0269-7491(02)00287-7

Wang XJ (2015) Carbon accumulation in arid croplands of Northwest China: pedogenic carbonate exceeding organic carbon. Sci Rep UK 5

Wang J, Xiong Z, Kuzyakov Y (2016) Biochar stability in soil: meta-analysis of decomposition and priming effects. Glob Change Biol Bioenergy 8:512–523. https://doi.org/10.1111/gcbb.12266

West TO, Post WM (2002) Soil organic carbon sequestration rates by tillage and crop rotation: a global data analysis. Soil Sci Soc Am J 66:1930–1946. https://doi.org/10.3334/CDIAC/tcm.002

Whipps JM, Lynch JW (1983) Substrate flow and utilization in rhizosphere of cereals. New Phytol 95:605–623. https://doi.org/10.1111/j.1469-8137.1983.tb03525.x

White PM, Rice CW (2009) Tillage effects on microbial and carbon dynamics during plant residue decomposition. Soil Sci Soc Am J 73(1). https://doi.org/10.2136/sssaj2007.0384

Williams A, Kane DA, Ewing PM, Atwood LW, Jilling A, Li M (2016) Soil functional zone management: a vehicle for enhancing production and soil ecosystem services in row-crop agroecosystems. Front Plant Sci 7:1–15. https://doi.org/10.3389/fpls.2016.00065

Yang LF, Li GT (2011) Advances in research of soil inorganic carbon. Chinese Journal of Soil Science 42:986–990

Zhao HL, Tian XH, Chen YL, Dong JJ, Shi JL (2017) Effect of exogenous substances on soil organic and inorganic carbon sequestration under maize stover addition. Soil Sci Plant Nutr 63:591–598. https://doi.org/10.1080/00380768.2017.1409603

Zimmerman AR, Gao B, Ahn MY (2011) Positive and negative carbon mineralization priming effects among a variety of biochar-amended soils. Soil Biol Biochem 43:11690–11179. https://doi.org/10.1016/j.soilbio.2011.02.005

Zink TA, Allen MF (1998) The effects of organic amendments on the restoration of a disturbed coastal sage scrub habitat. Restor Ecol 6:52–58. https://doi.org/10.1046/j.1526-100x.1998.00617.x

Chapter 7
Metal-Organic Frameworks for Carbon Dioxide Capture

Shivy Mangal and S. Shanmuga Priya

Abstract The ever increasing amounts of CO_2 in the atmosphere calls for immediate action and required remediation. A number of techniques for CO_2 capture have been used for quite some time including amine based scrubbing, absorption, membranes, zeolites etc. Metal organic frameworks (MOFs) have come across as worthy candidates for carbon dioxide capture, storage and transport on account of its large surface area, large pore volume and structures that can be tuned as per requirement. They exhibit properties superior to those of zeolites and other composites. This chapter deals with the use of metal organic frameworks (MOFs) as promising agents for carbon dioxide capture. A number of MOFs have been presented along with a description of their superior properties. Zirconium based MOF, it's characterization and demonstration of its high uptake capacity has been presented. Metal organic framework based derivatives obtained by pyrolysis and their performance as compared to the parent MOF is also presented.

Keywords Metal organic framework · CO_2 capture · Storage & transportation · Carbon sequestration · Clean energy · CO_2 remediation · Zirconium metal organic framework · Pyrolysis of metal organic framework

Abbreviations

BBC	4,4′,4″-(benzene-1,3,5triyl-tris(benzene-4,1 diyl))tribenzoate)
BDC	1, 4-Benzene dicarboxylic acid
BPEE	1,2 bipyridylethene)
BPTC	1,1′-biphenyl-3,3′,5,5′-tetracarboxylate

S. Mangal · S. Shanmuga Priya (✉)
Department of Chemical Engineering, Manipal Institute of Technology, Manipal Academy of Higher Education (MAHE), Manipal, Karnataka, India
e-mail: shan.priya@manipal.edu

Inamuddin et al. (eds.), *Sustainable Agriculture Reviews 38*, Sustainable Agriculture Reviews 38, https://doi.org/10.1007/978-3-030-29337-6_7

BTC	1,3,5-benzenetricarboxylate
BTE	4,4′,4″-(benzene-1,3,5-triyl-tris(ethyne-2,1-diyl)) tribenzoate
CNF	Carbon nano fiber
CNS	Carbon nitride sheets
DMF	N, N-dimethyl foramide,
DMS	Dimethyl Sulphate
DOBDC	2,5-dioxido-1,4-benzenedicarboxylate
GNP	Graphene nano plates
GO	Graphene oxide
HKUST	Hong Kong University of Science and Technology
MEA	Monoethanolamine
MIL	Materials of the Institut Lavoisier
MOF	Metal organic framework
PSA	Pressure swing adsorption
SBU	Secondary Building Units
TEA	triethylamine
TEOA	triethanolamine
TGA	Thermo-gravimetric analysis
UiO	University of Oslo
XRD	X-Ray Diffraction
ZIF	Zeolitic Imidazolate Frameworks

7.1 Introduction

It is a worldwide concern that the unprecedented increase in carbon dioxide (CO_2) levels in the atmosphere will cause global warming and become a threat to nature. This might lead to unpredictable consequences. It is also globally accepted that since the last century, excessive amounts of CO_2 have been generated by human activity, such as from large scale power plants and industries.

The levels of CO_2 emissions can be reduced by two ways, viz. by using alternative energy sources such as wind, solar, hydroelectric and nuclear energy sources, and by developing CO_2 capture technologies to be used with fossil fuel (coal, oil and gas) power plants. Even though other energy generating technologies are available, fossil fuel sources are the current main energy source and will still be in a foreseeable future. Therefore, the latter aspect which is to develop the carbon dioxide capture technology is urgently needed.

The fossil fuel utilization can be separated into four categories-industry, electricity & heat, transport and others (residence etc.). The former two, industry and electricity & heat, accounted for 63% of the CO_2 emission in 2012, with the part from industry being 20%. This ranges from small plants such as refineries and cement production to large scale plants like iron and steel industries. Electricity & heat are mainly supplied by large scale power plants which are responsible for the other 43% of the CO_2 emission. This points out the importance of using CO_2 capture technologies in fossil fuel combustion industries and large point sources (such as large scale power plants) (IEA Greenhouse Gas R&D Programme (IEA GHG) 2007).

7.2 Literature Survey

The recent use of metal organic frameworks in order to achieve the physisorption of CO_2 is more efficient than chemisorption as it uses less energy in order to carryout CO_2 adsorption. In the process of CO_2 capture most of the MOFs are made of ligands of organic compositions of carboxylates and poly pyridines and the activation metals used in most of the MOFs are from d and f block elements. The use of fluorinated MOF increases adsorption de-adsorption capacity and using amino functionalized aromatic compounds increased the strength of the activation sites and fuels the porosity of the MOF. Use of lightweight metal and use of carbon template MOFs or MOF-CNF(carbon nano fiber) increases the uptake drastically.

The quality of MOF adsorbent depends on various parameters such as high surface area, high selectivity, higher volumetric and porous capacity and a very good mechanical strength using minimal energy. Since amine scrubbers have been used in industries to capture CO_2, MOFs can be an alternative to this as is requires up to 30% less energy and has better mass and heat transfer conditions then most of the dry adsorbents.

Some of the MOF's with high adsorption capability are:

1. $Cu_3(BTC)_2$(where BTC = 1,3,5-benzenetricarboxylate).
2. $Zn_4O(BBC)_2(H_2O)$(where BBC = 4,4′,4″-(benzene-1,3,5triyl-tris(benzene-4,1 diyl))tribenzoate).
3. $Zn_4O(BTE)_{4/3}$(BDPC) where BTE = 4,4′,4″-(benzene-1,3,5-triyl-tris(ethyne-2,1-diyl)) tribenzoate and BPDC = biphenyl-4,4′-dicarboxylate).

Some of the MOF with high selectivity are:

1. Zn_2(BPDC)(BPEE) (where BPDC = 4,40-biphenyldicarboxylate and BPEE = 1,2 bipyridylethene).
2. (Ni_2L_2) (BPTC) (where BPTC =1,1′-biphenyl-3,3′,5,5′-tetracarboxylate).

Humidity facilitates the adsorption capacity of CO_2 but the presence of water is found to be damaging to Mo(DOBDC) (DOBDC = 2,5-dioxido-1,4-benzenedicarboxylate). Mg-MOF-74 binding sites have been studied and it is found that it has two activation sites which are amino and water bounded. Use of unsaturated metal centers and metal doping, using fluorinated MOFs and along with proper aligned ligand sites can increase adsorption capacity and can deliver output with less energy consumption.

NH_2-MIL-125 was experimentally concluded to be one of the most efficient sorbents at higher temperature Using this MOF which is functionalized by amino group the selectivity and permeability has shown good quality results. Usually this was combined with metal matrix membrane in order to increase the CO_2, H_2O adsorbing capacity in the pre combustion process in order to carry out CO_2 sequestration.

The use of multichannel monoliths for CO_2 adsorption using photo catalysis is proved to be more efficient than most of the chemisorption processes. Use of d & f block elements over extended surface of TiO_2, depositing these d & f block metals on monolithic structures of ceramic honeycomb with optic fiber for the process of vapor phase photo reduction of CO_2 has also been studied. The results achieved during this setup have proven to be more efficient for CO_2 adsorption than using chemisorption.

Metal organic frameworks (MOFs) not only finds application in the CO_2 storage and processing but also in various other fields such as drug delivery system, hydrogen, nitrogen, methane storage, various electrochemical, mechano-chemical sonochemical synthesis. By using lanthanide metal ions it can be also used in various sensors devices.

According to recent studies and experiments it is evident that CO_2 adsorption capacity of the MOF increases with increase with the increase in CO_2 pressure and decreases with increase in temperature. Adsorption capacity of a substance using MOF can be determined at various temperatures. MIL101(Cr) has a staggering adsorption capacity of 8.07 mmol/g and that of HKUST is said to 7.19 mmol/g.

Presence of heterogeneity in the adsorbing composite tend to increase the performance of MOF in adsorbing quality and regeneration. It has been experimentally shown that MOFs structures begin to slump on exposure to toxic gases and its perseverance indicated that these materials can be used again and again.

Thermo-gravimetric analysis (TGA), Fourier infrared spectroscopy, x-ray diffraction are some of the methods used in characterization of MOFs. Solvothermal reaction method to synthesize MOF is one of the most common methods of MOF synthesis. Certain MOF such as MIL-53(AL)/GNP (GNP = graphene nano plates) are produced using this method which show significant CO_2 adsorption capacities.

The separation and capture of CO_2 is said to better in ionic liquid when it is dispersed in a number of MOFs. On dispersion, properties of porous material change coulomb interaction which assist the frameworks and several topographic structures. The ionic liquid become more inclined in 2-D MOFs due to active metal ions and coulombic interactions are also improved. Injecting ionic liquids increases MOFs performance which has been proved through various simulation experiments.

The adsorption capacities of aqueous alkanol amine post combustion using wet scrubbing technology can be further more improved by using piperazine in addition to the process. The piperazine grafted CuBTC was inspected despite of reduced surface area along with the low availability of amine. The dominance of these materials at low pressure seems to be improved pertaining to the post combustion. On further comparison to other MOF such as mmen—CuBTC it didn't manifest the same improvement as shown in the chosen one. Presence of water vapor is a key feature in adsorption capabilities of adsorption. Due to presence of mmen in the second MOF it showed steric hindrance due to integration of secondary amine present in the MOFs by using post synthetic modification process it has proved to show better adsorption capabilities.

In various experiments performed it has been evident that application of sound shows improvement in gas solid contact as well as adsorption efficiency. The performance of MOFs such as HKUST-1 showed remarkably higher capacities at pressure from 0 to 15 bar and therefore an ex situ regeneration strategy must be applied. It is also seen that it is capable of 10 degenerations –generation cycles. It shows chemo-physical features in which water bed is used as a fluidized base.

At this level the prime challenges faced by MOF based photo catalysis is its feasibility in terms of economics and its shortcomings in terms of stability. Copper has proven to have good primary selectivity for photo reduction of CO_2 in methanol. MOFs have also been utilized as primary photo catalysts as they have the ability to harvest visible light and also has a huge number of active centers for CO_2 adsorption and the organic linkers can be adjusted for a more efficient light harvesting. Due to their versatility they can be coupled with semiconductors. It can also act as a co-catalyst to promote higher catalytic activity. Currently it is not industrially viable because its production volume is available in very less quantities along with its low turnover due to less electrical conductivity and lack to accessibility for active metal site. Ultimately, most of the MOFs application depend upon the use of TEA(triethylamine) as the sacrificial agent. As an alternate, water can be used as electron donor for its low cost accessibility and feasibility. Development of advanced MOFs appears to be a challenge and its economic sustainability is still a question in terms of performance.

MOFs can be synthesized using various methods such as slow diffusion methods and microwave irradiation method or solvothermal techniques however solvothermal and diffusion provide less surface area whereas microwave irradiation also saves us some minutes of reaction time in comparison to other methods along with better phase selectivity and facile morphology control. MIL-77, MIL 100 and MIL 101 were made and various tests were carried out on theses to concur these results. Physio-thermal properties of MOFs using microwaved MOF5 is similar to conventional hydrothermal heating.

Some studies have also proposed new metal organic frameworks based on the existing framework of $Cu_3(BTC)_2$. They were simulated with help of grand canonical Monte Carlo simulation in which three components were studied according to their performances:

1. $Cu_3(BTC)_2$.
2. $Sc_3(BTC)_2OH_3$.
3. $Sc_3(MeBTC)_2(OH)_3$.

In tests conducted $Sc_3(MeBTC)_2(OH)_3$ showed better selectivity of CO_2 with respect to other MOFs whereas selectivity didn't show any new result with Cu_3BTC_2 remaining the highest whereas working capacity of 2nd MOF was better than the rest in terms of PSA(pressure swing adsorption) which proved that the 3rd MOFs used was more efficient in CO_2 gas adsorption provided selectivity and work capacity were optimal.

MOFs have been proved to be superior in comparison to their porous solvents due to various geometric, facile and porosity factors. The versatility in the central metal frameworks of MOFs leads to adsorption of CO_2,SO_2 and other pollutants . Acid base properties play a dominant role in adsorption capacities as it improves the adsorption because of acid base interaction and formation of pi-complexes, in MOFs adsorption capacities have also improved due to exposure to electrostatic interactions which activate cations and anions present in the active metal sites. The hydrogen bonding between OH or O has been recommended for effective adsorption. However, due to pressure swinging adsorption, some MOFs tend to show difference in behaviors.

Recent studies have shown that modified compound HKUST-GO1 (GO= graphene oxide) instead of HKUST-1 showed better results in terms of adsorption due to its difference in porosity and the composite had a better adsorption capacity than that of the previous compound and showed a rise in adsorption blue up to 38% which was quite staggering.

MOFs are said to depict unique tunable functionalities which lead to various application including gas storage and adsorption. The ability of MOFs to be readily tuned by modifications of their organic linkers or metal centers has enabled their usage in a plethora of fields. For example, MOF-5 can be used as charge carrier for photo catalysis. It is composed of Zn_4O_{13} clusters surrounded by 6 terephthalate ligands and has shown effective results on analysis by spectroscopy. Various strategies can thus be applied for developing MOFs for photo catalysis. The photo catalytic properties can be enhanced by increasing the adsorption of light along with reducing the recombination rate of photo generated charge carriers, by modification and proper functionalization, the quality of ligands can be improved by decorating with co –catalyst. It has been proved in several experiments that how in MIL-15(Ti) modified by using NH_2 and using it as a ligand can help increase the CO_2 reduction capacity of MOFs, also metal doping NH_2-UiO-66(UiO = Universiteteti Oslo series) with platinum, or coupling with semiconductors. Such modifications of MOFs can help us in increasing CO_2 adsorption capacity.

Further studies have shown that use of mono substituted groups reduce the band gap for physio absorption. The introduction of functional groups o the organic linkers have shown a significant increase in absorbance. For example, UiO-66-NH_2 showed the highest absorbance among other fellow amino groups. Metal cations which are doped can also be performed with electronic mediation to aid change transfer process. The process of doping cerium to MIL101 using thermal treatment process along with adding Pd as nanoparticles showed improved results than the parent MOF.

It has also been seen in various experiments that the use of Zr instead of Ti ions have led to increment in photocatalytic activity in NH_2-UiO-66(Zr)which shows about 1.7 times greater capacity as Ti has a strong reducing capacity as compared to Zr. Also formation of titanium oxide cluster along with strong interactions with noble metals benefit the photocatalytic activity. Further if MOFs are coated with graphene oxide, it increases the conductivity of material for photo generation showing large surface area and higher absorption capacity. Use of sacrificial agent such as TEOA(triethanolamine) can increase CO_2 reduction over Uio-66/CNS(CNS = carbon nitride sheets) . The CO_2 yield obtained was about 59.4 pnumol/g it also showed reproducible photocatalytic activity. Use of carbon nitride sheet shows increased photocatalytic activity. Ag/AgCl@ZIF-8 showed better levels of activity and durability compared to other coupling of MOFs .

On comparison to other techniques absorption is more preferable in terms of cost reduction and energy saving for post combustion capture with respect to other technologies. Flow chemistry approach has been preferred in the recent years for synthesis of metal organic frameworks as it requires less time and offers a wide range of temperature for the process to be carried out. Carboxylates and nitriles are the most preferred linkers along with d&f block elements due to high affinity of lanthanide ions for donor atoms and ligands containing oxygen and nitrogen as hybrids. (DMF) i.e. N, N-dimethyl foramide, dimethyl sulfate (DMS) are some of the preferred solvents in the synthesis. Pore structure is an important factor to be considered in MOFs manufacturing which is formed after the guest molecule is evacuated. Reduction of pore size increase the CO_2 adsorption capacities of MOFs. Larger the specific surface area larger is the absorption of CO_2. MOFs with high content of Lewis acidity exhibit relatively high heats of CO_2 adsorption at a relatively lower pressure which leads to increase in enhanced selectivity for CO_2 capture. The quadruple moment is also responsible for greater polarizability of MOF for CO_2 capture.

MOFS like Mg-MOF -74, MOF-Mg-74, mmen-MG2(dobpdc, (Cu(Me-4py-trz-ia)), SNU-5 showed significant CO_2 uptake at .15,1.0 bar pressure compared to other conventional MOF. Spectroscopic and diffraction measurements along with density functional theory are few of the mechanism used for CO_2 capture. Chain model is preferred over pair models for CO_2 capture. MOFs like Porphyrin based MOF(Al)/(Cu) are preferred for photocatalytic conversion of CO_2 as they fall under visible light source forming methanol as the product and using water as proton donor wherein most of them used TEOA as their donors forming formic acid as the end product although providing a significant yield. Flexibility in design and proper synthesis of their frame work through appropriate choice of SBU (secondary building units) and functionalization of ligands can cause an exceptional amount of improvement for CO_2 capture. With development in new synthetic technique of manufacturing MOF material for the sole purpose of CO_2 adsorption, have opened a lot of doors for carbon capture and sequestration technologies with commercialization still being a challenge.

Among other applications MOFs can be used in pure form or materials can be derived from MOFs that can be used in electrolysis. Uniform pore structure, larger surface area along with tunable structure, existence of sites which are electroactive, Nano pore microenvironment which is distinctive in nature, appropriate thermal conductivity, chemical stability are the requirements of MOFs for its full potential application which will give us enhanced performance. The availability of MOFs at large scale has been hindered due to which it is not fulfilling the industrial requirements. Intrinsic conductivity of MOF significantly varies its electrolytic activity.

Single active sites of MOFs used for CO_2 photo reductions are OMS, metal-porphyrins and reactive functional group. Single active sites increase the potential and design opportunities of MOFs capable of forming highly selective products.

Zeolite containing MOFs are also quite common. They have two types of topologies namely rho and sod topologies. Studies have shown that zeolite containing MOFs exhibited better adsorption then most of the MOFs along with high heat of adsorption.

MOFs are capable of showing high porosity physio-absorption. It has the capacity of hosting a large number of guest species as photonic units based on their functions which can be stored in multiple units based on functional capabilities in luminescence sensing, white light emission photo catalysis, nonlinear optics, lasing devices, data storage and in biomedicine which have capabilities of high performance due to its photonic functions. Artificially, MOFs can provide us with a large surface area for operation and also a good template to carry out the porous catalysis.

7.2.1 Carbon Dioxide Capture and Storage

The complete process to capture the CO_2 in industry comprises of three stages, viz. CO_2 capture, transport and storage. Carbon capture technology has been used in the iron and steel industries for reuse since 1930s (Nonhebel 1936). However, it is still a new technology for fossil fuel combustion power plants. CCS systems were applied successfully to small scale coal and gas combustion plants in the 1980s, but there were no subsequent developments. Recently there has been a large increase in research into CCS systems to be used with large scale power plants. As there is the lack of CCS systems for most of the fossil fuel based power plants, studies are mainly focused on the feasibility of incorporating a CCS system into the current running power plants (IEA Greenhouse Gas R&D Programme 2005) and the engineering feasibility of building new large power plants with CCS (carbon capture and storage) systems in the near future (IEA Greenhouse Gas R&D Programme (IEA GHG) 2007).

7.2.1.1 Carbon Dioxide Capture

The carbon dioxide capture technology can be divided into three types based on the combustion type i.e. post-combustion, pre-combustion or oxy-combustion.

The post-combustion is the most common technology. It applies to the combustion system that burns fossil fuels in air. The post combustion product is a mixture of gases containing CO_2 and N_2 primarily. During the capture process, CO_2 is separated from the gas flow and captured.

The Pre-combustion process involves the chemical reaction before combustion. Before the combustion, oxygen and water steam are pumped in with the gas fuel to generate CO_2 and H_2 at a high temperature. Then CO_2 can be captured from the gas mixture and H_2 can be recycled as an energy source.

Oxy-combustion technology burns fuel with purge oxygen rather than air, in which CO_2 with high purity can be produced. Pure CO_2 can be easily recycled rather than going through other capture processes.

7.2.1.2 Carbon Dioxide Transport and Storage

After the successful capture of carbon dioxide the next hurdle is the transport and storage of the captured carbon dioxide. For the large scale transport of CO_2 captured at power plants, the most economic method is a pipeline to transport large amounts of CO_2 up to distances of 10 km. The other methods includes train and shipping, which are more feasible for smaller amounts of CO_2.

The final stage of a CCS system is the CO_2 storage. There are two main methods to store the gas. The first one is to bury the gas under the earth which has now been proven to be feasible. CO_2 can be compressed to 100–150 bar to be stored under 800 m depths of the previously-mined oil and gas fields. The second method is to compress the CO_2 to liquid and then inject it in the water near the seafloor. By doing so, the gas can be prevented from entering the atmosphere (IEA Greenhouse Gas R&D Programme 2005, IEA Greenhouse Gas R&D Programme (IEA GHG) 2007; Nonhebel 1936).

7.3 Current Carbon Dioxide Capture Technologies

7.3.1 Absorption- Amine Based Scrubbing

Amine based scrubbing has been used as an economically effective method to separate CO_2 from flue gas in industries for a long time (Rochelle 2009). In this process, amine solutions are used as solvents to capture CO_2 at low temperatures and then release the gas by heating. Amines are derivatives of ammonia, and can be obtained by replacing the ammonia hydrogen atom with alkyl or alryl groups. The most commonly used amine for CO_2 capture is MEA (monoethanolamine) .

First the gas is bubbled through amine solution to capture CO_2. Then the solution thus formed is heated up thereby releasing CO_2 which can then be compressed for further use and leaving behind the amine solution which can then be recycled for reuse.

Amine based scrubbing technologies are mostly likely to be incorporated into CCS systems in the near future because so far they are the best developed of the potential technologies, and because they have relatively lower costs compared with other proposed capture methods. However, there remain concerns for the environment due the volatility and toxicity of amines. The feasibility depends on the economic and environment concerns, and technology incorporation into large scale power plants has been studied by international organizationsand panels (Rochelle 2009; Rao and Rubin 2002; Shao and Stangeland 2009).

7.3.2 *Absorption: Aqueous Ammonium Absorption*

Aqueous ammonia solutions for CO_2 capture were first reported some decades ago (Yaghi and Li 1995) and have drawn considerable attention because they are less corrosive and more stable than amine solutions. However, since ammonia is a toxic gas, it is necessary to prevent ammonium entering into the atmosphere. Ammonium-based systems operate efficiently at lower temperature. The lower temperature also minimizes ammonium volatility and the possibility for its leakage.

The reversible reaction temperature for ammonia and CO_2 is lower than the interaction temperature for CO_2 and amine based solvents. Therefore the low quality waste heating equipment at power plants might be exploited to release captured CO_2 in amine based systems.

7.3.3 *Surface Adsorption*

Although the chemical absorption methods listed above have proved to be effective, they can be harmful to the environment on account of toxic nature of the chemicals used as absorbing agents. An alternative to such chemical absorption is the use of materials that adsorb CO_2 on internal surfaces. These materials, on account of having large surface areas can capture CO_2 molecules on their surface. One of such promising material is Metal organic frameworks (MOFs). MOFs have attracted a lot of attention as CO_2 capture candidates due their variety and tunable pore structure.

7.4 Introduction to Metal-Organic Frameworks

Metal organic frameworks (MOFs) are a novel class of crystalline materials possessing large surface areas and variable geometries. They were first reported in the early 1990s (Venkataraman et al. 1994), and were synthesized to by combining a metal center and organic ligands thereby producing cage like structure. Systematic MOF synthesis methods and characterization of properties have been reported by Yaghi and his group (Yaghi and Li 1995; Rowsell and Yaghi 2004). Due to their large surface areas and properties superior to zeolites, MOFs have gathered large attention and are considered to be worthy candidates to be used as agents in carbon dioxide capture technologies.

7.4.1 Structure of Metal Organic Frameworks

MOFs contain organic ligands linked to each other by metal-containing centres. The combination of large number of possible organic linkers and metal-containing centres such as zinc, aluminium, iron etc. increases the varieties of crystal structures. This has led to more than 20,000 MOFs reported so far. They have various well-defined porous framework structures and are robust enough to remain porous after guest species been removed, among which ZIF-8, Cu-BTC (also named HKUST-1) (Chui et al. 1999), MOF-5 (Li et al. 1999) and MIL-53 (Loiseau et al. 2004) etc. are commonly studied ones. The pore structure can be as simple as ZIF-8 and MOF-5 (Fig. 7.1a, b) with one kind of pore, or can be designed to be more complicated and containing different sizes of pores as in Cu-BTC (Fig. 7.1c). The pores can also be long channels as in MIL-53 (Fig. 7.1d). The diameter of the pores ranges from a few Angstrom to several tens of nanometers (mesopores) (Peng et al. 2014; Zhu and Xu 2014).

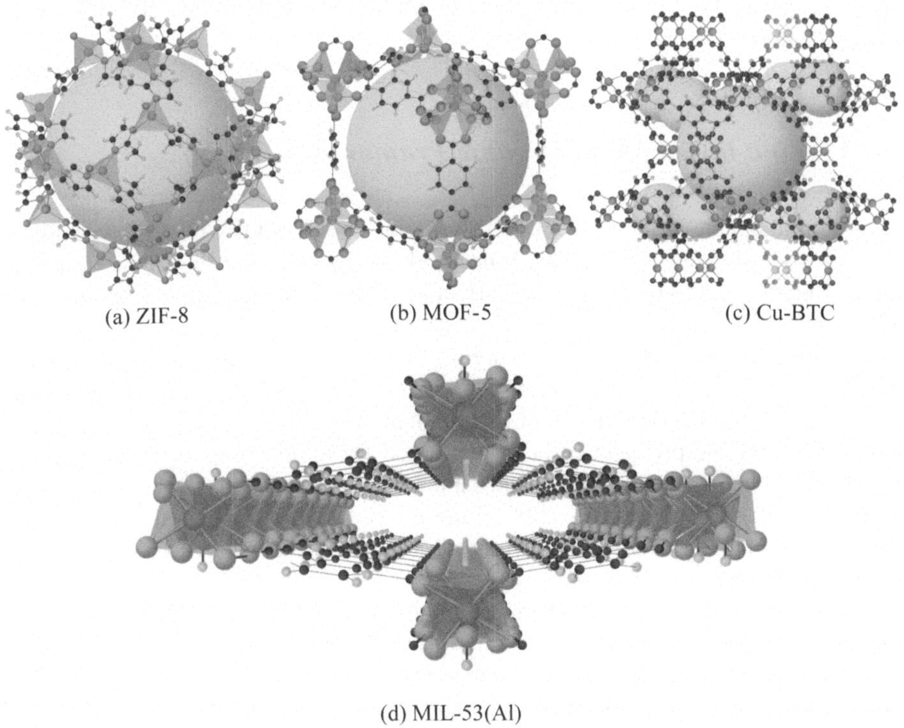

Fig. 7.1 Structures of some common MOFs. (**a**) Zeoliticimidazolate framework-8 and (**b**) MOF-5 have simple pore structures. (**c**) Cu-BTC contains pores of different sizes. (**d**) MIL-53 has pores in form of long elongated channels

7.4.2 *MOF Properties and Applications*

The large pore surface area gives the possibility of the gas absorption through the ability to adsorb gas molecules on the porous surfaces. Moreover, the various types of pore structure give the scope to tune molecule selectivity, and this makes it possible to design MOFs for particular target molecule selection and adsorption.

In addition to having a wide diversity of crystalline pore structures, many MOFs remain stable when heated to temperatures of a few hundred degree celsius (Li et al. 1999). However, there are also some flexible MOFs that react to the change of temperature and pressure quickly, and can exhibit phase transitions.

The studies that have been carried out in the last two decades indicate that the porous structure, stable chemical and thermal stability mean that MOFs may be excellent materials for CO_2 capture and gas separation for industry applications. Some MOFs like MOF-177 have shown higher CO_2 capacity than zeolites and active carbon (Millward and Yaghi 2005). The same is true for the possibility of using MOFs for H_2 gas storage, leading to application in industries for new clean energy resources.

7.4.3 *Zinc Based Metal Organic Frameworks (ZIFs)*

Zeoliticimidazolate frameworks (ZIFs) are zinc based MOFs that show similar structures to silicate zeolites. This framework is characterised by ZnN_4tetrahedra linked together through the ligands (Park et al. 2006). The metal-im-metal angle in ZIF (145°) is the same as the Si-O-Si angle in zeolites, as illustrated in Fig. 7.2. This material can be tailored by changing the metal cation and tuning the exact imidzolate linkage ligands (Banerjee et al. 2008). This has led to at least 100 types of ZIFs being found (Banerjee et al. 2008; Phan et al. 2010). Their various porous structures offer the prospect of being exploited for applications in areas such as catalysis, storing gases such as H_2, and capture of gases such as CO_2 (Li et al. 2010; Song et al. 2012; Lochan et al. 2008; Banerjee et al. 2008; Tran et al. 2011; Phan et al. 2010; Nguyen et al. 2012). Figure 7.3 shows the structures of some commonly studied ZIFs.

Most ZIFs have shown high chemical and thermal stabilities, which are required properties for applications for gas absorption and desorption. However, experiments have shown that some ZIFs are less stable. For example, ZIF-4 undergoes an irreversible structural phase transition to an amorphous phase and then to the ZIF-zni structure on heating to 300 °C (Bennett et al. 2010). ZIF-7 goes through a guest induced displacive phase transition (Zhao et al. 2014). ZIF-8 also shows a displacive phase transition induced by the guest argon gas molecules (Zhao et al. 2014).

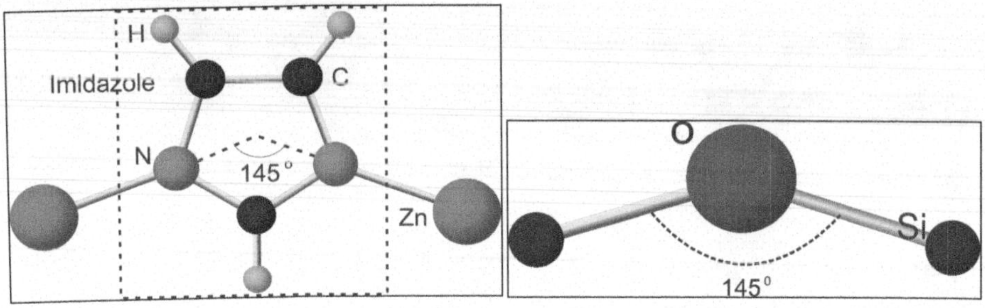

(a) Zn-Im-Zn (b) Si-O-Si

Fig. 7.2 Comparison of the metal-im-metal angle in ZIFs and the Si–O–Si angle in SiO_2. Both are 145° which confirms that ZIF structures are similar to silicate zeolites. (**a**) Zinc-imidazolate-zinc angle in ZIF. (**b**) Silicon-oxygen-silicon angle in silicon zeolites

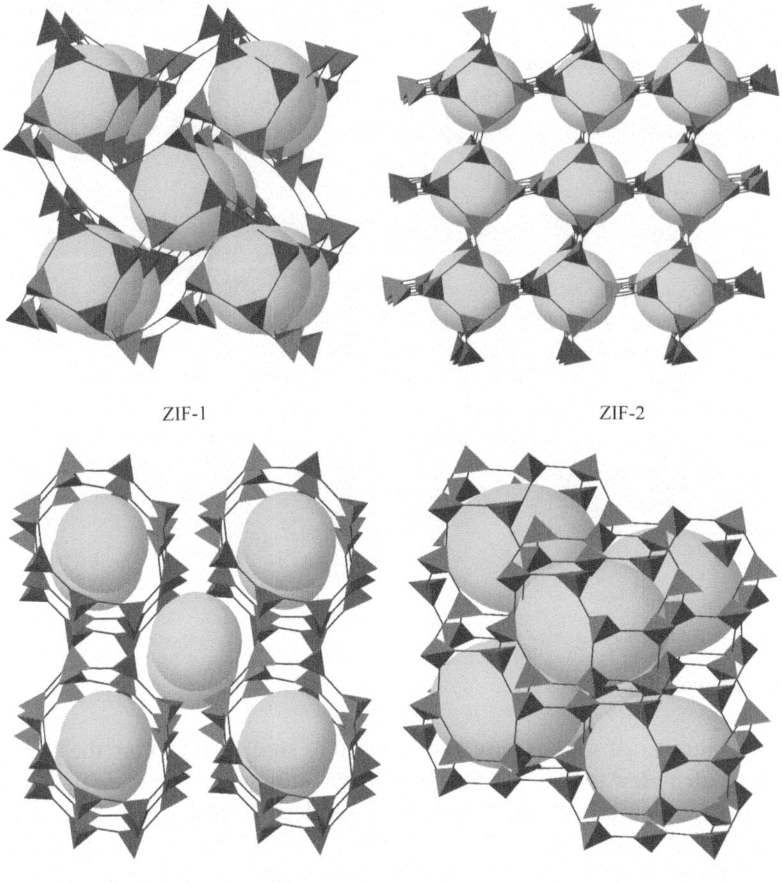

Fig. 7.3 Structures of some commonly studied zeolite-imidazolate frameowrks (ZIFs)

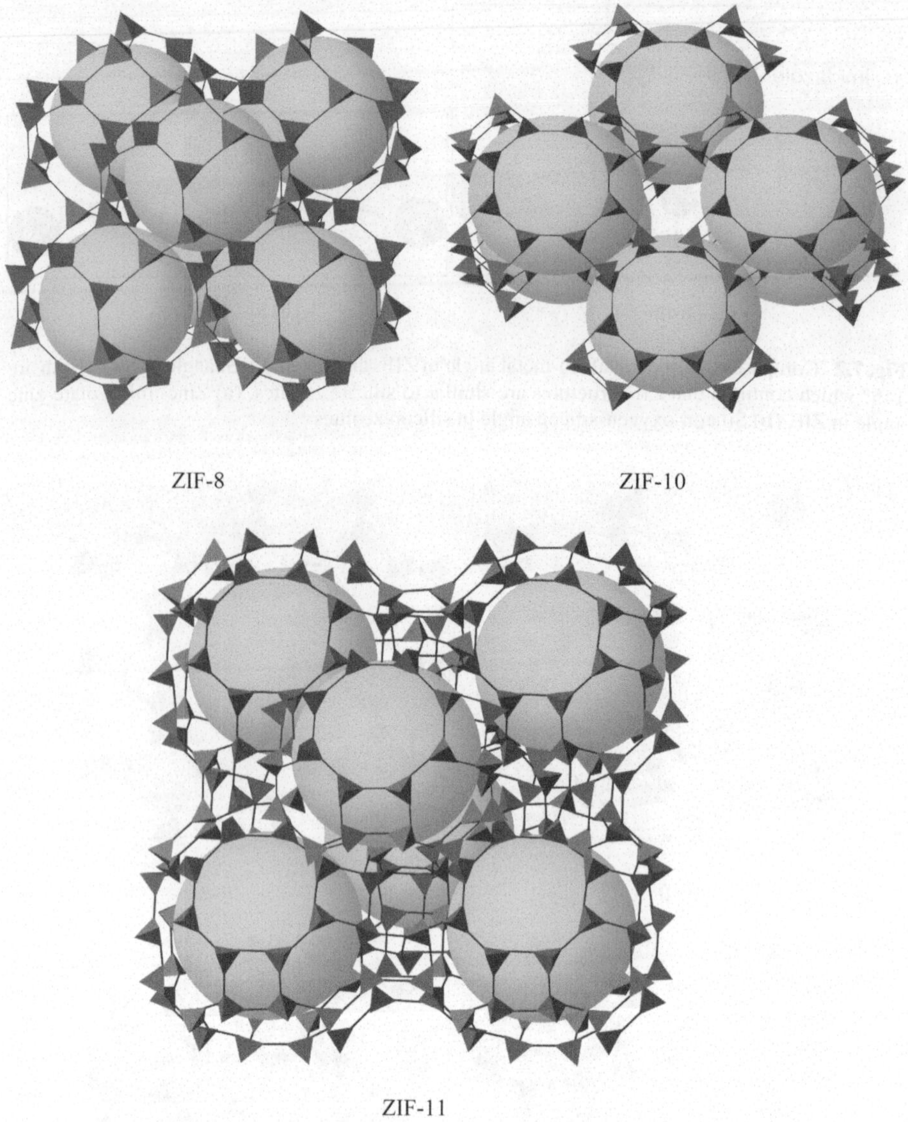

Fig. 7.3 (continued)

A large number of studies have been carried to study and explore the properties of MOFs with a view to their applications for gas storage and separation, still it is a tedious and long term research programme to collect the necessarily large quantity of experimental data due to the large number of the MOF structures. In the meantime, ever more MOF structures are being synthesized and reported quickly. On the other hand, the mechanism of the dynamics and interaction between the gas and

MOFs is very difficult to observe in experiments. Computer simulation is a technique that has the potential to rapidly screen many materials for their gas uptake capacity and selectivity.

7.5 Carbon Capture Using Metal Organic Frameworks

7.5.1 *Adsorption of Carbon Dioxide on a Zirconium Based Metal Organic Framework (MOF)*

Carbon dioxide capture at high pressure up to 34 atm has been extensively studied (Ravikovitch et al. 2000; Millward and Yaghi 2005). Still there has not been much research in the area of CO_2 capture at low pressures of the order of atmospheric pressure. The technologies that do exist are unsuitable for industrial applications (Wu et al. 2010; Britt et al. 2010).

MOFs are worthy candidates in CO_2 adsorption however, some MOFs have known to be thermally unstable. For this reason they become unsuitable to be used as adsorbents for high temperature CO_2 capture.

A zirconium based MOF known as Ui-O66 has shown good thermal stability at high temperatures of around 750 K. In addition, this particular MOF has also exhibited chemical resistance against some of the commonly used solvents.

Ui-O66 shows closely packed cubic structures with high symmetry as shown in Fig. 7.4 (Cavka et al. 2008; Valenzano et al. 2011). Various studies have also proven that Ui-O66 is a suitable material for CH_4 adsorption also (Yang et al. 2011). It has also been studied for H_2 adsorption at high pressure and its derivative Ui-O77 with longer linker has shown high capacity for H_2 adsorption, large surface area and large pore volume (Kandiah et al. 2010; Chavan et al. 2010, 2012). Some studies also suggest its potential application in catalysis (Gomes Silva et al. 2010; Vermoortele et al. 2011).

7.5.1.1 Synthesis of Zirconium Based MOF (Zr-MOF)

Synthesis of a modified form of Zr-MOF has been described by Abid et.al (2012). The method describes the solvothermal synthesis of Zr-MOF. According to the procedure, 2.27 mmol of BDC (1,4 Benzenedicarboxylic acid) was mixed with a 2.27 mmol solution of ZrCl4. This solution was further mixed with 400 mmol of DMF(N,N dimethylformamide). The prepared solution was then placed in an autoclave kept in a hot air oven for 24 h at 400 K temperature. The dried sample was then filtered under vaccum and finally white powder of Zr-MOF was collected and subjected to further drying.

Post synthesis, activation of the produced MOF needs to be carried out. A solvent exchange method was employed for the activation process. First chloroform was

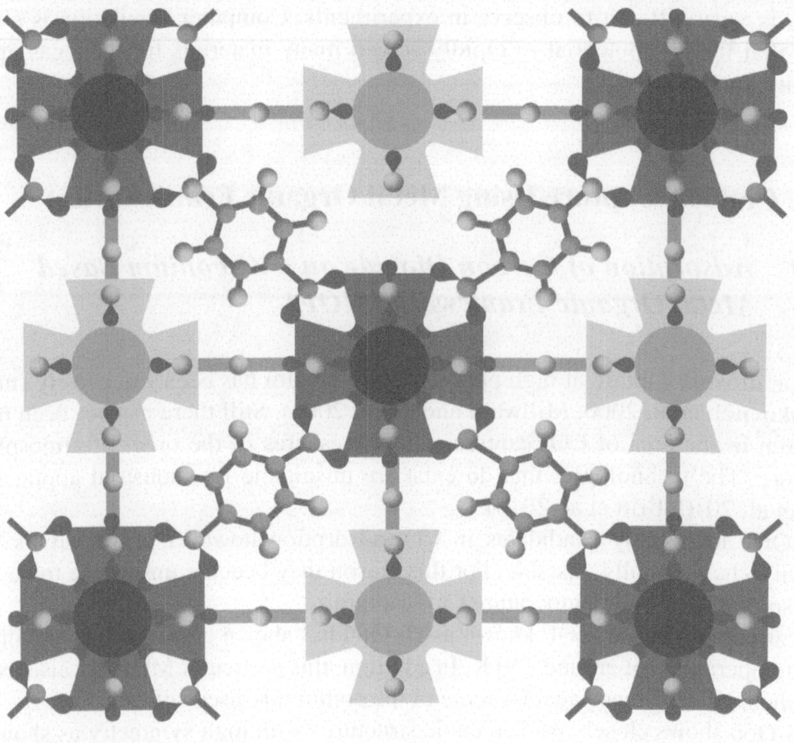

Fig. 7.4 Structure of UiO-66 showing closely packed cubic structures with high symmetry (Millward and Yaghi 2005)

used as a solvent wherein 0.3 g of synthesized MOF was mixed with 80 ml of chloroform. The solution was kept for about 5 days. After that, the solution was filtered and dried under vacuum. Further in order to remove the entrapped solvent, it was heated for 2 days under vacuum at 473 K. Secondly, methanol was used as a solvent to remove DMF and BDC from the sample.In order to do so, 25 ml of methanol was mixed with 0.5 g of prepared MOF and was placed in an autoclave which was then placed in a hot air oven for 24 h at 400 K. The sample was then filtered and dried. The obtained sample was subjected to further heating at 500 K for 2 days.

7.5.1.2 Characterization of Prepared MOF

XRD patterns of Zr-MOF present in the literature and activated samples of Zr-MOF show similar peak positions (Fig. 7.5). This implies that the prepared MOF has similar structure as UiO-66 but their crystallinities differ. Chloroform activated Zr-MOF exhibited higher crystallinity as compared to the methanol activated MOF. This can be attributed to the fact that chloroform is a better activating agent

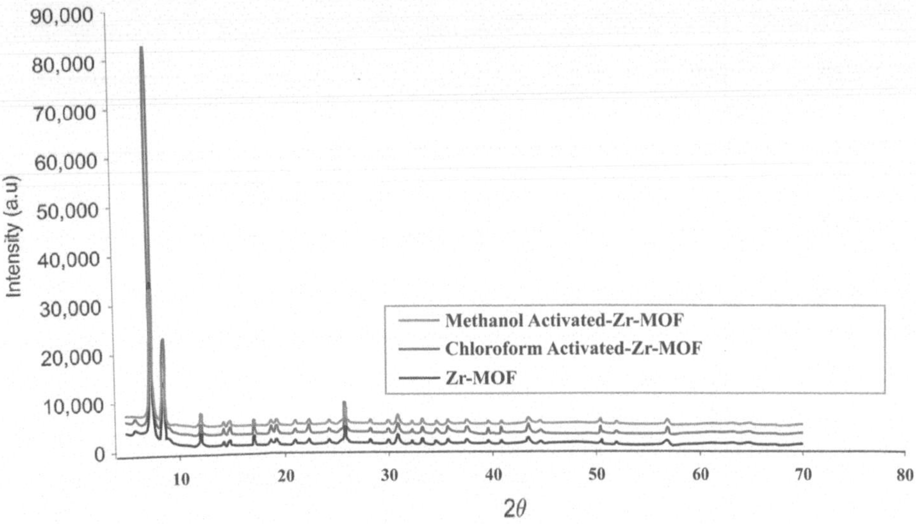

Fig. 7.5 XRD Patterns for Zirconium based MOF before and after activation compared with UiO-66 from literature. The prepared samples shows similar peaks as that of UiO-66 from literature. However, their crystallinities differ on account of using different activating agent viz. methanol and chloroform

than methanol although methanol is a better solvent for the purpose of unreacted solvent removal.. The prepared Zr-MOF exhibited great stability in presence of moisture. When exposed to ambient conditions for a long duration, this MOF retained its structure. When immersed in water for around 6 h, this MOF showed little to no change in its structure. This concludes that this MOF can be stored for extended durations without affecting its structure or crystallinity. Figure 7.6 shows the XRD pattern of the MOF when stored at ambient conditions and when immersed in water.

7.5.1.3 Results and Discussions

The samples activated by chloroform shows better CO2 adsorption capacity than methanol (Figs. 7.7 and 7.8). The methanol activated sample shows CO2 adsorption capacity of around 60 cc/g at 273 K and 35 cc/g at 298 K. The chloroform activated sample shows CO_2 adsorption capacity of around 80 cc/g at 273 K and 40 cc/g at 302 K. The chloroform activated sample has supposedly improved the surface area of MOF and has thus resulted in better adsorption of CO2. Figure 7.9 shows the superiority of chloroform activated sample over methanol activated sample at high temperature and high pressure.

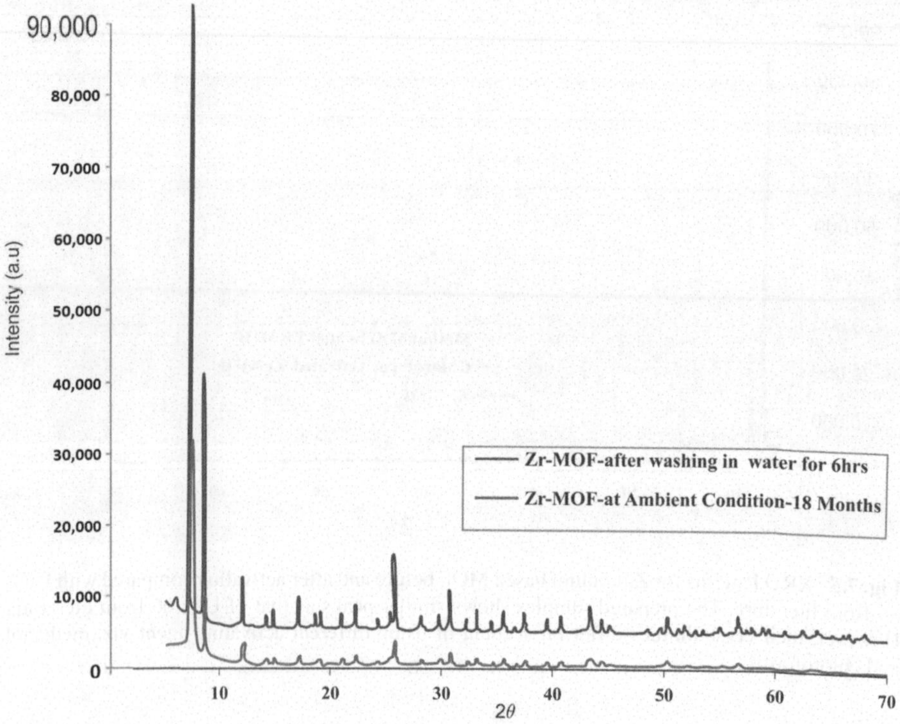

Fig. 7.6 XRD patterns for Zirconium based MOF under effect of presence of water. Blue curve which represents XRD pattern obtained after the sample is washed in water for 6 h. Red curve is the XRD pattern obtained after the samples is exposed to ambient conditions for 18 months. Similar peaks indicate that the sample is able to retain its structure and thus exhibits remarkable stability

7.5.2 *Adsorption of Carbon Dioxide on a Porous Carbon Material Derived from Zinc Based MOF*

MOFs have known to show instability in presence of moisture. This shortcoming of MOF is a great concern towards its applicability in CO2 capture from flue gas that contains moisture (water vapour). MOF-5 and MOF-177 are some of the such MOFs that have shown extreme instability in presence of moisture. The show a considerable reduction in CO2 adsorption capacity under humid conditions.

MOF derived carbon materials have been widely used in clean energy applications.These porous carbon materials are obtained by pyrolysis of MOFs. They remain stable when exposed to moisture owing to the hydrophobic property of carbon.

Zinc based MOFs such as MOF-5 or MOF-177 are pyrolyzed at 1000 Cfor 7–8 h. Under an atmosphere of argon to give porous carbon material. As a result of pyrolysis, zinc metal gets completely removed and porous carbon materials devoid

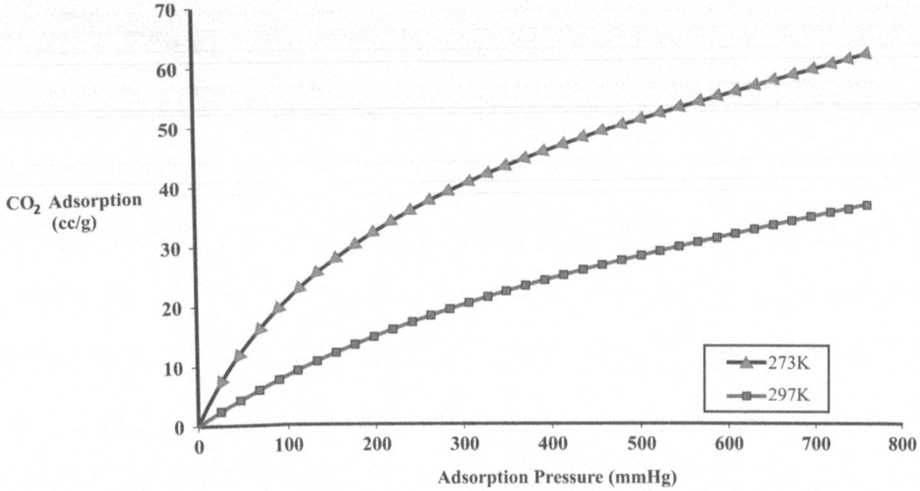

Fig. 7.7 CO_2 adsorption isotherms of methanol activated Zr-MOF. The CO_2 adsorption capacity increases with increasing absolute pressure. At high temperatures the graph becomes somewhat linear.Co2 adsorption capacity at 273 K is around 60 cc/g while that at 297 K is 35 cc/g. The curve follows a somewhat exponential pattern at lower temperatures

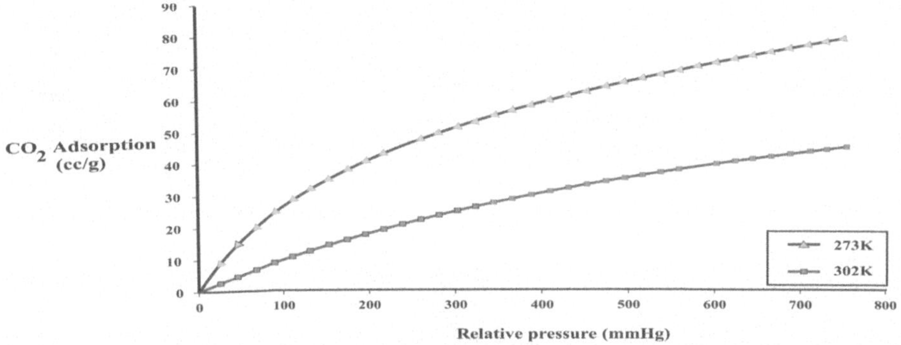

Fig. 7.8 CO_2 adsorption isotherms of chloroform activated MOF. CO_2 adsorption capacity at 273 K is 80 cc/g and at 302 K is 40 cc/g. The CO_2 adsorption capacity increases with increase in absolute pressure

of metal centers are formed. Studies have shown that porous carbon materials show superior CO2 adsorption capacities because pyrolysis has reduced the pore size of these materials thereby providing a confined space for CO2 adsorption. Fig. 7.10 shows a comparison between CO2 uptake capacity of MOF derived carbon materials and their parent MOFs (Kim et al. 2017).

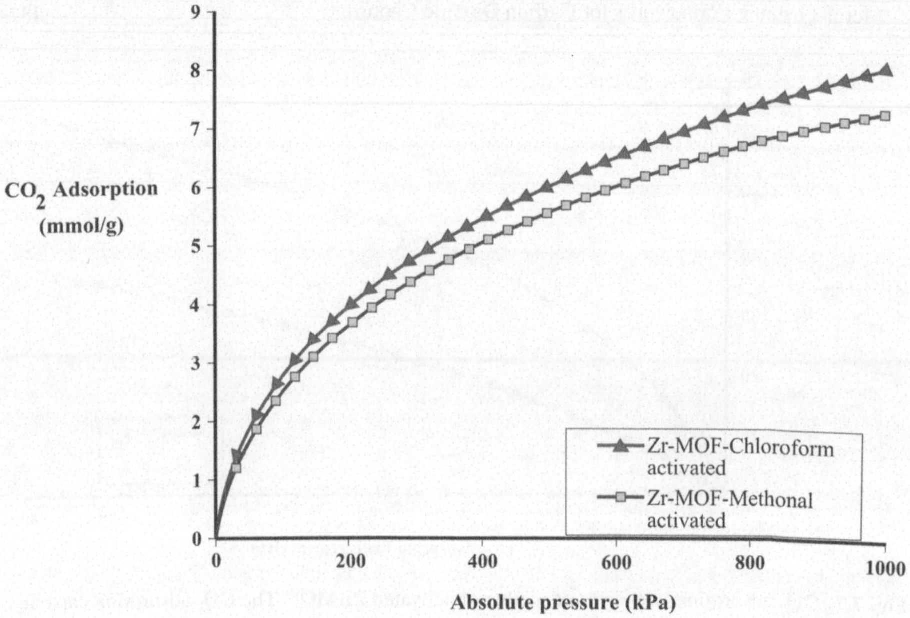

Fig. 7.9 CO_2 adsorption on modified Zr-MOFs at high pressure and 273 K. The two curves represent the two different solvents used for activation process. It can be clearly seen that activation by chloroform shows higher CO_2 adsorption capacity than methanol activated MOF. Hence, activation by chloroform is a better choice for CO_2 capture applications

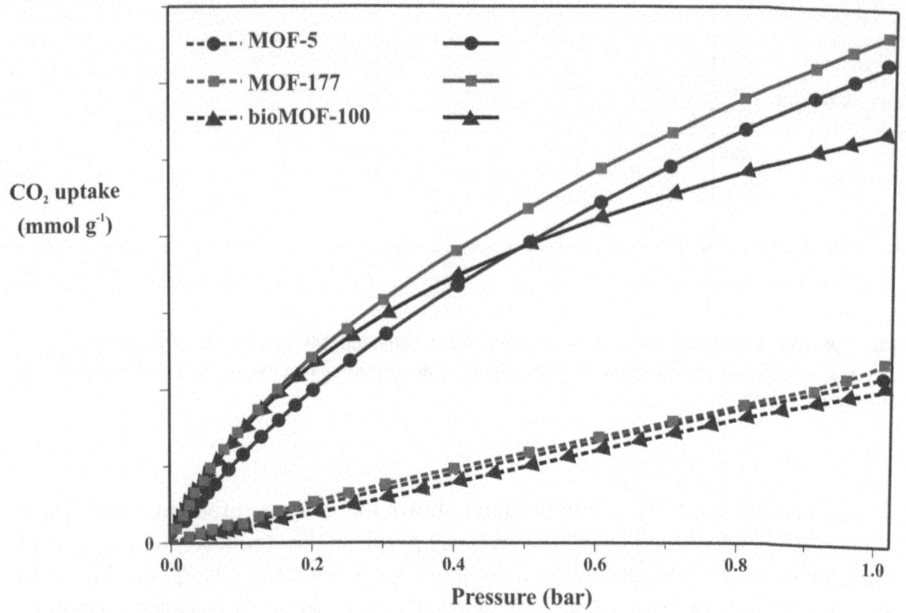

Fig. 7.10 Comparison of CO_2 uptake capacity between parent MOF and porous carbon material derived from them. The uptake capacity increases drastically for the derivatives derived from MOFs. This can be attributed to the fact that the absence of metal centers and pyrolysis has caused a reduction in pore volume thereby creating a confined space for efficient CO_2 capture

7.6 Conclusion

Metal Organic frameworks(MOF) find numerous application in the field of capture and storage owing to their high surface area. Due to the freedom that MOFs offer with respect to optimization as per requirement, they have proved to be worthy candidates in environmental applications. Since the structural and functional properties of MOFs can easily be modified a large number of derivatives and composites are frequently prepared that come out to be more efficient than their parent MOFs. Since their inception, MOFs have gone a long way and with time their applications are getting more diverse. The unprecedented rise in the atmospheric levels of CO_2 has made it clear that technologies to store and capture carbon dioxide are urgently needed if the catastrophe associated with excessive amounts of CO_2 is be prevented. MOFs might help us in solving this crisis that the world is currently facing and can turn out to be a boon for mankind.

References

Abid HR, Pham GH, Ang HM, Tade MO, Wang S (2012) Adsorption of CH4 and CO_2 on Zr-metal organic frameworks. J Colloid Interface Sci 366(1):120–124. https://doi.org/10.1016/j.jcis.2011.09.060. PMID: 22014395

Banerjee R, Phan A, Wang B, Knobler C, Furukawa H, O'Keeffe M, Yaghi OM (2008) High-throughput synthesis of zeoliticimidazolate frameworks and application to CO_2 capture. Science 319(5865):939–943

Bennett TD, Goodwin AL, Dove MT, Keen DA, Tucker MG, Barney ER, Soper AK, Bithell EG, Tan J-C, Cheetham AK (2010) Structure and properties of an amorphous metal-organic framework. Phys Rev Lett 104:115503

Britt D, Furukawa H, Wang B, Glover TG, Yaghi OM (2010) Highly efficient separation of carbon dioxide by a metal-organic framework replete with open metal sites. Proc Natl Acad Sci 106(49):20637–20640

Cavka JH, Jakobsen S, Olsbye U, Guillou N, Lamberti C, Bordiga S, Lillerud KP (2008) A new zirconium inorganic building brick forming metal organic frameworks with exceptional stability. J Am Chem Soc 130(42):13850–13851

Chavan S, Vitillo JG, Uddin MJ, Bonino F, Lamberti C, Groppo E et al (2010) Functionalization of UiO-66 metal-organic framework and highly cross-linked polystyrene with Cr(CO)(3): in situ formation, stability, and photoreactivity. Chem Mater 22(16):4602–4611

Chavan S, Vitillo JG, Gianolio D, Zavorotynska O, Civalleri B, Jakobsen S, Nilsen MH, Valenzano L, Lamberti C, Lillerud KP, Bordiga S (2012) H_2 storage in isostructural uio-67 and uio-66 mofs. Phys Chem Chem Phys 14(1614–1626)

Chui SSY, Lo SMF, Charmant JPH, Orpen AG, Williams ID (1999) A chemically functionalizablenanoporous material[$Cu_3(TMA)_2(H_2O)_3$]. Science 283:1148–1150

Gomes Silva C, Luz I, Llabrés i, Xamena FX, Corma A, García H (2010) Water stable Zr–benzenedicarboxylate metal–organic frameworks as photocatalysts for hydrogen generation. Chem Eur J 16:11133–11138

IEA Greenhouse Gas R&D Programme (2005) In: Dillon DJ, White V, Allam R, Wall RA, Gibbins J (eds) Oxy-combustion processes for CO_2 capture from power plant. IEA Greenhouse Gas R & D Programme, CheltenhamAll rights reserved to Mitsui Babcock Energy Limited

IEA Greenhouse Gas R&D Programme (IEA GHG) (2007) CO_2 capture ready plants, 2007/4, May 2007

Kandiah M, Nilsen MH, Usseglio S, SørenJakobsen UO, Tilset M, CherifLarabi EAQ, Bonino F, PetterLillerud K (2010) Synthesis and Stability of tagged UiO-66 Zr-MOFs. Chem Mater 22(24):6632–6640

Kim HR, Yoon TU, Kim SI, An J, Bae Y-S, Lee CY (2017) Beyond pristine MOFs: carbon dioxide capture by metal-organic frameworks (MOFs)-derived porous carbon materials. RSC Adv 7(3):1266–1270

Li H, Eddaoudi M, O'Keeffe M, Yaghi OM (1999) Design and synthesis of an exceptionally stable and highly porous metal-organic framework. Nature 402(6759):276–279

Li Y, Liang F, Bux H, Yang W, Caro J (2010) Zeoliticimidazolate framework ZIF-7 based molecular sieve membrane for hydrogen separation. J Membr Sci 354(1–2):48–54

Lochan RC, Khaliullin RZ, Head-Gordon M (2008) Interaction of molecular hydrogen with open transition metal centers for enhanced binding in metal-organic frameworks: a computational study. Inorg Chem 47(10):4032–4044

Loiseau T, Serre C, Huguenard C, Fink G, Taulelle F, Henry M, Bataille T, Férey G (2004) A rationale for the large breathing of the porous aluminum terephthalate (MIL-53) upon hydration. Chem Eur J 10:1373–1382

Millward AR, Yaghi OM (2005) Metal–organic frameworks with exceptionally high capacity for storage of carbon dioxide at room temperature. J Am Chem Soc 127:17998–17999

Nguyen LT, Ky KL, Nam TS (2012) A zeolite imidazolate framework ZIF-8 catalyst for friedel-crafts acylation. Chin J Catal 33:688–696

Nonhebel G (1936) A commercial plant for removal of smoke and oxides of sulphur from flue gases. Trans Faraday Soc 32:1291–1296, The Royal Society of Chemistry, 0014–7672

Park KS, Ni Z, Côté AP, Choi JY, Huang R, Uribe-Romo FJ, Chae HK, O'Keeffe M, Yaghi OM (2006) Exceptional chemical and thermal stability of zeoliticimidazolate frameworks. Proc Natl Acad Sci 103(27):10186–10191

Peng L et al (2014) Highly mesoporous metal-organic framework assembled in a switchable solvent. Nat Commun 5:4465

Phan A, Doonan CJ, Uribe-Romo FJ, Knobler CB, OKeeffe M, Yaghi OM (2010) Synthesis, structure, and carbon dioxide capture properties of zeoliticimidazolate frameworks. Acc Chem Res 43(1):58–67

Rao AB, Rubin ES (2002) A technical, economic, and environmental assessment of amine-based CO_2 capture technology for power plant greenhouse gas control. Environ Sci Technol 36(20):4467–4475

Ravikovitch PI, Vishnyakov A, Russo R, Neimark AV (2000) Unified approach to pore size characterization of microporous carbonaceous materials from N_2, Ar, and CO_2 adsorption isotherms. Langmuir 16(5):2311–2320

Rochelle GT (2009) Amine scrubbing for CO_2 capture. Science 325:1652–1654

Rowsell JLC, Yaghi OM (2004) Metal-organic frameworks: a new class of porous materials. Microporous Mesoporous Mater 73:3–14

R. Shao, A. Stangeland Amines used in CO_2 capture – health and environmental impacts., First online edition, 2009

Song Q, Nataraj SK, Roussenova MV, Tan JC, Hughes DJ, Li W, Bourgoin P, Alam MA, Cheetham AK, Al-Muhtaseb SA, Sivaniah E (2012) Zeoliticimidazolate framework (ZIF-8) based polymer nanocomposite membranes for gas separation. Energy Environ Sci 5:8359–8369

Tran UPN, Le KKA, Phan NTS (2011) Expanding applications of metal–organic frameworks: zeolite imidazolate framework ZIF-8 as an efficient heterogeneous catalyst for the Knoevenagel reaction. ACS Catal 1(2):120–127

Valenzano L, Civalleri B, Chavan S, Bordiga S, Lamberti C, Nilsen MH, Jakobsen S, Lillerud KP (2011) Disclosing the complex structure of UiO-66 metal organic framework: a synergic combination of experiment and theory. Chem Mater 23(7):1700–1718

Venkataraman D, Lee S, Zhang J, Moore JS (1994) An organic solid with wide channels based on hydrogen bonding between macrocycles. Nature 371(6498):591–593

Vermoortele F, Ameloot R, Vimont A, Serre C, De Vos D (2011) An amino-modified zrterephthalate metalorganic framework as an acidbase catalyst for cross-aldol condensation. Chem Commun 47:1521–1523

Wu H, Reali RS, Smith DA, Trachtenberg MC, Li J (2010) Highly selective CO2 capture by a flexible microporous metal–organic framework (MMOF) material. Chem Eur J 16:13951–13954

Yaghi OM, Li H (1995) Hydrothermal synthesis of a metal-organic framework containing large rectangular channels. J Am Chem Soc 117(41):10401–10402

Yang QY, Wiersum A, Hervé J, Guillerm V, Llewellyn PL, Maurin G (2011) Understanding the thermodynamic and kinetic behavior of the CO_2/CH_4 gas mixture within the porous zirconium terephthalate UiO-66(Zr): a joint experimental and modeling approach. J Phys Chem C 115(28):13768–13774

Zhao P, Lampronti GI, Lloyd GO, Wharmby MT, SébastienFacq AKC, Redfern SAT (2014) Phase transitions in zeolitic imidazolate framework 7: the importance of framework flexibility and guest-induced instability. Chem Mater 26(5):1767–1769

Zhu Q-L, Xu Q (2014) Metal–organic framework composites. Chem Soc Rev 43:5468–5512

Chapter 8
Ionic Liquids for Carbon Dioxide Capture

Maryam Raeisi, Amineh Keshavarz, and Mohammad Reza Rahimpour

Abstract The emissions of anthropogenic carbon dioxide have been considered as the noticeable contributor to the global warming. Many technologies are developed to mitigate the global warming issue through controling the emission of greenhous gasses into the environment. Among the existing methods, post-combustion process is a very promising approach for capturing carbon dioxide. In this regard, various solvents such as amine-based solvents have been employed. The ionic liquid has also shown a great potential as a solvent for capturing carbon dioxide, recently.

In this chapter, various properties of ionic liquids such as selectivity and solubility of carbon dioxide will be discussed. Besides, previous researches in this regards will be reviewed. Furthermore, various strategies for improving the performance of ionic liquid will be introduced. Finally, the economic aspects of carbon dioxide capture process through ionic liquids, which is of paramount importance in the industrial scale, will be evaluated.

Keywords Carbon dioxide capture · Post-combustion method · Solvent · Ionic liquid · Physiochemical properties of the ionic liquid · Solubility · Selectivity · Viscosity · Functionalized ionic liquid · Amine-ionic liquid

M. Raeisi
Department of Chemical Engineering, Shiraz University of Technology, Shiraz, Iran

A. Keshavarz · M. R. Rahimpour (✉)
Department of Chemical Engineering, Shiraz University, Shiraz, Iran
e-mail: rahimpor@shirazu.ac.ir

Inamuddin et al. (eds.), *Sustainable Agriculture Reviews 38*, Sustainable Agriculture Reviews 38, https://doi.org/10.1007/978-3-030-29337-6_8

Abbreviations

([hmim] [hFAP])	1-hexyl-3- methylimidazolium tris(heptafluoropropyl) trifluorophosphate
([P14666][Cl])	trihexyl(tetradecyl)phosphonium chloride
[aP_{4443}][AA]	dual aminofunctionalized phosphonium ionic liquids
[BF_4]	tetrafluoroborate
[bmim] [Br]	1-butyl-3-methylimidazolium bromide
[bmim] [DCA]	1-butyl-3-methylimidazolium dicyanamid
[bmim][BF_4]	1-butyl-3-methylimidazolium tetrafluoroborate
[bmim][PF_6]	1-butyl-3-methyl-imidazolium hexafluorophosphate
[bmim][TFO]	1-butyl-3-methylimidazolium trifluoromethanesulfonate
[C_2mim][CH_3CO_2]	1-ethyl-3-methyl-1H-imidazolium acetate
[C_2mim][NFO]	1-ethyl-3-methyl-1H-imidazolium nonafluorobutanesulfonate
[C_2mim][NO_2]	1-ethyl-3-methyl-1H-imidazolium nitrite
[C_2mim][NO_3]	1-ethyl-3-methyl-1H-imidazolium nitrate
[C_2mim][TFO]	1-ethyl-3-methyl-1H-imidazolium trifluoromethane sulfonate
[C_2mim][BF_4]	1-ethyl-3-methylimidazolium tetrafluoroborate
[C_2mim][CF_3SO_3]	1-ethyl-3-methylimidazolium trifluoromethanesulfonate
[C_2mim][CH3COO]	1-ethyl-3-methylimidazolium trifluoroacetate
[C_2mim][Cl]	1-ethyl-3-methylimidazolium chloride
[C_2mim][I]	1-ethyl-3-methylimidazolium iodide
[C_2mim][OTF]	1-ethyl-3-methylimidazolium triflate
[C_2mim][PF_6]	1-ethyl-3-methyl-1H-imidazolium hexafluorophosphate
[C_2mim][TF_2N]	1-ethyl-3-methyl-1H-imidazolium bis((trifluoromethyl)sulfonyl)imide
[C_3mim][Cl]	1-propyl-3-methylimidazolium chloride
[C_3mim][PF_6]	1- propyl −3-propylimidazolium hexafluorophosphate
[C_3mim][TF_2N]	1-propyl-3-methylimidazolium bis(trifluoromethyl)sulfonyl)imide
[C_4mim][BF_4]	1-butyl-3-methylimidazolium tetrafluoroborate
[C_4mim][NFO]	1-butyl-3-methylimidazolium nonafluorobutanesulfonate
[C_4mim][TFO]	1-butyl-3-methylimidazolium trifluoromethane sulfonate
[C_4mim][CF_3COO]	1-butyl-3-methylimidazolium trifluoroacetate
[C_4mim][Cl]	1-butyl-3-methylimidazolium chloride
[C_4mim][I]	1-butyl-3-methylimidazolium iodide
[C_4mim][PF_6]	1-butyl-3-methylimidazolium hexafluorophosphate
[C_4mim][TF_2N]	1-butyl-3-methylimidazolium bis((trifluoromethyl)sulfonyl)imide
[C_6mim][BF_4]	1-hexyl-3-methylimidazolium tetrafluoroborate
[C_6mim][Cl]	1-hexyl-3-methylimidazolium chloride
[C_6mim][PF_6]	1-hexyl-3-methylimidazolium hexafluorophosphate

$[C_8mim][BF_4]$	1-octyl-3-methylimidazolium tetrafluoroborate
$[C_8mim][Cl]$	1-octyl-3-methylimidazolium chloride
$[C_8mim][PF_6]$	1-octyl-3-methylimidazolium hexafluorophosphate
$[CH_3SO_4]$	methyl sulfate
$[emim][ESO_4]$	1-ethyl-3-methylimidazolium ethyl sulfate
[emim][TFO]	1-ethyl-3-methylimidazolium trifluoromethanesulfonate
[hmim] [TFO]	1-hexyl-3-methylimidazolium trifluoromethanesulfonate
$[hmim][BF_4]$	1-hexyl-3-methylimidazolium tetrafluoroborat
$[hmim][PF_6]$	1-hexyl-3-methylimidazolium hexafluorophosphate
$[N1888][Tf_2N]$	methyltrioctylammonium bis(trifluoromethylsulfonyl)imide
$[N\text{-}bupy][BF_4]$	N-butylpyridinium tetrafluoroborate
[NO3]	nitrate
[omim][Cl]	1-octyl-3-methylimidazolium chloride
[P(C4)4][AA]	tetrabutylphosphonium amino acid
[P(C4)4][OH]	tetrabutylphosphonium hydroxide
$[P14666][Tf_2N]$	trihexyl(tetradecyl)phosphonium bis(trifluoromethylsulfonyl) imide
[P5mim][nFAP]	1-pentyl-3 methylimidazolium tris(nonafluorobutyl) trifluorophosphate
APILs	aprotic ionic liquids
CCS	carbon capture and storage
ILs	ionic liquids
PILs	protic ionic liquids
PMG	pentamethylguanidinium
TGA	thermo gravimetric analysis
TMG	tetramethylguanidinium
TSIL	task specific ionic liquids

8.1 Introduction

In recent years, increasing attention has been paid to examining greenhouse gas emission and their negative impact on global climate change, in general, and the global warming, in particular. Among greenhouse gases, more attention has been paid to carbon dioxide due to higher generation and longer negative effect on the atmosphere. This mean carbon dioxide is the most determinant factor for the global warming as compared to other greenhouse gasses (Figueroa et al. 2008). In this condition, carbon capture and storage (CCS) is one of the pathways to lower the anthropogenic contribution of this gas to the global warming.

Separating, capturing, transporting and storing of carbon dioxide are involved in the CCS process. Carbon dioxide capture process can be conducted via a variety of methods (Shafeeyan et al. 2014). Unfortunately, each technology has some draw-backs such as high cost and energy consumption, hazardous secondary pollutions, which are big obstacles for the industrial application of CCS technologies (Boot-

Handford et al. 2014; Hu et al. 2016; Zhao et al. 2016; Liang et al. 2016). Definitely, designing high efficiency reversible processes to capture carbon dioxide has been a big challenge for centuries. In this regard, some efforts have been put by many researchers to address the challenges.

In general, CCS is categorized into three fundamental sections, pre-combustion, post-combustion, and oxy-combustion. In pre-combustion, hydrogen and carbon monoxide called syngas is produced from gasification of fuel (carbon based) (Bailera et al. 2017). In the water-gas shift reaction, carbon monoxide is converted to hydrogen and carbon dioxide which is used as a fuel in a gas turbine generator to produce electricity and captured from the outlet stream, respectively (Chen et al. 2012). Thus, the carbon dioxide capture stage is placed before it enters the combustion chamber to remove carbon dioxide. In the oxy-combustion and post-combustion, fuel is burned with oxygen and air, respectively. The difference between these two combustions is the flue gas composition and in both of them the carbon is captured after combustion. Although, more process units are needed to capture the carbon dioxide from the flue-gas in the post-combustion method. This technology offers the advantage of being easier to retrofit the existing plants compared to oxy-combustion and pre-combustion approaches. Therefore, pre/oxy- combustion methods can preferably be utilized in new plants (Sreedhar et al. 2017a).

As illustrated in (Fig. 8.1), five main subcategories (absorption, cryogenics, adsorption, membrane, and microbial (algae)) are defined for the post-combustion carbon capture technique (Aghaie et al. 2018). Among various post combustion

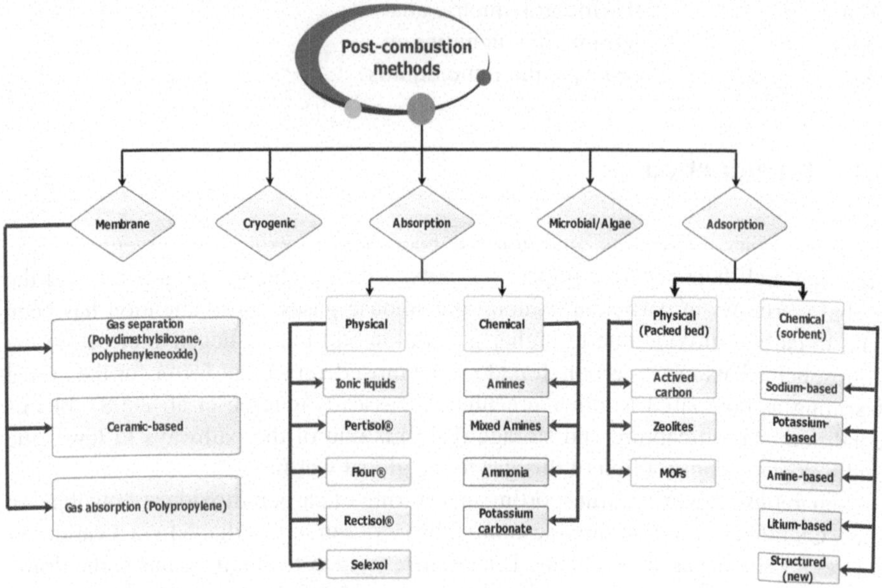

Fig. 8.1 Diagram of different categories of post combustion carbon dioxide methods. As can be seen, ionic liquids are one of the important physical absorber

methods, the absorption is more common to be employed in industrial plants for capturing the carbon dioxide. In this technology, the gases are contacted with solvents in absorption columns. In this stage, the rate of carbon dioxide absorption is important. Then the solvent is regenerated in the stripper (Freund et al. 2005; Sreedhar et al. 2017b).

As shown in (Fig. 8.1) amines, as the most prevalent commercial chemical absorbers among various absorbers components are used to capture the carbon dioxide. These absorbents are stood as a practical option for industrial applications due to the simplicity of their synthesis procedure, excellent performance, and economic feasibility (Anthony et al. 2001; Kazarian et al. 2002). These solvents owing to their absorption capacity and reactivity with carbon dioxide molecules are extensively utilized in the absorption process for carbon dioxide capturing (Riemer and Ormerod 1995). The monoethanolamine, methyl diethanolamine, aminomethyl propanol are three common primary, tertiary, and sterically- hindered amines, respectively (Kothandaraman 2010). The capturing carbon dioxide efficiency of these solvents decreases in the following order: monoethanolamine, diethanolamine, aminomethyl propanol, diisopropylamine and methyl diethanolamine (Rosli et al. 2017). Although, these absorbents have some advantages such as great purification of the gases, they face some momentous drawbacks such as high apparatus corrosion rate, construction cost, energy consumption in the regeneration step and a huge amount of absorber requirement (Yu et al. 2012). Amines have some properties such as high volatility, corrosive nature which causes these obstacles (Torralba-Calleja et al. 2013).

To replace the traditional chemical solvents with a new one for capturing and separating carbon dioxide, 100 ionic liquids with prominent properties, as potential compounds, have successfully been presented to the industries by researchers, but their effectiveness should be proved. Various researches carried out studies about the reliability of these materials for this purpose. Fortunately, the drawbacks associated with chemical solvent such as amine- based are addressed thanks to the ionic liquids (Anthony et al. 2002).

8.2 Ionic Liquids

A novel generation of solvents, namely ionic liquids (ILs), has received noticeable attentions as substitutes for traditional solvents. These types of solvent contain a large cation and a small anion. They are nonflammable and highly stable compounds. These organic salts have high boiling points whereas to their low vapor pressures are low (MacFarlane et al. 2014). The positive characteristics and properties i.e., thermal stability, high carbon dioxide solubility and selectivity, tunability of these materials have made them potential candidates for carbon dioxide capture and separation (Meine et al. 2010; Welton 1999; Revelli et al. 2010; Kenarsari et al. 2013; Zhang et al. 2012; Hayes et al. 2015; Zhao et al. 2011; Babamohammadi et al. 2015; Chiappe and Pomelli 2014; Ahrens et al. 2009). This type of solvent has been broadly investigated for this purpose since 1999. First of all, Blanchard et al. reported that solubility of ionic liquids in carbon dioxide can be ignored while the solubility of carbon dioxide in 1-butyl-3-methyl-imidazolium hexafluorophosphate ([bmim][PF6]) ionic liquid is significant (Blanchard et al. 1999). Their investigation was the turning point for further scientific research for capturing carbon dioxide with ionic liquids. Various kinds of ionic liquids have been synthesized through different combinations of anions and cations. To exemplify, imidazolium, pyridinium, pyrrolidinium, and ammonium are four common cations which are used to synthesize ionic liquids.

In comparison with imidazolium and ammonium based ionic liquids, pyrrolidiniumbased ionic liquids have attracted more attentions. Moreover, phosphonium-based ionic liquids have been used recently for capturing carbon dioxide (Harper et al. 2011). Also, iodide, bromide, chloride, tetrafluoroborate, hexafluorophosphate, trifluoromethylsulfonate and bis (trifluoromethanesulfonul) imide are conventional anions utilized in ionic liquid synthesis. Figure 8.2 shows the structure of common cations and anions used for synthesizing ionic liquids (Hayes et al. 2015).

In addition, they are categorized based on the structure of their molecules, dividing into protic ionic liquids (PILs) and aprotic ILs (APILs). In fact, the PILs and AILs are proton-donating and nonproton-donating, respectively. The utilized cations and anions can specify physicochemical properties of ionic liquids. Based on specific application, different ions are combined to produce appropriate ionic liquids.

For example, some ionic liquids are perfectly miscible with liquid water such as 1-alkyl-3-methylimidazolium nitrate ionic liquids; in contrast, 1-alkyl-3-methylimidazolium hexafluorophosphate ionic liquids are perfectly immiscible. Also, It is reported in a study conducted by Shariati et al. that the hydrophobicity of an ionic liquid does not decrease with increasing the length of their chains (Shariati et al. 2005). Generally, for evaluating ionic liquids for carbon dioxide capture and designing the capturing processes, the thermodynamic and physical properties of these organic salts such as melting point, critical properties, vapor pressure, thermal stability, solubility, density, viscosity and selectivity should necessarily be taken into account.

8.3 Characteristics of Ionic Liquids

8.3.1 Melting Point

Depending on kinds of cations and anions, melting points of ionic liquids varies, but most of them have melting point below100 °C. Melting points of 16 types of ionic liquids are shown in Table 8.1 (Huddleston et al. 2001; Bonhôte et al. 1996; Wilkes and Zaworotko 1992; Suarez et al. 1998; Ngo et al. 2000). The increase in cation and anion size can result in a deduction in the melting point of the ionic liquids.

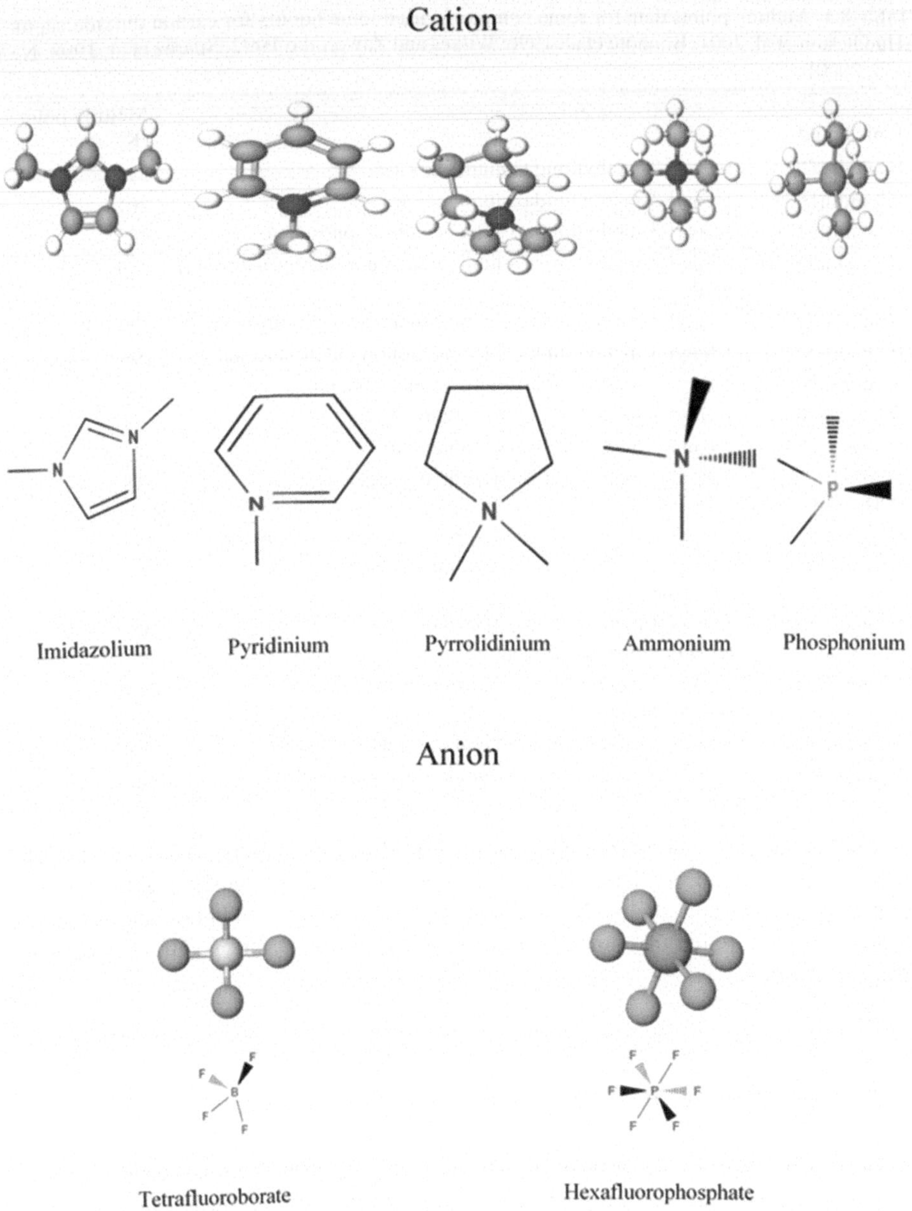

Fig. 8.2 Structure of most common cations and anions utilized in ionic liquid synthesis. Different kinds of ionic liquids have been synthesized through different combinations of anions and cations

Table 8.1 Melting points data for some commonly used ionic liquids for carbon dioxide capture (Huddleston et al. 2001; Bonhôte et al. 1996; Wilkes and Zaworotko 1992; Suarez et al. 1998; Ngo et al. 2000)

Ionic liquid		Melting point (K)
$[C_4mim][Cl]$	1-butyl-3-methylimidazolium chloride	314
$[C_4mim][I]$	1-butyl-3-methylimidazolium iodide	201
$[C_4mim][PF_6]$	1-butyl-3-methylimidazolium hexafluorophosphate	283
$[C_4mim][TF_2N]$	1-butyl-3-methylimidazolium bis((trifluoromethyl)sulfonyl) imide	269
$[C_4mim][TFO]$	1-butyl-3-methylimidazolium trifluoromethane sulfonate	289
$[C_4mim][NFO]$	1-butyl-3-methylimidazolium nonafluorobutanesulfonate	293
$[C_4mim][BF_4]$	1-butyl-3-methylimidazolium tetrafluoroborate	192
$[C_2mim][NO_3]$	1-ethyl-3-methyl-1H-imidazolium nitrate	311
$[C_2mim][NO_2]$	1-ethyl-3-methyl-1H-imidazolium nitrite	328
$[C_2mim][CH_3CO_2]$	1-ethyl-3-methyl-1H-imidazolium acetate	318
$[C_2mim][PF_6]$	1-ethyl-3-methyl-1H-imidazolium hexafluorophosphate	333
$[C_2mim][TFO]$	1-ethyl-3-methyl-1H-imidazolium trifluoromethane sulfonate	264
$[C_2mim][NFO]$	1-ethyl-3-methyl-1H-imidazolium nonafluorobutanesulfonate	301
$[C_2mim][TF_2N]$	1-ethyl-3-methyl-1H-imidazolium bis((trifluoromethyl) sulfonyl)imide	270
$[C_3mim][PF_6]$	1-propyl-3-propylimidazolium hexafluorophosphate	313
$[C_6mim][PF_6]$	1-hexyl-3-methylimidazolium hexafluorophosphate	212

Furthermore, the melting points of ionic liquids are extremely affected by length of alkyl chain of cations. The melting points of ionic liquids intensifies with increasing the length of their chains. In addition, the ionic liquids with halide anion have a higher melting point in comparison with other kinds of anion (Brennecke and Maginn 2001).

8.3.2 *Critical Properties*

Various techniques/methods have been reported by different scientists for determining the critical properties of ionic liquids. For example, group contribution approach was utilized to calculate critical temperature/pressure/volume, and also the acentric factor of some ionic liquids (Valderrama and Robles 2007; Maia et al. 2013; Valderrama and Rojas 2009; Valderrama et al. 2008). In Table 8.2, the acentric factors and critical temperature/pressure/volume of some imidazolium-based ionic liquids are tabulated.

Table 8.2 The critical properties of some commonly used ionic liquids for carbon dioxide capture (Valderrama et al. 2008; Valderrama and Robles 2007)

Ionic liquid		Critical pressure (MPa)	Critical temperature (K)	Critical volume (cm^3/mol)	Acentric factor
1-n-butyl-3-methylimidazolium hexafluorophosphate	[bmim] [PF_6]	1.73	708.9	779.5	0.7553
1-butyl-3-methylimidazolium tetrafluoroborate	[bmim] [BF_4]	2.04	623.3	672.0	0.8489
1-butyl-3-methylimidazolium trifluoromethanesulfonate	[bmim] [TFO]	2.94	1061.1	767.6	0.3677
1-butyl-3-methylimidazolium bromide	[bmim] [Br]	2.98	834.9	583.3	0.4891
1-butyl-3-methylimidazolium dicyanamid	[bmim] [DCA]	2.44	1035.8	712.0	0.8419
1-hexyl-3-methylimidazolium hexafluorophosphate	[hmim] [PF_6]	1.55	754.3	893.7	0.8352
1-hexyl-3-methylimidazolium tetrafluoroborat	[hmim] [BF_4]	1.79	679.1	786.2	0.9258
1-hexyl-3-methylimidazolium trifluoromethanesulfonate	[hmim] [TFO]	2.39	1287.3	1121.3	0.3539
1-ethyl-3-methylimidazolium ethyl sulfate	[emim] [ESO_4]	4.04	1061.1	676.8	0.3368
1-ethyl-3-methylimidazolium trifluoromethanesulfonate	[emim] [TFO]	3.58	985.2	653.4	0.2891
1-octyl-3-methylimidazolium chloride	[omim] [Cl]	2.03	860.1	814.2	0.6190
N-butylpyridinium tetrafluoroborate	[N-bupy] [BF_4]	2.03	597.6	648.1	0.8307

8.3.3 *Vapor Pressure*

To select the suitable solvent, design the separation process equipment and operation, vapor pressure of components are required. The ionic liquids have low vapor pressure, which is in favor of the separation process, due to their ionic structure. Regarding this specification, these solvents have low evaporation, and therefore their quantities are fixed during absorption and desorption process, unlike common industrial solvents e.g., Amin-based solvents.

There are low models to predict the vapor pressure of ionic liquids. Valderrama et al. reported equations for constant parameters of the Antoine equation (log $p^{sat} = A - B/(T + C)$) (Valderrama et al. 2008). In addition, the amount of saturation pressure of ionic liquids at various temperatures can be determined by Knudson (Paulechka et al. 2005) or transpiration (Emel'yanenko et al. 2007) methods. Furthermore, the thermogravimetric analysis (TGA) (Langmuir 1913) is another approach for determining vapor pressure of the ionic liquids.

8.3.4 *Thermal Stability*

The thermal stability is another factor for the industrial application of ionic liquids in different reactions and separation processes. Various studies have been performed to investigate the thermal stability of ionic liquids. Most of the ionic liquids are thermally stable below 450 °C for a limited period of time (Wasserscheid and Keim 2000). For instance, the research of Wilkes et al. showed that the decomposition temperature of 1-Butyl-3-methylimidazolium tetrafluoroborate is 445 °C (Wilkes and Zaworotko 1992). The maximum temperatures can be tolerated before decomposing of different ionic liquids are shown in Table 8.3.

Table 8.3 Decomposition temperature of some commonly used ionic liquids for carbon dioxide capture (Huddleston et al. 2001; Ngo et al. 2000; Bonhôte et al. 1996; Holbrey and Seddon 1999)

Ionic liquid		Decomposition temperature (K)
[C_2mim] [CF_3SO_3]	1-ethyl-3-methylimidazolium trifluoromethanesulfonate	713
[C_2mim][BF_4]	1-ethyl-3-methylimidazolium tetrafluoroborate	685
[C_2mim][PF_6]	1-ethyl-3-methylimidazolium hexafluorophosphate	648
[C_2mim][Cl]	1-ethyl-3-methylimidazolium chloride	558
[C_2mim][TF_2N]	1-ethyl-3-methylimidazolium bis((trifluoromethyl) sulfonyl)imide	728
[C_2mim][I]	1-ethyl-3-methylimidazolium iodide	576
[C_2mim] [CH3COO]	1-ethyl-3-methylimidazolium trifluoroacetate	423
[C_3mim][Cl]	1-propyl-3-methylimidazolium chloride	555
[C_3mim][PF_6]	1-propyl-3-methylimidazolium hexafluorophosphate	608
[C_3mim][TF_2N]	1-propyl-3-methylimidazolium bis(trifluoromethyl)sulfonyl)imide	725
[C_4mim][Cl]	1-butyl-3-methylimidazolium chloride	527
[C_4mim][I]	1-butyl-3-methylimidazolium iodide	538
[C_4mim][BF_4]	1-butyl-3-methylimidazolium tetrafluoroborate	676
[C_4mim][PF_6]	1-butyl-3-methylimidazolium hexafluorophosphate	622
[C_4mim][TF_2N]	1-butyl-3-methylimidazolium bis((trifluoromethyl) sulfonyl)imide	712
[C_6mim][PF_6]	1-hexyl-3-methylimidazolium hexafluorophosphate	690
[C_6mim][Cl]	1-hexyl-3-methylimidazolium chloride	253
[C_8mim][PF_6]	1-octyl-3-methylimidazolium hexafluorophosphate	376
[C_8mim][Cl]	1-octyl-3-methylimidazolium chloride	243

Table 8.4 Densities of some commonly used ionic liquids for carbon dioxide capture (Bonhôte et al. 1996; Seddon et al. 2000; Suarez et al. 1998; Huddleston et al. 2001; Blanchard et al. 2001; Dullius et al. 1998)

Ionic liquid		Temperature (K)	Density(g/cm^3)
$[C_2mim][CF_3SO_3]$	1-ethyl-3-methylimidazolium trifluoromethanesulfonate	298	1.29
$[C_2mim][OTF]$	1-ethyl-3-methylimidazolium triflate	295	1.39
$[C_4mim][PF_6]$	1-butyl-3-methylimidazolium hexafluorophosphate	293,298	1.37,1.36
$[C_4mim][Cl]$	1-butyl-3-methylimidazolium chloride	298	1.08
$[C_4mim][BF_4]$	1-butyl-3-methylimidazolium tetrafluoroborate	303,298	1.12,1.17
$[C_4mim][I]$	1-butyl-3-methylimidazolium iodide	298	1.44
$[C_4mim][TF_2N]$	1-butyl-3-methylimidazolium bis(trifluoromethyl)sulfonyl)imide	298	1.43
$[C_4mim][CF_3COO]$	1-butyl-3-methylimidazolium trifluoroacetate	298	1.21
$[C_6mim][PF_6]$	1-hexyl-3-methylimidazolium hexafluorophosphate	298	1.29
$[C_6mim][Cl]$	1-hexyl-3-methylimidazolium chloride	293	1.03
$[C_6mim][BF_4]$	1-hexyl-3-methylimidazolium tetrafluoroborate	293	1.15
$[C_8mim][Cl]$	1-octyl-3-methylimidazolium chloride	298	1.00
$[C_8mim][BF_4]$	1-octyl-3-methylimidazolium tetrafluoroborate	313	1.08
$[C_8mim][PF_6]$	1-octyl-3-methylimidazolium hexafluorophosphate	298	1.22

8.3.5 *Density*

Based on kinds of cations/anions and temperature, broad ranges of densities are available for ionic liquids. Although most of the ionic liquids have the density of more than 1 g/cm3, densities of imidazolium and phosphonium - based ionic liquids are less than this at the specific temperatures. In addition, solubility of carbon dioxide lead to increase the density of ionic liquids (Seddon et al. 2000; Cadena et al. 2004). Furthermore, the density of ionic liquids is extremely affected by number of carbons in alkyl chain of cations. The densities of ionic liquids are adversely related to their chains (Marsh et al. 2004).

To compare densities of various ionic liquids with different cations and anions, these data are tabulated in Table 8.4.

8.3.6 *Viscosity*

One obstacle to utilize ionic liquids for absorbing carbon dioxide is their great viscosity owing to existing impurities and high hydrogen bonding/van der Walls interactions. In comparison with common solvents, viscosities of ionic liquids are high, which decreases the rate of carbon dioxide capture, and thus it is inappropriate for industrial application (Ramdin et al. 2012). Although high temperature can solve this problem, components with low viscosities are more favorable to be used in the separation processes since cost of apparatuses is lower and rates of mass transfer is higher (Kirchner 2008).

At room temperature, viscosities of ionic liquids can change between 10 mPa.s to approximately 500 mPa.s, suggesting that there are more viscous in comparison with common solvents. The viscosity can be modified by changing cations and anions to the appropriate one. The number of carbon atoms in alkyl chain of cations as well as type of anions and their symmetry can influence the viscosity. Actually, the viscosities of ionic liquids increase with increasing the length of their cation chains by enhancing the molecular interactions. However, the impact of the anion type is greater than the length of chain (Krupiczka et al. 2015). In order to compare viscosities of various ionic liquids with different cations and anions, these data are tabulated in Table 8.5.

8.3.7 *Biodegradability*

Biodegradability properties consider the effect of ionic liquids on the environment. Biodegradability is defined as complete and fast mineralization. Thus, the ultimate potential of biodegradability is seen as broad molecular cleavage (Petkovic et al. 2011).

Same as other properties, types of cations and anions play a significant role in biodegradability. The previous investigations showed that biodegradability properties can be improved by choosing proper ester groups for pyridinium or imidazolium ionic liquids (Gathergood et al. 2004; Harjani et al. 2009). In addition, imidazolium ionic liquids have lower biodegradability potential in comparison to pyridinium ones (Grabińska-Sota and Kalka 2003). Furthermore, ionic liquids with phosphonium cations are identified to persist in the microbial attack (Atefi et al. 2009).

Due to lack of knowledge, further investigations should be conducted to evaluate the biodegradability behaviors of various ionic liquids in the environment.

8.3.8 *Solubility*

Various researches have been done to investigate the solubility of different gases in common ionic liquids. As mentioned before, for the first time, that solubility of carbon dioxide in 1-Butyl-3-methylimidazolium hexafluorophosphate ionic liquid had been measured by Blanchard et al. (Blanchard et al. 1999). Furthermore, the

Table 8.5 Viscosities of some commonly used ionic liquids for carbon dioxide capture (Seddon et al. 2000; McEwen et al. 1999; Suarez et al. 1998; Huddleston et al. 2001; Bonhôte et al. 1996; Tokuda et al. 2006)

Ionic liquid		Temperature(K)	Viscosity(mPa.S)
[C_2mim] [CF_3SO_3]	1-ethyl-3-methylimidazolium trifluoromethanesulfonate	298	90
[C_2mim][BF_4]	1-ethyl-3-methylimidazolium tetrafluoroborate	293,298	66.5,43
[C_2mim][PF_6]	1-ethyl-3-methylimidazolium hexafluorophosphate	343	23.4
[C_2mim][OTf]	1-ethyl-3-methylimidazolium triflate	293	50
[C_2mim] [TF_2N]	1-ethyl-3-methylimidazolium bis((trifluoromethyl)sulfonyl)imide	298	28
[C_4mim][Cl]	1-butyl-3-methylimidazolium chloride	323	1534
[C_4mim][I]	1-butyl-3-methylimidazolium iodide	298	1110
[C_4mim][BF_4]	1-butyl-3-methylimidazolium tetrafluoroborate	293,298	154,219
[C_4mim][PF_6]	1-butyl-3-methylimidazolium hexafluorophosphate	293,298	371,450
[C_4mim] [TF_2N]	1-butyl-3-methylimidazolium bis((trifluoromethyl)sulfonyl)imide	298	52
[C_4mim] [CF_3COO]	1-butyl-3-methylimidazolium trifluoroacetate	298	73
[C_6mim][Cl]	1-hexyl-3-methylimidazolium chloride	323,298	1124,716
[C_6mim][BF_4]	1-hexyl-3-methylimidazolium tetrafluoroborate	293	314
[C_6mim][PF_6]	1-hexyl-3-methylimidazolium hexafluorophosphate	273,298	690,585
[C_8mim][Cl]	1-octyl-3-methylimidazolium chloride	323,298	1930,337
[C_8mim][BF_4]	1-octyl-3-methylimidazolium tetrafluoroborate	298	682
[C_8mim][PF_6]	1-octyl-3-methylimidazolium hexafluorophosphate	293	439

solubility of carbon dioxide and eight different gases in 1-Butyl-3-methylimidazolium hexafluorophosphate ionic liquid were reported Anthony et al. The results showed that carbon dioxide has the highest solubility in 1-Butyl-3-methylimidazolium hexafluorophosphate ionic liquid, in comparison to methane, nitrogen, hydrogen, ethane, carbon monoxide, ethylene and oxygen (Anthony et al. 2002). In addition, not only 1-Butyl-3-methylimidazolium hexafluorophosphate ionic liquid but also imidazolium-based tetrafluoroborate ([bmim][BF4]) ionic liquid can solve high amount of carbon dioxide (Chen et al. 2006). Totally, various researches have been done to explore and understand the solubility of different gases in common imidazolium-based ionic liquids (Ramdin et al. 2012; Kirchner 2008; Krupiczka et al. 2015).

In addition, Kroon et al. succeeded to predict the solubility of carbon dioxide in some kinds of ionic liquids with PC-SAFT equation of state. Results demonstrated that modeling and real data agrees very well (Kroon et al. 2006). Furthermore, the ability of various equation of state such as PC-SAFT, RK, SRK, vdW, and PR for predicting solubility of various gases in ionic liquids have been examined in many research work. The results showed that there is a good agreements between modeling and experimental data (Yokozeki and Shiflett 2010; Shiflett and Yokozeki 2010; Andreu and Vega 2008; Chen et al. 2012; Yokozeki et al. 2008; Kroon et al. 2006).

Furthermore, the cations have a lower impact on the solubility of gas, especially carbon dioxide in ionic liquids, compared to anions. Researchers proved that the solubility of carbon dioxide depends on the characteristics of anions. In addition, it is affected by number of carbons in alkyl chain of cations. Due to high dispersion forces, the alkyl group of cations can interact better with carbon dioxide. Thus, the solubility of gases increases with increasing the length of the ionic liquid chain (Chen et al. 2006; Zhang et al. 2005; Baltus et al. 2004). For instance, the solubilities of carbon dioxide in eight traditional hydroxyl ammonium (imidazolium – based) ionic liquids were determined (Yuan et al. 2007). Results showed that the ionic liquids with the long alkyl chain in the cation ring dissolves a high amount of carbon dioxide. However, in comparison with the long alkyl chain in the cation ring, the effect of type of anions on the solubility of carbon dioxide is greater due to strong interaction with carbon dioxide and anions.

Furthermore, to evaluate the influence of the anions, the seven different ionic liquids with the same cations were studied by Aki et al. Among 1-butyl-3-methylimidazolium nitrate, 1-butyl-3-methylimidazolium dicyanamide, 1-butyl-3-methylimidazolium tetrafluoroborate, 1-butyl-3-methylimidazolium hexafluorophosphate, the1-butyl-3-methylimidazolium bis(trifluoromethylsulfonyl) imide dissolved high amount of carbon dioxide. Thus, the solubility of carbon dioxide in the ionic liquids containing fluoroalkyl groups was higher than others (Aki et al. 2004). The fact that the nature of the anion and cation have first and secondary roles, respectively, in managing the solubility of carbon dioxide was represented by Canda et al. In other words, they confirmed that solubility of this gas is more influenced by the type of anion in comparison with the cation (Cadena et al. 2004). Moreover, the strongest interactions between hexafluorophosphate anion and carbon dioxide were reported (Anthony et al. 2002). Not only hexafluorophosphate but also bis (trifluoromethylsulfonyl)imide made a strong association with carbon dioxide (Cadena et al. 2004). Among ordinarily alkylimidazolium based ionic liquids, the alkylimidazolium bis(trifluoromethylsulfonyl)imide exhibits the greatest carbon dioxide solubility and affinity (Anthony et al. 2002). Nonetheless, in comparison with other anions, the bis(trifluoromethylsulfonyl)imide does not adequately boost volumetric carbon dioxide solubility (Bara et al. 2009a).

Although the solubility of carbon dioxide can be improved by designing ionic liquids with appropriate cations/onions, the carbon dioxide capture capacity using ionic liquid is lower than current post-combustion methods such as chemical solvent such as amine-based.

8.3.9 *Selectivity*

Besides carbon dioxide, various gases such as methane, ethane, nitrogen, hydrogen, and impurities are presented in the natural/flue gas. Hence the selectivity of carbon dioxide in the ionic liquids is an important factor as its solubility that affects the performance of carbon dioxide capture with ionic liquids. Fortunately, ionic liquids selectivity for carbon dioxide is superlative in comparison to other gases because of great quadrupole moments of this gas (Anderson et al. 2007; Anthony et al. 2002; Zhao et al. 2011; Bara et al. 2009b). Zhang et al. reported that among carbon monoxide, hydrogen, carbon dioxide, and oxygen, the selectivity of carbon dioxide in conventional ionic liquids was the highest (Zhang et al. 2012). Therefore, researchers try to synthesize an ionic liquid to be more selective toward carbon dioxide than other gases.

8.3.10 *Conductivity*

The size of cations and anions and molecular weight determine the conductivity of ionic liquids which are related to accessible charge carriers, and their mobility. The mobility of ions in aggregation condition limits protic ionic liquids conductivity (Greaves and Drummond 2008; Greaves et al. 2006). Thus, low interactions between ions results in high conductivity (Davoodnia et al. 2010; Wang et al. 2008). The aprotic ionic liquids have lower conductivity as compared to protic ionic liquids. In addition, the conductivity of ionic liquids is affected by size and number of carbons in alkyl chain of cations (Markusson et al. 2007). The big size and long alkyl chain cations decrease the conductivity of them (Greaves and Drummond 2008). Although the types of anions can effect on these properties, obvious trend has not been found for these ions (Belieres and Angell 2007).

8.4 Improving Ionic Liquids Performance

The researchers have tried to formulate appropriate solvents with both good features of amines and advantageous properties of ionic liquids for absorbing carbon dioxide. Synthesizing task specific ionic liquids (TSIL) by changing the structure of the ionic liquid with an amine is one approach to meet this target, and thereby improve the rate of absorption of carbon dioxide (Sánchez et al. 2011).

The performance of ionic liquids in capturing carbon dioxide can be considerably improved by adding an amine to the structure of ionic liquid which is known as functionalized ionic liquids (Sánchez et al. 2011). First of all, Bates et al. synthesized amine functionalized ionic liquid and investigated the mechanism of reaction with carbon dioxide. They reported that this new ionic liquid reacts with carbon

Table 8.6 Comparison of different properties of chemical and physical absorber (Zhang et al. 2012; Xie et al. 2014)

	Chemical absorber		Physical absorber		
Type	Functionalized ionic liquids	Monoethanolamine (30%wt)	Conventional ionic liquids	DEPG	pc
Viscosity(mPa.s)	50–2000	18.98	20–2000	5.8	3
Vapor pressure (Pa) T = 298 K	1.33×10^{-4}	85	1.33×10^{-4}	9.73×10^{-2}	11.3
Solubility of CO2 (mol/mol) P = 100,000 Pa, T = 293-313 K	1.6	50–85	>2.51	3.63	3.41
Selectivity(CO_2/ CH_4)			8–35	15	26
Selectivity(H_2S/ CO_2)		1–2	2–10	8.8	3.3
Selectivity(CO_2/H_2)			50–150	77	126
Selectivity(CO_2/N_2)			30–100	50	117
Price	80/g	370$/L	40$/gr	80$/L	117$/L

dioxide the same as an organic amine. Furthermore, this amine functionalized ionic liquid can also be regenerated more efficiently because of its low volatility (Bates et al. 2002). Subsequently, other researchers have followed this approach to design and synthesize novel ionic liquids.

Table 8.6 illustrates that pure ionic liquid as a physical absorber is not a good option for carbon dioxide absorption whereas the functionalized ionic liquids have the ability to absorb carbon dioxide chemically and consequently overcome drawbacks of utilizing pure ionic liquids. The mechanism of reactions between carbon dioxide and functionalized ionic liquids is the same as that of carbon dioxide and organic amine's mechanism (Yu et al. 2008).

The cation/anion/dual amino functionalized ionic liquids are three kinds of these task specific ionic liquids (Yu et al. 2008). For instance, the natural amino acids and alkaline groups were utilized to functionalize both ions (cation and anion) of ionic liquids to obtain dual aminofunctionalized phosphonium ionic liquids ([aP_{4443}][AA]) in order to enhance the capacity of carbon dioxide absorption. Actually, for a period of 80 min, one mole of this ionic liquid can absorb one mole carbon dioxide (Zhang et al. 2009). The molecular simulations showed that the hydrogen bond that exist between two anions, and also between cation and anion causes the low viscosity. The viscosity of dual aminofunctionalized phosphonium ionic liquids ranges between 758 mPa.s and 1985 mPa.s (temperature = 298 K) (Liu et al. 2009).

In another research, the tetrabutylphosphonium hydroxide [P(C4)4][OH] reacted with five different amino acids to synthesize tetrabutylphosphonium amino acid [P(C4)4][AA] for capturing carbon dioxide. This new solvent had a good carbon dioxide absorption capacity. Although the partial pressure of carbon dioxide is low, the amino- functionalized ionic liquids are chemically able to capture carbon dioxide up to approximately 30 mole% (Bates et al. 2002). Recently, new kinds of functionalized ionic liquids are synthesized. In this super basic amino acid ionic liquids, there are two functional (amino and carboxylic acid) groups, providing beneficial characteristics for capturing carbon dioxide. Furthermore, this new generation of ionic liquids are cheap, and it is not harmful for the environment (Fukumoto et al. 2005).

It should be noted that slow diffusion of carbon dioxide arising from the high viscosity of functionalized ionic liquids in addition to high cost attributing to complicated synthesis and purification processes are the two major weaknesses of utilization of such ionic liquids to capture carbon dioxide. In addition, they are not economically profitable as compared to conventional alkanolamines.

Furthermore, mixing ionic liquids with amine or water is another strategy to improve the performance of ionic liquids and alleviate their faults such as high cost and viscosity. Through this strategy the favorable features of both solvents are combined, leading to higher absorption efficiency and capacity carbon dioxide capture (Zhang et al. 2014). Camper et al. utilized the mixture of monoethanolamine and diethanolamine with 1-Ethyl-3-methylimidazolium bis(trifluoromethylsulfonyl) imide ionic liquids to enhance carbon dioxide absorption process. The results indicated that the mixture of ionic liquids and amines provides higher absorption performance with low energy consumption for desorption. Also, the shortcomings of using functionalized ionic liquids are eliminated in this mixture (Camper et al. 2008). Some recent researches dealt with absorption of carbon dioxide from gas streams (e.g., flue gas) by using a mixture of ionic liquid and methyl diethanolamine/monoethanolamine. Undeniably, the physicochemical properties information such as the solubility of special gases in the specific solvents is required to design processes and equipment (Camper et al. 2008; Feng et al. 2010).

Different absorption behaviors are observed for a mixture solution of ionic liquids and amines at various pressures. In fact, more carbon dioxide is absorbed by solution of methyl diethanolamine and lysine based ionic liquid than the mixture of methyl diethanolamine and glycine based ionic liquid at a pressure more than 25 KPa while the amount of absorption is the same for both liquid mixtures at a pressure below 25 KPa. Furthermore, the researchers studied regeneration behaviors of a variety of mixtures of ionic liquid and amine, demonstrating that the efficiency of regeneration was not the same for various aqueous solutions (Feng et al. 2010).

These new ionic liquid solvents have low volatility, and also low amount of energy is needed to regenerate the solvent. Besides, the performance of this mixture is equal to the amine aqueous solutions for carbon dioxide capture (Camper et al. 2008). In this mixture, water causes a decrease in the viscosity and an increase in the solubility properties.

Most recent strategy is the application of an ionic liquid in the membrane to separate carbon dioxide. The previous studies illustrated that during the carbon dioxide absorption process with the combination of the membrane with the ionic liquid, the viscosity is reduced and the selectivity and permeability of carbon dioxide increase significantly (Zeng et al. 2017). Regarding stabling composition of the membrane is essential for keeping their structure; the ionic liquid with low vapor pressure is good option for this usage. The low volatility of the ionic liquid solvents make them a proper candidate to be used in capture processes through membrane. Indeed, it can pave the way for the membrane structure to remain unchanged.

Furthermore, the functionalized ionic liquids can participate in the composition of membranes. In such cases, carbon dioxide capture rate can be intensified. Many scientists have investigated the potential applications of membranes which are supported with ionic liquid for the gas separation process. Noble and Gin and Bare et al. published the review articles about the usage of membranes supported by ionic liquid being used for separation of various gases (Bara et al. 2010; Noble and Gin 2011).

8.5 Previous Research

Analyzing the relevant research papers of using ionic liquid in capturing of carbon dioxide, the research interests have been mainly in the following aspects:

1. Designing and synthesizing new ionic liquids, especially functionalized ones, to enhance carbon dioxide absorption performance by introducing the functional groups to their ions (MacDowell et al. 2010; Wappel et al. 2010; Karadas et al. 2010; Hasib-ur-Rahman et al. 2010).
2. Blending ionic liquid with membrane composites to form a unique ionic liquid based membrane to improve carbon dioxide selectivity and capacity, and also decreasing the viscosity of the ionic liquid to accelerate carbon dioxide mass transfer in solvent required in industrial application (Bara et al. 2010; Adibi et al. 2011; Bara et al. 2009a, b; Bara et al. 2007; Barghi et al. 2010; Gonzalez-Miquel et al. 2011; Hong et al. 2009).
3. The prediction and estimation of thermo physical properties of the ionic liquid and other components of the system, and developing proper thermodynamic models (Breure et al. 2007; Domańska et al. 2007; Vega et al. 2010; Andreu and Vega 2008).
4. Investigating kinetics of absorption and industrial design of the process using ionic liquids as well as evaluating energy consumption and other economic issues (Blanchard et al. 1999; Brennecke and Gurkan 2010; Wang et al. 2010; MacDowell et al. 2010; Wappel et al. 2010; Karadas et al. 2010).
5. Investigating the environmental effects of different ionic liquids (Zhang et al. 2012; Zhang et al. 2008; Petkovic et al. 2011; Tian et al. 2011)

8.6 Important Factors for Choosing Suitable Ionic Liquid for Carbon Dioxide Capturing

Various ionic liquids are synthesized by adjusting cations and anions. The tunability of ionic liquids is resulted in various possible structures (Wappel et al. 2010). Definitely, it is hard to choose appropriate ionic liquid among these varieties for capturing carbon dioxide. Different factors should be taken into account to determine a suitable one for this purpose. The proper ionic liquids should have the ability to absorb a high amount of carbon dioxide in the presence of different gases. In addition, the biodegradability, being economical, and toxicity are the other important factors that should be considered.

The information about the solubility of carbon dioxide in various ionic liquids is available, which are important to be considered to assess the ability of ionic liquids for carbon dioxide capture. As discussed before, the type of anion employed, play a significant role in enhancing carbon dioxide solubility. In addition, a high quantity of carbon dioxide is dissolved in the ionic liquids when the flora ion is used in the anion part of their structure. Recently, the functionalized ionic liquids are proposed as a promising solvent to dissolve higher amounts of carbon dioxide.

The previous researches proved that the methyltrioctylammonium bis(trifluoromethylsulfonyl)imide ([N1888][Tf2N]), 1-pentyl-3 methylimidazolium tris(nonafluorobutyl)trifluorophosphate ([P5mim][nFAP]), trihexyl(tetradecyl) phosphonium bis(trifluoromethylsulfonyl) imide ([P14666][Tf2N]), trihexyl(tetradecyl)phosphonium chloride ([P14666][Cl]) and 1-hexyl-3- methylimidazolium tris(heptafluoropropyl)trifluorophosphate ([hmim] [FAP]) can solve high amount of carbon dioxide, so they are known as the best ionic liquids (Oliferenko et al. 2011). Furthermore, the Henry's law constant is an essential parameter that determines the physical absorption of carbon dioxide by ionic liquid. However, for functionalized ionic liquids which have chemical absorption, this parameter does not matter.

The ionic liquid which contains tetrafluoroborate [BF4], nitrate [NO3] or methyl sulfate [CH3SO4] anion and tetramethylguanidinium (TMG), tetrabutylammonium or pentamethylguanidinium (PMG) cation, have superlative carbon dioxide selectivity (Mortazavi-Manesh et al. 2013).

From the point of industrialization, the cost of ionic liquid plays a critical role. Due to lack of large scale production, this solvent is currently expensive. As a result, one of the main shortcomings for utilizing ionic liquid to capture carbon dioxide at industrial scale is the high cost of process. This is why it is not easy to utilize ionic liquids in the industrial scale owing to the fact that it is only produced in small quantity in the laboratory. The Sigma-Aldrich is a well-known ionic liquids supplier which provides and defines prices of a majority of ionic liquids. Aghaie et al. compared different prices of ionic liquid in different scales (Aghaie et al. 2018).

8.7 Economic View

In comparison with amine solvents, ionic liquids are not widely used in industry for carbon dioxide capture process. The operating cost, and the efficiency of ionic liquids in the lab-scale are very different from those conducted in large scale in the industry. The prices of ionic liquids are between 1$ and 10$ per gram. These prices are high for industrial usage and should be decreased to be comparable with conventional amines (Arshad 2009). Ramdin et al. investigated the price of them which are about 1000$ per kilogram ionic liquids, which is 100–1000 times higher than that of conventional solvents. The BASF company believed that this price will be dropped to 40$–100$ per kilogram ionic liquid if they are produced in large scale.

Actually, the high cost of utilizing ionic liquids can be attributed to the complexity of synthesis (Ramdin et al. 2012). The research of Aghaie et al. compared the prices of conventional ionic liquids and functionalized ones. In general, conventional ionic liquids are five times cheaper than the tailored ones (Aghaie et al. 2018). Furthermore, the price of ionic liquids with hydrogen sulfate anion and protic ones are comparable with acetone as organic solvent (Chen et al. 2014).

The cost of plant for capturing carbon dioxide is divided into two parts: operating and capital cost (Yang et al. 2014). When using ionic liquid instead of the amine solvent, some amounts of the capital cost are decreased. For instance, ionic liquid can be regenerated completely in the stripper because the vapor pressure is so low. In contrast, some amounts of amines are vaporized due to their high vapor pressure. Thus, a percentage of the amines can be lost in the stripper (Chen et al. 2014). Moreover, the energy consumptions which are required for the regeneration step is decreased 37% when a mixture of amine and ionic liquids are used instead of pure amine.

Furthermore, nowadays, the applications of ionic liquids in different field have increased significantly. It seems that producing ionic liquid has become commercialized. Therefore, low-cost ionic liquid will be available. Furthermore, the technology may help to synthesize them simply.

8.8 Challenge

As discussed before, ionic liquids have high viscosity. This property is one of the important challenges for using them because of influences on other properties including the solubility of gas (carbon dioxide) in this material. Although employing a mixture of amine, water, and ionic liquid is a solution for this problem, further investigation should be performed to determine optimum conditions and composition of the process.

Another drawback of utilizing ionic liquids for capturing carbon dioxide is the cost. Further research should be done to find an economical and easy approach for synthesizing cheaper ionic liquids solutions. Furthermore, the design of system and operating conditions should be optimized to be suitable for using ionic liquids in the industrial scale.

In addition, some critical properties data of ionic liquid are difficult to be found that are essential for thermodynamic modeling. Furthermore, selecting an appropriate equation of state can provide correct prediction of thermodynamic characteristics.

8.9 Conclusion

The ionic liquid is a novel solvent for the absorption of carbon dioxide. The outstanding features of these new generations of solvents make them suitable for usage in the carbon dioxide capturing process. In this Chapter, the physiochemical properties of ionic liquids are briefly explained. In addition, the impacts of ions including anions and cations and length (size) of the chain on determining various properties including solubility, density, viscosity are discussed. The functionalized ionic liquids, amine- ionic liquid solvents and ionic liquid- membranes are investigated as innovative approaches to improve their performance for separation of carbon dioxide. The characteristics of an appropriate ionic liquid for this application are defined. Furthermore, the economic issues of using ionic liquid for separation process are discussed.

References

Adibi M, Barghi SH, Rashtchian D (2011) Predictive models for permeability and diffusivity of CH4 through Imidazolium-based supported ionic liquid membranes. J Membr Sci 371(1–2):127–133. https://doi.org/10.1016/j.memsci.2011.01.024

Aghaie M, Rezaei N, Zendehboudi S (2018) A systematic review on CO_2 capture with ionic liquids: current status and future prospects. Renew Sust Energ Rev 96:502–525. https://doi.org/10.1016/j.rser.2018.07.004

Ahrens S, Peritz A, Strassner T (2009) Tunable aryl alkyl ionic liquids (TAAILs): the next generation of ionic liquids. Angew Chem Int Ed 48(42):7908–7910. https://doi.org/10.1002/anie.200903399

Aki SN, Mellein BR, Saurer EM, Brennecke JF (2004) High-pressure phase behavior of carbon dioxide with imidazolium-based ionic liquids. J Phys Chem B 108:20355–20365. https://doi.org/10.1021/jp046895+

Anderson JL, Dixon JK, Brennecke JF (2007) Solubility of CO 2 , CH 4 , C 2 H 6 , C 2 H 4 , O 2 , and N 2 in 1-hexyl-3-methylpyridinium bis(trifluoromethylsulfonyl)imide: comparison to other ionic liquids. Acc Chem Res 40(11):1208–1216. https://doi.org/10.1021/ar7001649

Andreu JS, Vega LF (2008) Modeling the solubility behavior of CO 2 , H 2 , and Xe in [C n-Mim] [Tf 2 N] ionic liquids. J Phys Chem B 112(48):15398–15406. https://doi.org/10.1021/jp807484g

Anthony JL, Maginn EJ, Brennecke JF (2001) Solution thermodynamics of Imidazolium-based ionic liquids and water. J Phys Chem B 105(44):10942–10949. https://doi.org/10.1021/jp0112368

Anthony JL, Maginn EJ, Brennecke JF (2002) Solubilities and thermodynamic properties of gases in the ionic liquid 1- n -butyl-3-methylimidazolium hexafluorophosphate. J Phys Chem B 106(29):7315–7320. https://doi.org/10.1021/jp020631a

Arshad MW (2009) CO2 capture using ionic liquids. Technical University of Denmark, DTU, DK-2800 Kgs, Lyngby

Atefi F, Garcia MT, Singer RD, Scammells PJ (2009) Phosphonium ionic liquids: design, synthesis and evaluation of biodegradability. Green Chem 11(10):1595. https://doi.org/10.1039/b913057h

Babamohammadi S, Shamiri A, Aroua MK (2015) A review of CO2 capture by absorption in ionic liquid-based solvents. Rev Chem Eng 31(4). https://doi.org/10.1515/revce-2014-0032

Bailera M, Lisbona P, Romeo LM, Espatolero S (2017) Power to gas projects review: lab, pilot and demo plants for storing renewable energy and CO_2. Renew Sust Energ Rev 69:292–312. https://doi.org/10.1016/j.rser.2016.11.130

Baltus RE, Culbertson BH, Dai S, Luo H, DePaoli DW (2004) Low-pressure solubility of carbon dioxide in room-temperature ionic liquids measured with a quartz crystal microbalance. J Phys Chem B 108(2):721–727. https://doi.org/10.1021/jp036051a

Bara JE, Lessmann S, Gabriel CJ, Hatakeyama ES, Noble RD, Gin DL (2007) Synthesis and performance of polymerizable room-temperature ionic liquids as gas separation membranes. Ind Eng Chem Res 46(16):5397–5404. https://doi.org/10.1021/ie0704492

Bara JE, Carlisle TK, Gabriel CJ, Camper D, Finotello A, Gin DL, Noble RD (2009a) Guide to CO 2 separations in imidazolium-based room-temperature ionic liquids. Ind Eng Chem Res 48(6):2739–2751. https://doi.org/10.1021/ie8016237

Bara JE, Gabriel CJ, Carlisle TK, Camper DE, Finotello A, Gin DL, Noble RD (2009b) Gas separations in fluoroalkyl-functionalized room-temperature ionic liquids using supported liquid membranes. Chem Eng J 147(1):43–50. https://doi.org/10.1016/j.cej.2008.11.021

Bara JE, Camper DE, Gin DL, Noble RD (2010) Room-temperature ionic liquids and composite materials: platform technologies for CO 2 capture. Acc Chem Res 43(1):152–159. https://doi.org/10.1021/ar9001747

Barghi SH, Adibi M, Rashtchian D (2010) An experimental study on permeability, diffusivity, and selectivity of CO2 and CH4 through [Bmim][PF6] ionic liquid supported on an alumina membrane: investigation of temperature fluctuations effects. J Membr Sci 362(1–2):346–352. https://doi.org/10.1016/j.memsci.2010.06.047

Bates ED, Mayton RD, Ntai I, Davis JH (2002) CO 2 capture by a task-specific ionic liquid. J Am Chem Soc 124(6):926–927. https://doi.org/10.1021/ja017593d

Belieres JP, Angell CA (2007) Protic ionic liquids: preparation, characterization, and proton free energy level representation. J Phys Chem B 111(18):4926–4937. https://doi.org/10.1021/jp067589u

Blanchard LA, Hancu D, Beckman EJ, Brennecke JF (1999) Green processing using ionic liquids and CO2. Nature 399(6731):28–29. https://doi.org/10.1038/19887

Blanchard LA, Gu Z, Brennecke JF (2001) High-pressure phase behavior of ionic liquid/CO 2 systems. J Phys Chem B 105(12):2437–2444. https://doi.org/10.1021/jp003309d

Bonhôte P, Dias AP, Papageorgiou N, Kalyanasundaram K, Grätzel M (1996) Hydrophobic, highly conductive ambient-temperature molten salts. Inorg Chem 35(5):1168–1178. https://doi.org/10.1021/ic951325x

Boot-Handford ME, Abanades JC, Anthony EJ, Blunt MJ, Brandani S, Mac Dowell N, Fernández JR, Ferrari MC, Gross R, Hallett JP, Haszeldine RS (2014) Carbon capture and storage update. Energy Environ Sci 7(1):130–189. https://doi.org/10.1039/C3EE42350F

Brennecke JF, Gurkan BE (2010) Ionic liquids for CO 2 capture and emission reduction. J Phys Chem Lett 1(24):3459–3464. https://doi.org/10.1021/jz1014828

Brennecke JF, Maginn EJ (2001) Ionic liquids: innovative fluids for chemical processing. AICHE J 47(11):2384–2389. https://doi.org/10.1002/aic.690471102

Breure B, Bottini SB, Witkamp GJ, Peters CJ (2007) Thermodynamic modeling of the phase behavior of binary systems of ionic liquids and carbon dioxide with the group contribution equation of state. J Phys Chem B 111(51):14265–14270. https://doi.org/10.1021/jp0776098

Cadena C, Anthony JL, Shah JK, Morrow TI, Brennecke JF, Maginn EJ (2004) Why is CO_2 So soluble in imidazolium-based ionic liquids? J Am Chem Soc 126(16):5300–5308. https://doi.org/10.1021/ja039615x

Camper D, Bara JE, Gin DL, Noble RD (2008) Room-temperature ionic liquid−amine solutions: tunable solvents for efficient and reversible capture of CO_2. Ind Eng Chem Res 47(21):8496–8498. https://doi.org/10.1021/ie801002m

Chen Y, Zhang S, Yuan X, Zhang Y, Zhang X, Dai W, Mori R (2006) Solubility of CO_2 in Imidazolium-based tetrafluoroborate ionic liquids. Thermochim Acta 441(1):42–44. https://doi.org/10.1016/j.tca.2005.11.023

Chen Y, Mutelet F, Jaubert JN (2012) Modeling the solubility of carbon dioxide in imidazolium-based ionic liquids with the PC-SAFT equation of state. J Phys Chem B 116(49):14375–14388. https://doi.org/10.1021/jp309944t
Chen L, Sharifzadeh M, Mac Dowell N, Welton T, Shah N, Hallett JP (2014) Inexpensive ionic liquids: [HSO 4] – -based solvent production at bulk scale. Green Chem 16(6):3098–3106. https://doi.org/10.1039/C4GC00016A
Chiappe C, Pomelli CS (2014) Point-functionalization of ionic liquids: an overview of synthesis and applications. Eur J Org Chem 2014(28):6120–6139. https://doi.org/10.1002/ejoc.201402093
Davoodnia A, Bakavoli M, Moloudi R, Tavakoli-Hoseini N, Khashi M (2010) Highly efficient, one-pot, solvent-free synthesis of 2,4,6-Triarylpyridines using a Brønsted-acidic ionic liquid as reusable catalyst. Monatsh Chem 141(8):867–870. https://doi.org/10.1007/s00706-010-0329-x
Domańska U, Żołek-Tryznowska Z, Królikowski M (2007) Thermodynamic phase behavior of ionic liquids. J Chem Eng Data 52(5):1872–1880. https://doi.org/10.1021/je700204d
Dullius JE, Suarez PA, Einloft S, de Souza RF, Dupont J, Fischer J, De Cian A (1998) Selective catalytic Hydrodimerization of 1,3-butadiene by palladium compounds dissolved in ionic liquids. Organometallics 17(5):815–819. https://doi.org/10.1021/om970982p
Emel'yanenko VN, Verevkin SP, Heintz A (2007) The gaseous enthalpy of formation of the ionic liquid 1-butyl-3-methylimidazolium dicyanamide from combustion calorimetry, vapor pressure measurements, and Ab initio calculations. J Am Chem Soc 129(13):3930–3937. https://doi.org/10.1021/ja0679174
Feng Z, Cheng-Gang F, You-Ting W, Yuan-Tao W, Ai-Min L, Zhi-Bing Z (2010) Absorption of CO2 in the aqueous solutions of functionalized ionic liquids and MDEA. Chem Eng J 160(2):691–697. https://doi.org/10.1016/j.cej.2010.04.013
Figueroa JD, Fout T, Plasynski S, McIlvried H, Srivastava RD (2008) Advances in CO2 capture technology—the U.S. Department of Energy's Carbon Sequestration Program. Int J Greenhouse Gas Control 2(1):9–20. https://doi.org/10.1016/S1750-5836(07)00094-1
Freund H, Bauer J, Zeiser T, Emig G (2005) Detailed simulation of transport processes in fixed-beds. Ind Eng Chem Res 44(16):6423–6434. https://doi.org/10.1021/ie0489453
Fukumoto K, Yoshizawa M, Ohno H (2005) Room temperature ionic liquids from 20 natural amino acids. J Am Chem Soc 127(8):2398–2399. https://doi.org/10.1021/ja043451i
Gathergood N, Garcia MT, Scammells PJ (2004) Biodegradable ionic liquids: part I. concept, preliminary targets and evaluation. Green Chem 6(3):166. https://doi.org/10.1039/b315270g
Gonzalez-Miquel M, Palomar J, Omar S, Rodriguez F (2011) CO_2/N_2 selectivity prediction in supported ionic liquid membranes (SILMs) by COSMO-RS. Ind Eng Chem Res 50(9):5739–5748. https://doi.org/10.1021/ie102450x
Grabińska-Sota E, Kalka J (2003) An assessment of the toxicity of Pyridinium chlorides and their biodegradation intermediates. Environ Int 28(8):687–690. https://doi.org/10.1016/S0160-4120(02)00112-5
Greaves TL, Drummond CJ (2008) Protic ionic liquids: properties and applications. Chem Rev 108(1):206–237. https://doi.org/10.1021/cr068040u
Greaves TL, Weerawardena A, Fong C, Krodkiewska I, Drummond CJ (2006) Protic ionic liquids: solvents with tunable phase behavior and physicochemical properties. J Phys Chem B 110(45):22479–22487. https://doi.org/10.1021/jp0634048
Harjani JR, Singer RD, Garcia MT, Scammells PJ (2009) Biodegradable pyridinium ionic liquids: design, synthesis and evaluation. Green Chem 11(1):83–90. https://doi.org/10.1039/B811814K
Harper ND, Nizio KD, Hendsbee AD, Masuda JD, Robertson KN, Murphy LJ, Johnson MB, Pye CC, Clyburne JA (2011) Survey of carbon dioxide capture in phosphonium-based ionic liquids and end-capped polyethylene glycol using DETA (DETA = Diethylenetriamine) as a model absorbent. Ind Eng Chem Res 50(5):2822–2830. https://doi.org/10.1021/ie101734h
Hasib-ur-Rahman M, Siaj M, Larachi F (2010) Ionic liquids for CO2 capture—development and Progress. Chem Eng Process 49(4):313–322. https://doi.org/10.1016/j.cep.2010.03.008
Hayes R, Warr GG, Atkin R (2015) Structure and nanostructure in ionic liquids. Chem Rev 115(13):6357–6426. https://doi.org/10.1021/cr500411q

Holbrey JD, Seddon KR (1999) The phase behaviour of 1-Alkyl-3-methylimidazolium tetrafluoroborates; ionic liquids and ionic liquid crystals. J Chem Soc 13:2133–2140. https://doi.org/10.1039/a902818h

Hong SU, Park D, Ko Y, Baek I (2009) Polymer-ionic liquid gels for enhanced gas transport. Chem Commun 46:7227. https://doi.org/10.1039/b913746g

Hu G, Nicholas NJ, Smith KH, Mumford KA, Kentish SE, Stevens GW (2016) Carbon dioxide absorption into promoted potassium carbonate solutions: a review. Int J Greenhouse Gas Control 53:28–40. https://doi.org/10.1016/j.ijggc.2016.07.020

Huddleston JG, Visser AE, Reichert WM, Willauer HD, Broker GA, Rogers RD (2001) Characterization and comparison of hydrophilic and hydrophobic room temperature ionic liquids incorporating the Imidazolium Cation. Green Chem 3(4):156–164. https://doi.org/10.1039/b103275p

Karadas F, Atilhan M, Aparicio S (2010) Review on the use of ionic liquids (ILs) as alternative fluids for CO 2 capture and natural gas sweetening. Energy Fuel 24(11):5817–5828. https://doi.org/10.1021/ef1011337

Kazarian SG, Sakellarios N, Gordon CM (2002) High-pressure CO2-induced reduction of the melting temperature of ionic liquids. Electronic supplementary information (ESI) available: Fig. S1: ATR-IR Spectrum of [C16mim][PF6] after it has been subjected to solution of Ferrocene in CO2 at 50 °C. Chem Commun 12:1314–1315. https://doi.org/10.1039/b202759c

Kenarsari SD, Yang D, Jiang G, Zhang S, Wang J, Russell AG, Wei Q, Fan M (2013) Review of recent advances in carbon dioxide separation and capture. RSC Adv 3(45):22739. https://doi.org/10.1039/c3ra43965h

Kirchner B (2008) Ionic liquids from theoretical investigations. InIonic liquids:213–262. https://doi.org/10.1007/128_2008_36

Kothandaraman A (2010) Carbon dioxide capture by chemical absorption: a solvent comparison study 72

Kroon MC, Karakatsani EK, Economou IG, Witkamp GJ, Peters CJ (2006) Modeling of the carbon dioxide solubility in Imidazolium-based ionic liquids with the TPC-PSAFT equation of state. J Phys Chem B 110(18):9262–9269. https://doi.org/10.1021/jp060300o

Krupiczka R, Rotkegel A, Ziobrowski Z (2015) Comparative study of CO2 absorption in packed column using Imidazolium based ionic liquids and MEA solution. Sep Purif Technol 149:228–236. https://doi.org/10.1016/j.seppur.2015.05.026

Langmuir I (1913) The vapor pressure of metallic tungsten. Phys Rev 2(5):329–342. https://doi.org/10.1103/PhysRev.2.329

Liang Z, Fu K, Idem R, Tontiwachwuthikul P (2016) Review on current advances, future challenges and consideration issues for post-combustion CO2 capture using amine-based absorbents. Chin J Chem Eng 24(2):278–288. https://doi.org/10.1016/j.cjche.2015.06.013

Liu X, Zhou G, Zhang S, Yao X (2009) Molecular dynamics simulation of dual amino-functionalized imidazolium-based ionic liquids. Fluid Phase Equilib 284(1):44–49. https://doi.org/10.1016/j.fluid.2009.06.004

MacDowell N, Florin N, Buchard A, Hallett J, Galindo A, Jackson G, Adjiman CS, Williams CK, Shah N, Fennell P (2010) An overview of CO2 capture technologies. Energy Environ. Sci 3(11):1645. https://doi.org/10.1039/c004106h

MacFarlane DR, Tachikawa N, Forsyth M, Pringle JM, Howlett PC, Elliott GD, Davis JH, Watanabe M, Simon P, Angell CA (2014) Energy applications of ionic liquids. Energy Environ Sci 7(1):232–250. https://doi.org/10.1039/C3EE42099J

Maia FM, Calvar N, González EJ, Carneiro AP, Rodriguez O, Macedo EA (2013) Modeling of ionic liquid systems: phase equilibria and physical properties. In: Ionic liquids – new aspects for the future. InTech. https://doi.org/10.5772/51812

Markusson H, Belieres JP, Johansson P, Angell CA, Jacobsson P (2007) Prediction of macroscopic properties of protic ionic liquids by Ab initio calculations. J Phys Chem A 111(35):8717–8723. https://doi.org/10.1021/jp072036k

Marsh KN, Boxall JA, Lichtenthaler R (2004) Room temperature ionic liquids and their mixtures—a review. Fluid Phase Equilib 219(1):93–98. https://doi.org/10.1016/j.fluid.2004.02.003

McEwen AB, Ngo HL, LeCompte K, Goldman JL (1999) Electrochemical properties of imidazolium salt electrolytes for electrochemical capacitor applications. J Electrochem Soc 146(5):1687. https://doi.org/10.1149/1.1391827

Meine N, Benedito F, Rinaldi R (2010) Thermal stability of ionic liquids assessed by potentiometric titration. Green Chem 12(10):1711. https://doi.org/10.1039/c0gc00091d

Mortazavi-Manesh S, Satyro MA, Marriott RA (2013) Screening ionic liquids as candidates for separation of acid gases: solubility of hydrogen sulfide, methane, and ethane. AICHE J 59(8):2993–3005. https://doi.org/10.1002/aic.14081

Ngo HL, LeCompte K, Hargens L, McEwen AB (2000) Thermal properties of imidazolium ionic liquids. Thermochim Acta 357–358:97–102. https://doi.org/10.1016/S0040-6031(00)00373-7

Noble RD, Gin DL (2011) Perspective on ionic liquids and ionic liquid membranes. J Membr Sci 369(1–2):1–4. https://doi.org/10.1016/j.memsci.2010.11.075

Oliferenko AA, Oliferenko PV, Seddon KR, Torrecilla JS (2011) Prediction of gas Solubilities in ionic liquids. Phys Chem Chem Phys 13(38):17262–17272. https://doi.org/10.1039/c1cp20336c

Paulechka YU, Zaitsau DH, Kabo GJ, Strechan AA (2005) Vapor pressure and thermal stability of ionic liquid 1-Butyl-3-methylimidazolium bis(trifluoromethylsulfonyl)amide. Thermochim Acta 439(1–2):158–160. https://doi.org/10.1016/j.tca.2005.08.035

Petkovic M, Seddon KR, Rebelo LP, Pereira CS (2011) Ionic liquids: a pathway to environmental acceptability. Chem Soc Rev 40(3):1383–1403. https://doi.org/10.1039/C004968A

Ramdin M, de Loos TW, Vlugt TJ (2012) State-of-the-art of CO 2 capture with ionic liquids. Ind Eng Chem Res 51(24):8149–8177. https://doi.org/10.1021/ie3003705

Revelli AL, Mutelet F, Jaubert JN (2010) High carbon dioxide solubilities in imidazolium-based ionic liquids and in poly(ethylene glycol) dimethyl ether. J Phys Chem B 114(40):12908–12913. https://doi.org/10.1021/jp1057989

Riemer PW, Ormerod WG (1995) International perspectives and the results of carbon dioxide capture disposal and utilisation studies. Energy Convers Manage 36(6–9):813–818. https://doi.org/10.1016/0196-8904(95)00128-Z

Rosli A, Ahmad AL, Lim JK, Low SC (2017) Advances in liquid absorbents for CO2 capture: a review. J Phys Sci 28:121–144. https://doi.org/10.21315/jps2017.28.s1.8

Sánchez LG, Meindersma GW, De Haan AB (2011) Kinetics of absorption of CO2 in amino-functionalized ionic liquids. Chem Eng J 166(3):1104–1115. https://doi.org/10.1016/j.cej.2010.12.016

Seddon KR, Stark A, Torres MJ (2000) Influence of chloride, water, and organic solvents on the physical properties of ionic liquids. Pure Appl Chem 72(12):2275–2287. https://doi.org/10.1351/pac200072122275

Shafeeyan MS, Daud WM, Shamiri A (2014) A review of mathematical modeling of fixed-bed columns for carbon dioxide adsorption. Chem Eng Res Des 92(5):961–988. https://doi.org/10.1016/j.cherd.2013.08.018

Shariati A, Gutkowski K, Peters CJ (2005) Comparison of the phase behavior of some selected binary systems with ionic liquids. AICHE J 51(5):1532–1540. https://doi.org/10.1002/aic.10384

Shiflett MB, Yokozeki A (2010) Separation of carbon dioxide and sulfur dioxide using room-temperature ionic liquid [Bmim][$MeSO_4$]. Energy Fuel 24(2):1001–1008. https://doi.org/10.1021/ef900997b

Sreedhar I, Vaidhiswaran R, Kamani BM, Venugopal A (2017a) Process and engineering trends in membrane based carbon capture. Renew Sust Energ Rev 68:659–684. https://doi.org/10.1016/j.rser.2016.10.025

Sreedhar I, Nahar T, Venugopal A, Srinivas B (2017b) Carbon capture by absorption – path covered and ahead. Renew Sust Energ Rev 76:1080–1107. https://doi.org/10.1016/j.rser.2017.03.109

Suarez PA, Einloft S, Dullius JE, De Souza RF, Dupont J (1998) Synthesis and physical-chemical properties of ionic liquids based on 1-n-Butyl-3-methylimidazolium cation. J Chim Phys Phys Chim Biol 95(7):1626–1639. https://doi.org/10.1051/jcp:1998103

Tian X, Zhang X, Zeng S, Xu Y, Yao Y, Chen Y, Huang L, Zhao Y, Zhang S (2011) Process analysis and multi-objective optimization of ionic liquid-containing acetonitrile process to produce 1,3-butadiene. Chem Eng Technol 34(6):927–936. https://doi.org/10.1002/ceat.201000426

Tokuda H, Tsuzuki S, Susan MA, Hayamizu K, Watanabe M (2006) How ionic are room-temperature ionic liquids? An indicator of the physicochemical properties. J Phys Chem B 110(39):19593–19600. https://doi.org/10.1021/jp064159v

Torralba-Calleja E, Skinner J, Gutiérrez-Tauste D (2013) CO2 capture in ionic liquids: a review of solubilities and experimental methods. J Chem 2013:1–16. https://doi.org/10.1155/2013/473584

Valderrama JO, Robles PA (2007) Critical properties, normal boiling temperatures, and acentric factors of fifty ionic liquids. Ind Eng Chem Res 46(4):1338–1344. https://doi.org/10.1021/ie0603058

Valderrama JO, Rojas RE (2009) Critical properties of ionic liquids. Revisited Ind Eng Chem Res 48(14):6890–6900. https://doi.org/10.1021/ie900250g

Valderrama JO, Sanga WW, Lazzús JA (2008) Critical properties, normal boiling temperature, and acentric factor of another 200 ionic liquids. Ind Eng Chem Res 47(4):1318–1330. https://doi.org/10.1021/ie071055d

Vega LF, Vilaseca O, Llovell F, Andreu JS (2010) Modeling ionic liquids and the solubility of gases in them: recent advances and perspectives. Fluid Phase Equilib 294(1–2):15–30. https://doi.org/10.1016/j.fluid.2010.02.006

Wang W, Shao L, Cheng W, Yang J, He M (2008) Brønsted acidic ionic liquids as novel catalysts for Prins reaction. Catal Commun 9(3):337–341. https://doi.org/10.1016/j.catcom.2007.07.006

Wang C, Luo H, Luo X, Li H, Dai S (2010) Equimolar CO2 capture by imidazolium-based ionic liquids and superbase systems. Green Chem 12(11):2019. https://doi.org/10.1039/c0gc00070a

Wappel D, Gronald G, Kalb R, Draxler J (2010) Ionic liquids for post-combustion CO2 absorption. Int J Greenhouse Gas Control 4(3):486–494. https://doi.org/10.1016/j.ijggc.2009.11.012

Wasserscheid P, Keim W (2000) Ionic liquids—new 'solutions' for transition metal catalysis. Angew Chem 39(21):3772–3789. https://doi.org/10.1002/1521-3773(20001103)39:21<3772::AID-ANIE3772>3.0.CO;2-5

Welton T (1999) Room-temperature ionic liquids. Solvents for synthesis and catalysis. Chem Rev 99(8):2071–2084. https://doi.org/10.1021/cr980032t

Wilkes JS, Zaworotko MJ (1992) Air and water stable 1-ethyl-3-methylimidazolium based ionic liquids. J Chem Soc Chem Commun 13, 965. https://doi.org/10.1039/c39920000965

Xie Y, Zhang Y, Lu X, Ji X (2014) Energy consumption analysis for CO 2 separation using imidazolium-based ionic liquids. Appl Energy 136:325–335. https://doi.org/10.1016/j.apenergy.2014.09.046

Yang J, Yu X, Yan J, Tu ST (2014) CO2 capture using amine solution mixed with ionic liquid. Ind Eng Chem Res 53(7):2790–2799. https://doi.org/10.1021/ie4040658

Yokozeki A, Shiflett MB (2010) Gas Solubilities in ionic liquids using a generic van Der Waals equation of state. J Supercrit Fluids 55(2):846–851. https://doi.org/10.1016/j.supflu.2010.09.015

Yokozeki A, Shiflett MB, Junk CP, Grieco LM, Foo T (2008) Physical and chemical absorptions of carbon dioxide in room-temperature ionic liquids. J Phys Chem B 112(51):16654–16663. https://doi.org/10.1021/jp805784u

Yu KM, Curcic I, Gabriel J, Tsang SC (2008) Recent advances in CO 2 capture and utilization. ChemSusChem 1(11):893–899. https://doi.org/10.1002/cssc.200800169

Yu CH, Huang CH, Tan C (2012) A review of CO2 capture by absorption and adsorption. Aerosol Air Qual Res 12(5):745–769. https://doi.org/10.4209/aaqr.2012.05.0132

Yuan X, Zhang S, Liu J, Lu X (2007) Solubilities of CO2 in hydroxyl ammonium ionic liquids at elevated pressures. Fluid Phase Equilib 257(2):195–200. https://doi.org/10.1016/j.fluid.2007.01.031

Zeng S, Zhang X, Bai L, Zhang X, Wang H, Wang J, Bao D, Li M, Liu X, Zhang S (2017) Ionic-liquid-based CO 2 capture systems: structure, interaction and process. Chem Rev 117(14):9625–9673. https://doi.org/10.1021/acs.chemrev.7b00072

Zhang S, Yuan X, Chen Y, Zhang X (2005) Solubilities of CO 2 in 1-Butyl-3-methylimidazolium hexafluorophosphate and 1,1,3,3-tetramethylguanidium lactate at elevated pressures. J Chem Eng Data 50(5):1582–1585. https://doi.org/10.1021/je050046d

Zhang Y, Bakshi BR, Demessie ES (2008) Life cycle assessment of an ionic liquid versus molecular solvents and their applications. Environ Sci Technol 42(5):1724–1730. https://doi.org/10.1021/es0713983

Zhang Y, Zhang S, Lu X, Zhou Q, Fan W, Zhang X (2009) Dual amino-functionalised phosphonium ionic liquids for CO_2 capture. Chem Eur J 15(12):3003–3011. https://doi.org/10.1002/chem.200801184

Zhang X, Zhang X, Dong H, Zhao Z, Zhang S, Huang Y (2012) Carbon capture with ionic liquids: overview and progress. Energy Environ Sci 5(5):6668. https://doi.org/10.1039/c2ee21152a

Zhang X, Bao D, Huang Y, Dong H, Zhang X, Zhang S (2014) Gas-liquid mass-transfer properties in CO_2 absorption system with ionic liquids. AICHE J 60(8):2929–2939. https://doi.org/10.1002/aic.14507

Zhao Y, Zhang X, Dong H, Zhen Y, Li G, Zeng S, Zhang S (2011) Solubilities of gases in novel alcamines ionic liquid 2-[2-hydroxyethyl (methyl) amino] ethanol chloride. Fluid Phase Equilib 302(1–2):60–64. https://doi.org/10.1016/j.fluid.2010.08.017

Zhao B, Tao W, Zhong M, Su Y, Cui G (2016) Process, performance and modeling of CO 2 capture by chemical absorption using high gravity: a review. Renew Sust Energ Rev 65:44–56. https://doi.org/10.1016/j.rser.2016.06.059

Chapter 9
Methods for the Recovery of CO_2 from Chemical Solvents

Maryam Ebrahimzadeh Sarvestani, Maryam Raeisi, and Mohammad Reza Rahimpour

Abstract Regarding the increasing rate of greenhouse gases emission such as CO_2, and subsequently the global warming issue, recovery of CO_2 exhausted from industry, and also its utilization is of paramount importance. There are several practical methods to capture CO_2 involving physical and chemical absorption, adsorption, membrane and cryogenic technologies, among which the chemical absorption through solvent commonly amines; e.g., primary, secondary and tertiary amines is considered to be most efficient method due to high CO_2 capacity arising from the reaction taking place between the solvent and CO_2. However, two major obstacles encounter this method including high energy requirement for solvent regeneration and developing novel solvent to enhance CO_2 absorption capacity, when conventional absorption/desorption process is applied.

To address the mentioned obstacles, one can either employ membrane technology or develop more suitable chemical solvents. Many investigators have dealt with developing novel solvents including blending of amin-based solvents, potassium carbonate, metal hydroxide solutions e.g., Sodium hydroxide, ammonia aqueous solution. In addition, membrane technology can either be applied for both absorption and desorption process simultaneously, or it can be joined to the conventional process. If it is aimed to join membrane technology with the conventional process, some novel processes including membrane flash, electrodialysis membrane and electrolysis membrane are substituted with the desorption section of the conventional processes.

Keywords Greenhouse gases · CO_2 capture · CO_2 recovery processes · Chemical solvents · Solvent regeneration · Membrane technology · Absorption/desorption process · CO_2 absorption capacity · Flash membrane · Electrodialysis membrane · Electrolysis membrane

M. Ebrahimzadeh Sarvestani · M. R. Rahimpour (✉)
Department of Chemical Engineering, Shiraz University, Shiraz, Iran
e-mail: rahimpor@shirazu.ac.ir

M. Raeisi
Department of Chemical Engineering, Shiraz University of Technology, Shiraz, Iran

Inamuddin et al. (eds.), *Sustainable Agriculture Reviews 38*, Sustainable Agriculture Reviews 38, https://doi.org/10.1007/978-3-030-29337-6_9

Abbreviations

AEM	Anion exchange membrane
AHPD	2-amino-2-hydroxymethyl-1,3-propanediol
AMP	2-amino-2-methyl-1-propanol
BPM	Bipolar membrane
C_2H_4O	Ethylene oxide
$C_8H_{23}N_5$	Tetraethylenepentamine
CEM	Cation exchange membrane
CFC	Chlorofluorocarbons
CH_3	Methyl
CH_4	Methane
CO_2	Carbon dioxide
DEPG	Dimethyl ether of polyethylene glycol
ETA	Ethanolamine
H_3AsO_3	Arsenious acid
H_3BO_3	Boric acid
IPCC	Intergovernmental Panel on Climate Change
K_2CO_3	potassium carbonate
MDEA	Methyldiethanolamine
MEA	Monoethanolamine
MeOH	Methanol
$NaHCO_3$	Sodium bicarbonate
NaOH	Sodium hydroxide
NH_3	Ammonia
NMP	Normal methyl pyrrolidone
NO	Nitrogen Oxide
O_2	Oxygen
PC	Propylene carbonate
PE	polyethylene
PFA	Poly(tetrafluoroethylen-co-perfluorovinylether)
PM	2-Piperadine methanol
PP	Polypropylene
PTFE	Polytetrafluoroethylene
PVDF	Polyvinylidene fluoride
PZ	Piperazine
PZEA	2-(1-piperazinyl)-ethylamine
SF_6	Sulfur hexafluoride
SO_2	Sulfur dioxide
SOx	Sulfur Oxides
TBP	Tributyl phosphate
T-S1	Titanium silicate
UOP LLC	Universal Oil Products

9.1 Introduction

From the beginning of the twentieth century, with the progress in technology, the global warming issue and greenhouse gases have come into sharp focus of scientists and governments. The most important factors contributing to the global warming are human activities, specifically burning of fossil fuels such as oil and coal in the industry (Abu-Zahra et al. 2007; Wall 2007).

According to the reports of The Intergovernmental Panel on Climate Change (IPCC), considering the existence of greenhouse gases, the global warming issue has increased up to 37% from 1990 to 2015 (Change 2015). It is obvious that the concentration of greenhouse gases such as Carbon dioxide (CO_2), Chlorofluorocarbons (CFC), Nitrogen Oxide (NO), Methane (CH_4), Sulfur Oxides (SOx) and Sulfur hexafluoride (SF_6) in the atmosphere have had ascending trend owing to the human activities (Metz et al. 2005b). Figure 9.1 illustrates the percentage of major greenhouse gases including CO_2, N_2O and CH_4 that contribute to the global warming, according to IPCC reports in 2014 (Edenhofer et al. 2014).

Among the aforementioned greenhouse gases, many investigators believe that the most influential one is CO_2, and thus they have concentrated their attention on limiting its emission into the environment (Frumhoff et al. 2015; Heede and Oreske 2016, Heede 2014).

Generally, CO_2 can be produced in two ways including direct one such as energy production and indirect one e.g., power production in the industry among which energy production is performed through burning of fossil fuels. Therefore, as the demand for energy increases in the world, the consumption of fossil fuels levels up which can undeniably induce more CO_2 production and its emission into the environment. Unfortunately, according to reported statistics in recent years it can be speculated that until 2025 the total consumption of energy all over the world increases up to 75%, resulting in higher concentration of CO_2 in the atmosphere (Bao et al. 2018).

Regarding the noted issues, from 1980, scientists have conducted comprehensive studies to control emission of CO_2 as well as its separation and utilization (Farla et al. 1995; Khanipour et al. 2017). By way of illustration, Moomaw introduces efficient and proper utilization of fossil fuels in addition to designing novel industrial processes as the most effective way to restrict CO_2 emission from 1990 to 2020 (Moomaw 1996).

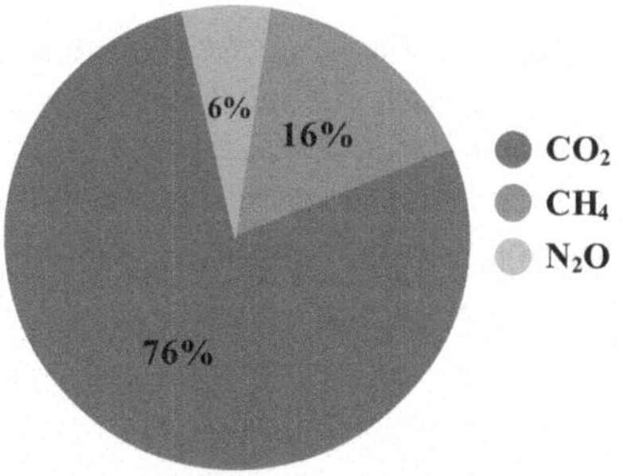

Fig. 9.1 Percentage of major greenhouse gases was reported by IPCC in 2014

The applied methods for this purpose are generally classified into physical and chemical absorption, adsorption, membrane and cryogenic technologies, which are strongly in favor of large industrial units such as chemical industries in terms of using economical instruments and thereby reducing costs, in addition to improving the efficiency of production (Gielen 2003; Göttlicher and Pruschek 1997; McKee 2002).

Among these methods, chemical absorption through solvent can be regarded as the best choice owing to its high reaction rate, providing higher CO_2 absorption capacity, especially when the partial pressure of CO_2 is low e.g., CO_2 which is produced in power plants or large industrial units consuming fossil fuels. This is why this method has been studied extensively by the researchers (Abu Zahra and Jansens 2009; Aroonwilas and Veawab 2004; Mattisson et al. 2001; Zhao et al. 2007).

Nevertheless, applying this method is accompanied by some challenges including corrosion, decomposition of the solvents and high required thermal energy for regeneration of the solvents. By way of illustration, Singh et al. reported that for removal of 90% CO_2, around 15–30% of the energy which is generated through power plant is consumed to regenerate the solvent (Singh and Versteeg 2008).

As a result, scientists try to address the problem of high energy requirement for regeneration processes through developing novel solvents, designing processes with high efficiency as well as optimization of the processes through simulation and modeling.

Therefore, in this chapter, some conventional and innovative chemical solvents are introduced; in addition, influential and efficient methods are proposed all with the purpose of minimizing the stated issues, and the proposed methods are evaluated from different aspects. Firstly, we have outlined different CO_2 recovery processes. Secondly, the chemical absorption process is evaluated and the ractical applied solvents are compared. Thirdly, the application of membrane technology for the absorption/desorption process is reviewed and compared with the conventional absorption/desorption process. Also, various types of membranes are presented and compared with each other. Finally, different membrane methods i.e., flash membrane, electrodialysis membrane and electrolysis membrane which can be joined with the conventional process to enhance the regeneration process are introduced.

9.2 Chemical and Physical Absorption

Generally, according to condition e.g., concentration, pressure and solubility of CO_2, physical, chemical and hybrid solvents are utilized for the absorption process (Raksajati et al. 2013, 2016).

In the first place, in physical absorption which is carried out without chemical reaction, the main driving force to absorb CO_2 is the solubility (Ban et al. 2014). In general, solubility increases with increase in pressure and decrease in temperature, and thus these two variables are introduced as the most important factors to enhance CO_2 absorption processes (Henni et al. 2005). Besides, if the concentration of CO_2 is high enough in the entrance feed, it is preferable to apply physical absorption rather than chemical one since not only it reduces energy demand of regeneration significantly, but also the size of utilized instruments becomes smaller (Henni et al. 2005).

Some solvents which are applied in the physical absorption are as the following (Ban et al. 2014; Burr and Lyddon 2008):

1. Conventional solvents: Tributyl phosphate (TBP), Dimethyl ether of polyethylene glycol (DEPG), Methanol (MeOH), Normal methyl pyrrolidone (NMP), propylene carbonate (PC).
2. Novel solvents: Mixture of polyethylene glycol dialkyl ethers, Diethylene Glycol Monobutyl Ether Acetate.

In the second place, in chemical absorption, chemical reaction takes place between solvent and CO_2, and thereby the driving force of the mass transfer improves, leading to increase in mass transfer. This is why in contrast with the physical absorption, the chemical absorption is strongly dependent on the temperature.

9.3 Characteristics of Chemical Solvents

In Fig. 9.2, the common chemical solvents are classified and explained briefly as the following.

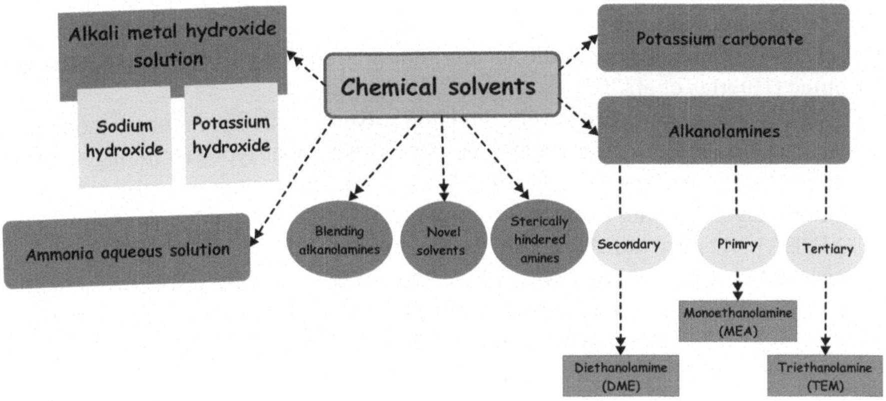

Fig. 9.2 Classification of chemical solvents were introduced in CO_2 capturing process

9.3.1 Alkanolamines

One of the most common chemical solvents in absorption and recovery of CO_2 processes is alkanolamine which has been used from 89 years ago in the industry (Booth 2005). In general, the applied alkanolamines are classified into three types including primary, secondary and tertiary in which the difference comes from the degree of substitution on nitrogen atom, meaning that one, two and three alkanol chain can be attached to the atoms of nitrogen producing primary, secondary and tertiary alkanolamines, respectively (Kohl and Nielsen 1997).

To explain the primary type alkanolamine, the most prevalent primary alkanolamin is monoethanolamine (MEA) which due to its low viscosity and density as well as high rate of reaction with CO_2 is known as a suitable solvent in CO_2 absorption processes. This chemical substance is produced through reaction of ammonia (NH_3) and ethylene oxide (C_2H_4O). This reaction is shown in Fig. 9.3, which is known with the chemical formula of ($HOCH_2CH_2$) NH_2 and the commercial name of MEA or ethanolamine (ETA) (Vaidya and Kenig 2007).

Nevertheless, there are several faults in utilization of this type of solvent MEA, which are listed below, concisely:

1. Degradation of the solvent in oxygen-rich environments-Since in the flue gas some oxygen-containing substances such as O_2 and SO_2 exist, it can result in degradation of the solvent through oxidizing, and consequently it is not advised to use such solvent in oxygen-rich environments (Davis and Rochelle 2009; Davison 2007).
2. High thermal energy requirement for regeneration of the solvent- It is noted that the CO_2 capture process through the monoethanolamine is a costly one, because approximately 36% of whole operating cost is used to regenerate the solvent, i.e., around 4.2 GJ/ton CO_2 is needed for the regeneration process (Chapel et al. 1999; Resnik et al. 2004).
3. Low absorption capacity of the solvent- it is around 720 g CO_2/Kg monoethanolamine (Huertas et al. 2015).
4. High corrosion of instruments through the solvent- Actually, monoethanolamine can cause more corrosion in comparison with the other types including secondary e.g., diethanolamime, tertiary i.g., triethanolamine amines and the mixture amines e.g., 2-amino-2-methyl-1-propanol (AMP) and methyldiethanolamine (MDEA). Results indicated that the maximum and minimum corrosion, was occurred in the processes with MEA and MDEA amines, respectively, while processes with AMP and diethanolamime (DEA) showed the moderate corrosion (Davidson 2007; Kittel et al. 2009).

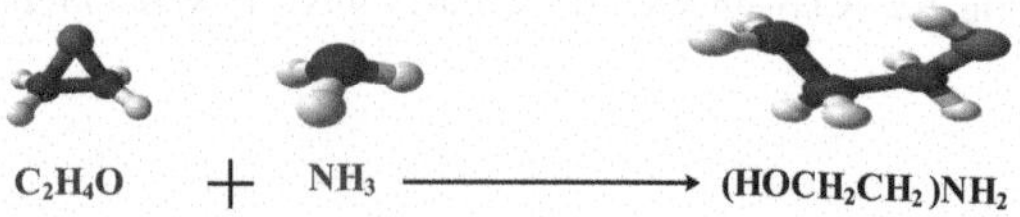

Fig. 9.3 Production reaction of monoethanolamine

Considering the stated faults above, it can be concluded that the overall operating cost of CO_2 absorption processes is relatively high when the monoethanolamine solvent is employed.

To tackle the aforementioned issue, many scientists have dealt with lowering the thermal energy requirement noticeably through modeling of the CO_2 absorption (Alie et al. 2005; Chang and Shih 2005; Rao et al. 2004; Sander and Mariz 1992; Singh et al. 2003) and recovery processes (Barchas and Davis 1992; Chapel et al. 1999; Mariz 1998; Suda et al. 1992).

The secondary alkanolamine e.g., DEA differs from the primary one in one hydrogen atom or one alkanol chain, meaning that the chemical formula of DEA comprises of two alkanol chains and a hydrogen atom bonded to a nitrogen atom $(CH_2CH_2OH)_2NH$.

Likewise, the tertiary type of alkanolamine e.g., triethanolamine (TEA) consists of three alkanol chains $(CH2OHCH2)_3$ bonded to a nitrogen atom, resulting in the molecular formula of $(CH_2OHCH_2)_3N$. In addition to triethanolamine, to exemplify another tertiary type, if one methyl (CH_3) group substitutes the alkanol chain, MDEA being formed with the molecular formula of $(CH_2OHCH_2)_2N(CH_3)$ (Li and Chang 1994).

It should also be noted that as a characteristic of the tertiary type, which can regarded as a superiority over other types, is the lack of nitrogen bonded to hydrogen, which can be in favor of CO_2 hydrolysis reaction to produce bicarbonate. In fact, the tertiary type has higher cyclic capacity, low volatility and degradation in comparison with the other types owing to the bicarbonate reaction pathway (Aroonwilas and Veawab 2007).

9.3.2 *Blend of Alkanolamines*

Today, due to high operating cost which mainly arises from the regeneration process, utilizing a mixture of amines as the solvents has come into sharp focus. In this regard, Dow Chemical Company and Universal Oil Products (UOP LLC) are two large companies producing amine blends (Tontiwachwuthikul et al. 1992).

By way of illustration, park et al. introduced AMP and MDEA as two examples of amine blends which are suitable alternatives for single amines since thermal energy requirement for regeneration of the solvents reduces when these blends are used, and this is why these blends have been used extensively, recently (Park et al. 2006; Vaidya and Kenig 2007).

In another research, MEA and MDEA are mixed, and results showed that the thermal energy for regeneration decreases when this blend is applied in a specific ratio instead of single monoethanolamine as the solvent to capture CO_2. Actually, it reduces the energy demand for regeneration up to the point that the thermal efficiency improves around 3% when compared to the choice where single monoethanolamine is used as the solvent (Aroonwilas and Veawab 2007).

In conclusion, in order to select a suitable blend, two major factors should be taken into account including:

1. Absorption performance: it can improve when primary and secondary amines are used as the blend components.
2. Energy requirement for solvent generation: it can be reduced when tertiary amines are applied as the blend component.

To enhance the solvent characteristics in both aspects, one should use a blend which contains all three types of amines (Adeosun and Abu-Zahra 2013).

9.3.3 *Sterically Hindered Amines*

This type of amines are mainly produced by large companies involving Kansai Electric Power Co and Mitsubishi (Mangalapally and Hasse 2010; Reddy et al. 2003).

To explain about the structures of these amines, they are formed from primary and secondary amines in which the amino groups are attached to secondary or tertiary carbon atoms (Vega et al. 2018). The different forms of the sterically hindered amines are listed in Table 9.1.

The benefit of using this type of chemical solvent in CO_2 capture includes reducing the required energy of solvent regeneration as well as increasing CO_2 loading in the solvent.

For instance, in a research studied by park et al. it is reported that the regeneration energy of KS that a type of sterically hindered amines is approximately 3GJ/tonCO$_2$ which is nearly 20% lower than that of MEA. Moreover, Park et al. examined another sterically hindered amines i.e., AHPD. It is stated in their work that if the partial pressure of CO_2 is in the high range e.g., above 4 kPa, the loading of CO_2 in the absorption solution of AHPD is much greater than that of monoethanolamine (Park et al. 2003).

9.3.4 *Novel Amine-Based Solvents*

Aside from the blending method mixing alkanolamines; e.g., AMP/MEA, DEA/MDEA, MEA/MDEA to enhance solvent characteristics in the absorption process, comprehensive studies have been carried out to develop innovative solvents (Choi et al. 2012; Dugas and Rochelle 2011; Pacheco et al. 2000; Seo and Hong 1996; Zoghi et al. 2012).

The main objective of developing such solvents is reducing thermal energy for solvent regeneration, increase in the CO_2 recovery rate, decrease in enthalpy of the absorption process, improving CO_2 absorption capacity and accelerating reaction rate of in solvents with CO_2 with the purpose of increasing absorption kinetics.

Table 9.1 Different types of sterically hindered amines

Name & abbreviation	Chemical structure	Chemical formulae
2-Amino-2-methyl-1-propanol(AMP)	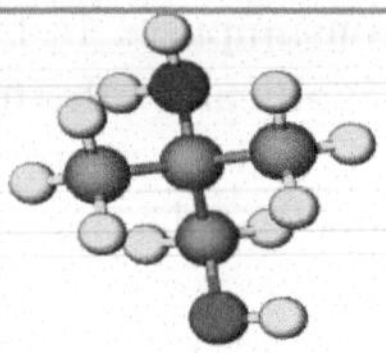	$C_4H_{11}NO$
2-Amino-2-hydroxymethyl-1,3-propanediol (AHPD)	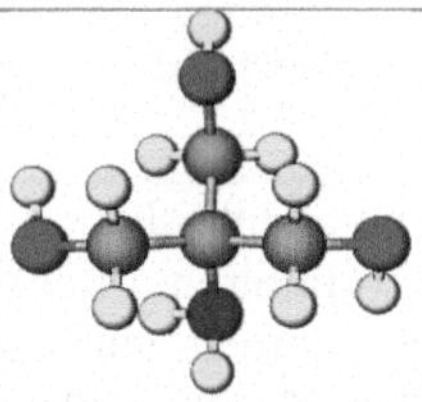	$C_4H_{11}NO_3$
2-piperidineethanol(PE)	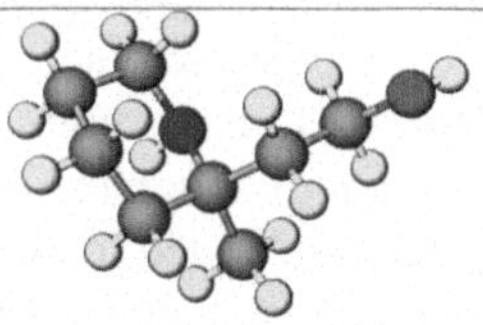	$C_7H_{15}NO$
2-Piperadine methanol(PM)	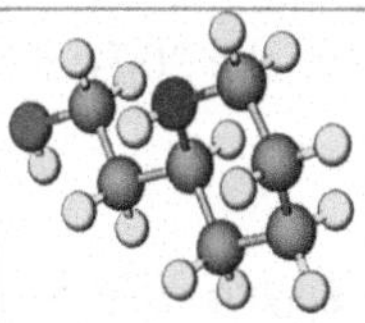	$C_6H_{13}NO$
N-methylalanine	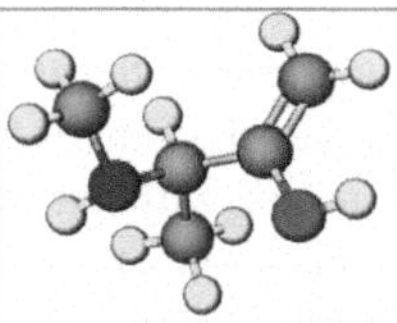	$C_4H_9NO_2$

Carbon atom = Gray; Nitrogen atom = blue; Hydrogen atom = white; Oxygen atom = Red

Some novel solvents which have been developed in recent years are as the following (Bishnoi and Rochelle 2000; Bruder et al. 2011; Chakravarty et al. 1985; Freeman et al. 2010; Xu et al. 1998):

1. Diamine: e.g., aqueous piperazine(PZ)
2. Cyclic monoamine: e.g., piperidine
3. Derivatives of alkanolamines: e.g., mixed alkanolamines with enzymes such as carbonic anhydrase

To exemplify, a concentrated aqueous PZ can manifest superior performance than alkanoamines such as monoethanolamine in terms of CO_2 absorption rate, thermal degradation and energy consumption. In fact, when PZ is used as the solvent, CO_2 absorption rate enhances significantly, thermal degradation becomes negligible for the temperature of up to 150 °C and the energy consumption can lower around 10–20%, all in relation to using monoethanolamine as the solvent (Freeman et al. 2010).

9.3.5 *Potassium Carbonate*

This type of solvent has come into sharp focus of many industrialists due to low cost and volatility in CO_2 absorption in different processes such as ammonia synthesis, crude hydrogen, natural gas treatment and power plant (Borhani et al. 2015; Isa et al. 2018; Kothandaraman 2010). In this regard, positive sides and negative sides of using this type of solvent are listed in the Table 9.2.

Despite attractive characteristics of potassium carbonate, it has also the weakness of low reaction rate, leading to high instrument cost which makes the process uneconomic. Thus, to solve this problem in the industry and accelerate CO_2 absorption rate, some promoters are added to potassium carbonate including amines, carbonate anhydrase enzyme, boric acid (H_3BO_3), piperazine ($C_4H_{10}N_2$), arsenious acid (H_3AsO_3), tetraethylenepentamine ($C_8H_{23}N_5$) and 2-(1-piperazinyl)-ethylamine (PZEA) (Cullinane and Rochelle 2004; Kim et al. 2012; Smith et al. 2012).

9.3.6 *Alkali Metal Hydroxide Solution*

From 1940, the CO_2 capture process through alkali metal hydroxide solutions, specifically Sodium hydroxide, has been studied (Nguyen et al. 2010; Tepe and Dodge 1943).

To compare such solutions with amine-based solvents e.g., primary alkanolamine, on one hand, regeneration of sodium hydroxide aqueous solution is more difficult than that of monoethanolamine since in the sodium hydroxide case when CO_2 is absorbed in the solution, the product is sodium bicarbonate ($NaHCO_3$) which has high solubility in aqueous solutions, leading to high operating cost. On the other hand, not only CO_2 absorption capacity of sodium hydroxide is greater than that of

Table 9.2 Advantages and disadvantages of potassium carbonate solvent

	Potassium carbonate solution
Advantages	Low toxicity, high volatility, low consumption energy, high
Disadvantages	Low reaction rate, poor kinetic reaction
CO_2 recovery efficiency	95%

monoethanolamine, but also Sodium hydroxide is more plentiful, cheaper, more well-known and a less volatile substance with lower vapor pressure (Heda et al. 1995; Siriwardane et al. 2007).

9.3.7 Ammonia Aqueous Solution

In comparison with amine-based solvents e.g., MEA, ammonia aqueous solution is superior in terms of CO_2 absorption capacity and regeneration process, which can provide the higher CO_2 absorption efficiency. By way of illustration, if one compares aqueous ammonia with a amine-based solvent e.g., MEA according to literature, firstly, absorption capacity of aqueous ammonia is reported to be above 1 kg CO_2/kg ammonia, while that of monoethanolamine is around 0.36 kg CO_2/kg MEA (Bai and Yeh 1997). Secondly, the regeneration temperature of aqueous ammonia is 85–95 °C whereas that of monoethanolamine is 120 °C, leading to low thermal energy requirement as well as less solvent degradation.

In conclusion, the two mentioned superiority of aqueous ammonia over monoethanolamine result in higher absorption efficiency, i.e., using ammonia can obtain CO_2 removal efficiency of 95–100%, whereas that of monoethanolamine is 90% (Darde et al. 2009; Zhang 1996). In Table 9.3, the above mentioned chemical solvents are compared in terms of operating conditions (Temperature& pressure), advantages and disadvantages, and CO_2 recovery Efficiency.

9.4 Methods of CO_2 Recovery

9.4.1 Absorption/Desorption Process Through Packed Columns

One of the well-known methods to recover CO_2 is the desorption/absorption process, which has been used comprehensively (Veawab et al. 2001). One of advantages of this method is that suitable automation and monitoring of the process can minimize the need for human operators. However, it is reported that the total cost of this process including utilization of new solvents, maintenance and operating costs is high around 40–70 $/ton CO_2 (Chakma and Tontiwachwuthikul 1999).

The schematic diagram of absorption/desorption process is depicted in Fig. 9.4. The selected column for absorption/desorption process is a packed column owing to its high stability and efficiency. The absorption and desorption columns are named absorber and stripper, respectively (Rao and Rubin 2002). Firstly, the mixture of gas is entered to the absorption column from the bottom while the pure solvent enters from the top in order to contact counter-currently, and the reaction between CO_2 and

Table 9.3 Comparison between different chemical solvents for CO_2 capture process

Types of chemical solvents	MEA	DEA	AMP	MDEA	NH3	NaOH	PZ
Temperature &Pressure operation (ºC -bar)	40–65, 1.3–1.7	40–80, 1	25–45, 1	40, 1	27, 1	27, 1	25–70, 1–8
Advantages	Very fast reaction with CO_2			High equilibrium capacity	Saving for regeneration energy (62%)		
	Low cost	Less corrosion	Low thermal energy	Low degradation		Lower vapor pressure	High capacity
	High capacity for absorption CO_2	Less degradation	High capacity	High loading capacity		Less water evaporation	Favour rapid formation
Disadvantages	High thermal energy						
	High degradation	Energy intensive	Low absorption rate	Lower CO_2 absorption rate(compared MEA)			High thermal energy
	High corrosion				High toxicity	High cost operation	
	Limited loading capacity	Limited loading capacity	Do not form stable carbonate				Limited solvent solubility
	Unstable (e.g.; Volatile and chemically)						
CO_2 recovery efficiency	90%	NR	NR	NR	99%	NR	NR
References	Bello and Idem (2005), Figueroa et al. (2008), and Kothandaraman (2010)	Kim et al. (2013)	Saha et al. (1995)	(Mandal et al. (2001)	Yeh et al. (2005)	(Lin et al. (2003)	(Bishnoi and Rochelle (2000)

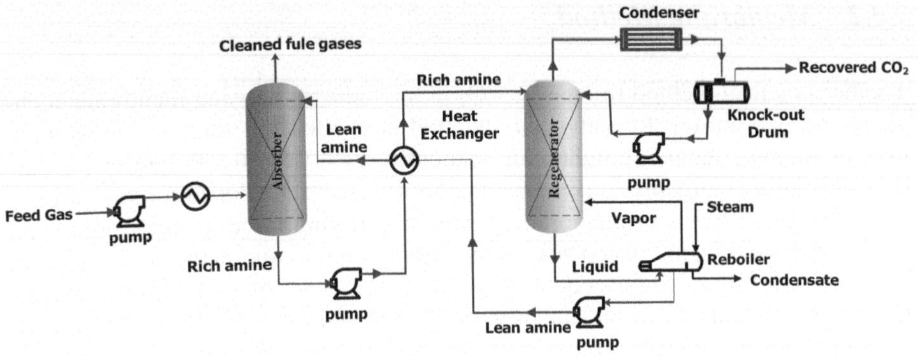

Fig. 9.4 Schematic diagram of absorption/desorption process for CO_2 recovery

the solvent e.g., primary alkanolamine takes place, and thereby some proportion of CO_2 is solved in the absorbent. The absorber operates at the temperature of below 60 °C around 38–60 °C and ambient pressure.

The rich solvent then exits from the bottom of the absorber column and after passing a heat exchanger, which is heated through the lean solvent regenerated from the desorption column, it is delivered to the top of the desorption column to be regenerated at the temperature of around 100 °C to 120 °C and pressure of 1.5 atm to 2 atm. Then, the reboiler provides the required energy for regeneration of the solvent through evaporating the lean solvent and returning it back to the stripper column, and the nonvolatile waste is purged from the system. After that, the regenerated solvent exits from the bottom of stripper and returns back to the absorber column, after passing the heat exchanger to decrease the temperature. The pure CO_2 which exits from the top of the stripper column is compressed to the pressure of 148 atm or 15 MPa, and then entered to the knock-out drum to remove the remaining impurities. The outlet stream of the drum is divided into two parts; i.e., the first is the recovered CO_2 that exits from the top and the remaining is refluxed to the stripper column (Ciferno et al. 2009; Wang et al. 2011; Yu et al. 2009).

Calculation of required energy for CO_2 recovery illustrates that theoretical amount of energy is 0.396 GJ/ton CO_2, whereas as the real amount of energy is 0.72 GJ/ton CO_2 (Rochelle 2009).

Ogawa et al. studied and compared three different amine solvents in respect of regeneration energy demand through thermodynamic simulation in the conventional CO_2 recovery processes. Results showed that for the 30 wt % MEA aqueous solution the minimum energy consumption is around 4.4 GJ/ton CO_2, while it is approximately 15% and 37% lower than MEA for AMP and titanium silicate (TS-1), respectively. After finding the TS-1 as the best choice for absorption/desorption process, they launched a pilot plant to assess the recovery percentage of CO_2 as well as the energy heat consumption for the regeneration process. Results revealed that CO_2 recovery could reach 94% and heat consumption was 2.9 GJ/ton CO_2 (Ogawa et al. 2009).

9.4.2 Membrane Method

Another practical method to recover CO_2 in the industry is using membrane technology for separation. Recently, this method is applied in order to remove CO_2 from other gases including light hydrocarbons more efficiently. Polymeric, metallic, solid-liquid, inorganic and ceramic membranes have developed for this purpose until now. In all forms of membranes, the major driving force for permeation can be provided through the sweep gas, sweep liquid or a vacuum on the permeate side. The selectivity of membranes depends on their ability to interact with the desired molecules. The interacted molecules then are diffused through two mechanism of solution-diffusion or absorption-diffusion across the membrane (Baltus and DePaoli 2002).

Figure 9.5 shows the overall process of CO_2 absorption/desorption through gas-liquid membrane contactor system. In this system, the gas and liquid phases are separated through the membrane, with a specific pressure difference between the two phases in order to inhibit diffusion of the liquid phase into the membrane. In fact, the membrane system should have the quality of high wetting resistance and gas diffusion rate with high CO_2 selectivity. In order to achieve the high wetting resistance feature, the applied liquid absorbent must have the qualities of high surface tension, low vapor pressure, high chemical stability, and low corrosiveness (Ma'mun et al. 2007).

In order to choose an ideal membrane for CO_2 separation specifically, some characteristics should be taken into account including (Li and Keener 2016; Yamasaki 2003):

- High mechanical strength
- High chemical and mechanical stability
- High CO_2 permeation rate
- High CO_2 selectivity
- High hydrophobicity

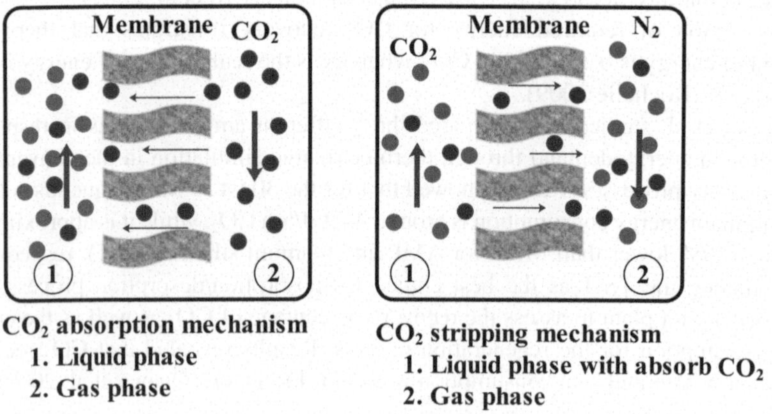

Fig. 9.5 CO_2 absorption/desorption mechanism through gas-liquid membrane contactor system

It should also be mentioned that in cases at which the membrane itself cannot provide the required selectivity for CO_2, choosing a proper absorbent solvent can compensate for lack of selectivity of the membrane. To exemplify, alkanolamines e.g., MEA are good candidates to capture CO_2 with high selectivity owing to their high CO_2 absorption capacity (Khalilpour et al. 2015). In the following, different types of membranes are explained, briefly.

9.4.2.1 Polymeric Based Membranes

Common polymeric membranes which are used for CO_2 separation processes includes polytetrafluoroethylene (PTFE), polyethylene (PE), polypropylene (PP), polyvinylidene fluoride (PVDF) and poly tetrafluoroethylen-co-perfluorovinylether (PFA). One of the most beneficial features of polymeric type membranes is that they are generally hydrophobic, and thereby penetration of absorbent liquid to the membrane is prohibited, except PP; actually, although the PP type membrane is cheap and abundant, it is not hydrophobic, and thus providing weak performance in comparison with other types (Khalilpour et al. 2015; Koonaphapdeelert et al. 2008; Yeon et al. 2003).

In this regard Takahashi et al. studied and examined CO_2 absorption through utilizing different types of polymeric membranes involving PTFE, PP and PE with hollow fibers in different sizes. The absorbent liquid was 30 wt% MEA aqueous solution. Results showed that using the drawn PE membrane can obtain high CO_2 recovery rate, suggesting that using hollow fiber membrane can effectively boost mass transfer (Takahashi et al. 2008).

In another work, PTFE and PVDF hollow fiber membranes were compared, and the results demonstrated that the PVDF exhibited higher CO_2 separation performance than PTFE since PVDF was more resistant to pore wetting (Yeon et al. 2003). Different types of polymeric membranes examined by investigators previously are classified and compared in aspects of utilized solvents, separated gases and challenges in Table 9.4.

9.4.2.2 Ceramic and Metallic Membranes

The mechanism of permeation in this type of membrane, which is porous, is based on size of gas molecules. In fact, they works like a sieve in which the targeted gas i.g., CO_2 can be separated from large gas molecules. In comparison with the polymeric type, this type possesses higher chemical and thermal stability, making it a more suitable choice for severe chemical conditions and high temperatures, while it can also be modified to hydrophobic materials (Koonaphapdeelert et al. 2008) (Table 9.5).

Table 9.4 Comparison between various types of polymer membranes

Polymeric membranes	Used Solvents (liquid phase)	Gases separated	Challenges	References
Polyvinylidene fluoride	Diethanolamine	N_2/CO_2	Rate controlled e.g.; absorbent flow rate membrane wetting	Mansourizadeh and Ismail (2009) and Zhang et al. (2008)
Polytetrafuloroethylene	Monoethanolamine Potassium carbonate AMP	CO_2/CO, $H_2/N_2/CH_4$	Pore wetting observed with MEA, but not potassium carbonate	Kentish et al. (2008) and Kim and Yang (2000)
Polypropylene	MEA, DEA, MDEA, MEA/MDEA, DEA/AMP, and MDEA/PZ	N_2/CO_2	Pore wetting observed with MEA, but not potassium carbonate	Ozturk and Hughes (2012) and Wang et al. (2013)

Table 9.5 Merits & demerits of polymer and ceramic membranes

	Merits	Demerits
Polymeric membranes	No need for regeneration	Pretreatment may be required
	Long life e.g.; 7 year	Pretreatment may be required
	Capability separate CO_2	Energy requirement for compressors
	Low cost compared ceramic& metallic	
Ceramic membranes	High resistance	High cost than polymer membrane
	Thermally stability	
	High a stability	High face velocities required
	Backwashable	

9.4.2.3 Microporous Solid Membranes

In this type, a microporous solid membrane is impregnated in a liquid absorbent. The targeted gas e.g., CO_2 diffuses through the membrane selectively, and subsequently absorbed through the liquid absorbent.

One of the most essential parameters that influences the performance of such membranes is its thickness, meaning that a thin membrane can be much more permeable than a thick one (Kovvali and Sirkar 2002). For instance, a 10-mm thick membrane can permeates 20 times more than a 200-thick one (Meisen and Shuai 1997).

9.4.3 Superiority of Membrane Method over Absorption/Desorption through Packed Columns Method

A general comparison between the conventional and membrane methods are outlined below. Firstly, while there are drawbacks of flooding and channeling in the conventional method, it is eliminated in the membrane method.

Secondly, since using thin hollow fibers can remarkably increase the contact area between the gas and liquid phases, the size of utilized equipment and instruments becomes smaller in comparison to packed columns, leading to lower operating cost.

Nevertheless, in the membrane method there is additional mass transfer resistance, and therefore the features of hollow fiber membrane should be optimized to improve CO_2 separation performance.

9.4.4 Membrane Flash Process

As it was stated before, one of the most challenging issues of the conventional absorption/desorption process is providing the required thermal energy for regeneration of the solvent through the stripping column. Therefore, in order to tackle this issue, membrane flash process is proposed, which can be regarded as an energy-saving alternative for the stripping section of the conventional process (Okabe et al. 2008, 2009).

It is possible to use the waste energy of some processes i.e., processes in which CO_2 is exhausted, especially in chemical industries and iron manufacturing industries, to lower CO_2 recovery energy requirement (Okabe et al. 2010). The waste energy can either be used to supply the stripper column heat in conventional process or be employed to increase the temperature of the flash process. However, energy consumption calculations reveal that not only the amount of waste energy that can be employed for the membrane flash process is greater than that of the conventional process, but also the amount of electrical energy required for the flash process is much lower than that of conventional one (Okabe et al. 2010). As a result, the membrane flash technique surpasses the conventional one in the matter of energy consumption (Okabe et al. 2010).

In the membrane flash process, the rich solvent exiting the absorption column enters the tube side of microporous hollow fibers where the solvent permeates to the membrane provided that the shell side pressure is decreased by applying vacuum as shown in Fig. 9.6. Then, CO_2 is separated from the rich solvent while the solvents permeates through the membrane to the shell side.

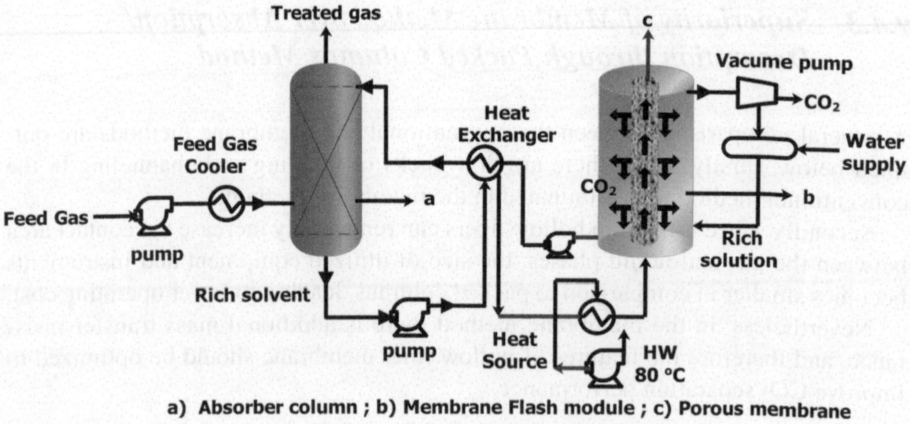

Fig. 9.6 Schematic diagram of absorption/desorption process through membrane flash method

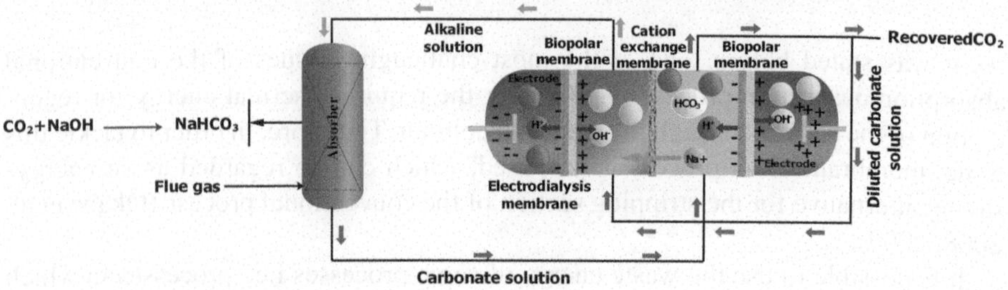

Fig. 9.7 Schematic diagram of absorption/desorption process using electro dialysis membrane method

9.4.5 Eelectro Dialysis Membrane Process

As explained before, the major drawback of the conventional CO_2 recovery process is the high energy requirement for regeneration through the stripping column; in other words, almost all the energy consumption of the CO_2 capture process is consumed for the regeneration process through the stripping column (Metz et al. 2005a). To address the mentioned issue, aside from the membrane flash process to substitute the stripping column which is described in Sect. 9.4.4. Another practical alternative is the electro dialysis membrane process (Eisaman et al. 2011a, b; Nagasawa et al. 2009). In fact, the electro dialysis process is employed to recover CO_2 from the rich solvent exiting the absorption column. Figure 9.7 illustrate absorption/desorption process using electro dialysis membrane method. This process operates through utilization of membrane in the atmospheric pressure and room temperature.

The overall CO_2 recovery process through electrochemical method consists of two major parts as the following:

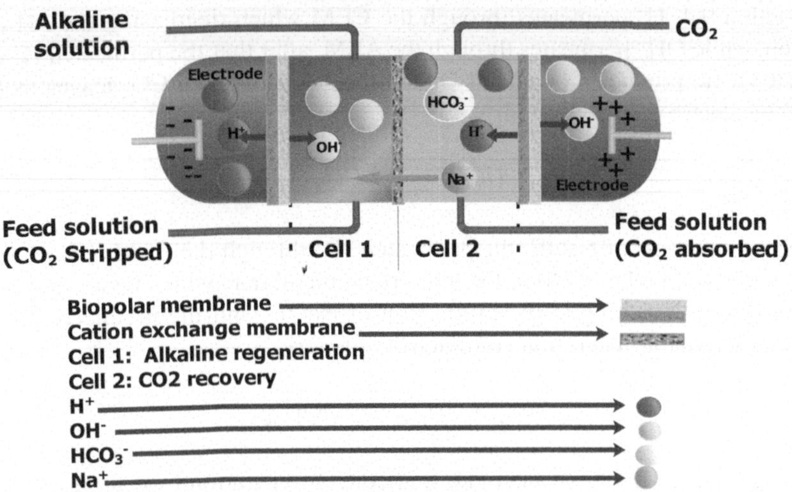

Fig. 9.8 Mechanism of electro dialysis membrane process

1. CO_2 absorption in an absorption column: The gas stream containing CO_2 enters the absorption column from the bellow to be absorbed chemically, via Reactions 9.1 and 9.2, to the NaOH aqueous solution which enters the column from above.

$$CO_2 + \mathrm{NaOH} \rightarrow \mathrm{NaHCO_3} \tag{9.1}$$

$$\mathrm{NaHCO_3} \rightarrow Na^+ + HCO_3^- \tag{9.2}$$

2. CO_2 recovery and NaOH regeneration by electro dialysis process: The rich aqueous solution $NaHCO_3$ leaves the absorption column then enters the electro dialysis system in which CO_2 is recovered and NaOH is regenerated, simultaneously.

Actually, the electro dialysis system is comprised of two sections; the left section is responsible for regeneration of NaOH solution whereas the right section is responsible for recovery of CO_2 which illustrate in Fig. 9.8. There are two types of membranes in the electrolysis system including bipolar membrane (BPM) and cation exchange membrane (CEM), which simultaneously contribute to the recovery of CO_2 and the regeneration of NaOH.

The BPM possesses both anion exchange membrane (AEM) that in which positive ions are fixed and negative ions are mobile, and CEM which in which negative ions are fixed and positive ions are mobile.

Through applying a voltage higher than what is required for dissociation of H_2O molecules, H_2O converts to H^+ and OH^- ions. This reaction is shown in Reaction 9.3.

$$\mathrm{H_2O} \xrightarrow{\text{electrical power}} \mathrm{H^+} + OH^- \tag{9.3}$$

In Reaction 9.4, H^+ permeates through the CEM which decreasing the PH of the solution, while OH^- permeates through the AEM, after that the permeated H^+ reacts with HCO_3^- to produce CO_2 and H_2O, and then the produced CO_2 leaves from the top of the electro dialysis system.

$$\mathrm{H}^+_{(aq)} + HCO3^-_{(aq)} \rightarrow CO_{2(\mathrm{g})} + \mathrm{H_2O} \quad (9.4)$$

For the regeneration of NaOH, the permeated Na^+ through the CEM reacts with the OH^- Reaction 9.5 released from the left bipolar membrane which increasing the PH of the solution, resulting in the production of NaOH, and the produced NaOH is then returned to the absorption column from the top.

$$Na^+_{(aq)} + OH^-_{(aq)} \rightarrow \mathrm{NaOH}_{(aq)} \quad (9.5)$$

By way of illustration, Eismana et al. studied CO_2 separation from atmospheric air by employing bipolar membrane electro dialysis. The energy consumption for CO_2 recovery from carbonate solution was calculated to be around 100 kJ/mole CO_2. Furthermore, they examined applying the pressure of up to 20 atm, finding that such pressures can lower resistance, applied voltage, and the overall energy required for the recovery process (Eisaman et al. 2011a).

Despite the advantageous features of the electro dialysis membrane process, it should be noted that according to literature, for a 400 MW coal-fired power plant, the minimum operating cost for CO_2 recovery process through this method is approximately US$180/ton CO_2, which is higher than that of chemical absorption through amine, implying that in order to make the electrochemical method a good alternative for the conventional process, the membrane costs should be reduced (Iizuka et al. 2012).

9.4.6 *Electrolysis Membrane Process*

This method is another substitute for the desorption process, in which membrane is utilized similar the electro dialysis and membrane flash processes. However, this method is applied when the absorption solvent is K_2CO_3. Actually, CO_2 capture process through K_2CO_3 has the drawback of low CO_2 absorption rate, high capital cost and energy consumption. To deal with the mentioned faults, Martin and Kubic proposed a new technology for stripping process e.g., recover CO_2 from rich K_2CO_3 solution which works through ion-exchange membrane electrolysis (Martin and Kubic 2007). They also stated that this method can lower the energy requirement dramatically around up to 96% in relation to the conventional heating process for stripping; however, it is not proved yet that the process is economical owing to lack of sufficient information. In addition, Zhao et al. investigated the mechanism and energy consumption for this process. They concluded that energy requirement can decrease around 16.6% and 25.8% for regeneration process and total CO_2 capture and compression process, respectively (Zhao et al. 2011).

The CO_2 recovery process through membrane electrolysis is described briefly in the following.

Identical to other recovery processes, in this process, firstly, CO_2 is absorbed through solvent. The absorption reaction is as follows in Reaction 9.6.

$$K_2CO_3 + CO_2 \rightarrow 2KHCO_3 \tag{9.6}$$

The rich solution contains both HCO_3^- and CO_3^{2-}.

After the absorption process, the electrolysis process is conducted in which CO_2 is released in three steps as the following (Zhao et al. 2011):

1. No CO_2 release step- in this step Reaction 9.7 takes place, and its details are demonstrated in Fig. 9.9.

$$2K_2CO_3 + 3H_2O \rightarrow 2KHCO_3 + 2KOH + H_2 + O_2 \tag{9.7}$$

2. Enhanced CO_2 release step- in this step Reaction 9.8 occurs to produced CO_2, its details are depicted in Fig. 9.10.

$$2KHCO_3 + H_2O \rightarrow 2CO_2 + H_2 + 2KOH + 1/2O_2 \tag{9.8}$$

3. Steady CO_2 release step- CO_2 is released from this step through Reaction 9.9, the detailed process is illustrated in Fig. 9.11.

$$(2-y)K_2CO_3 + (3-y)H_2O \rightarrow yCO_2 + 2(1-y)KHCO_3 + 2KOH + H_2 + 1/2O_2 \tag{9.9}$$

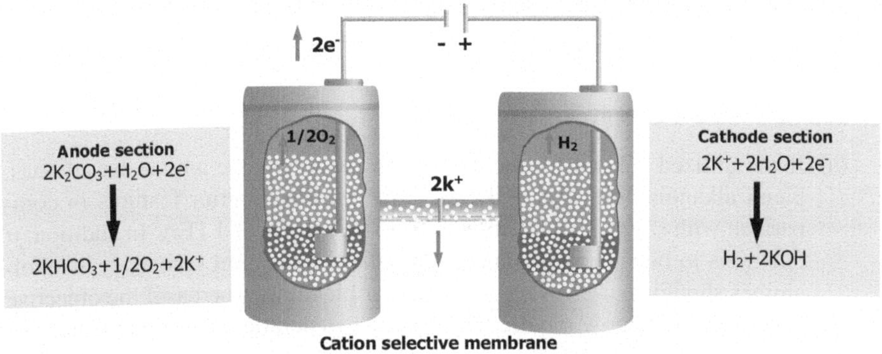

Fig. 9.9 The first step mechanism. No carbon dioxide gas is released at this stage

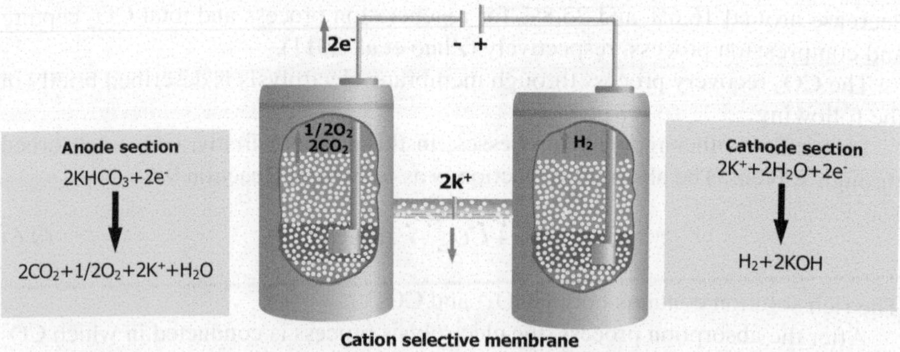

Fig. 9.10 The second step mechanism with enhanced CO_2 release step

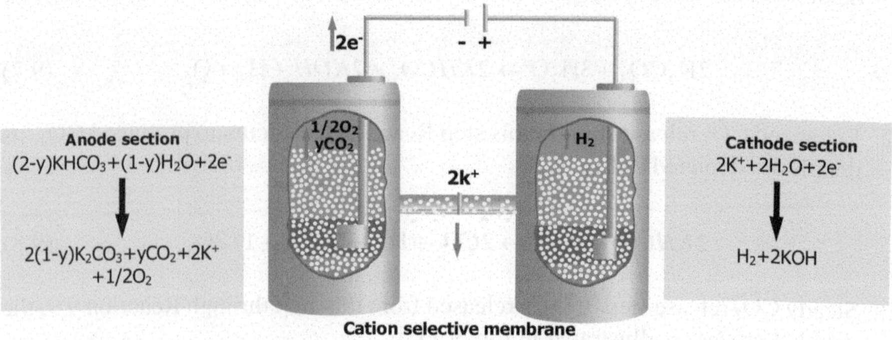

Fig. 9.11 The third step mechanism. At this step, carbon dioxide gas is released

9.5 Conclusion

In this chapter, CO_2 recovery processes were evaluated and recent studies with this purpose were reviewed. According to literature, the most proper method is the chemical absorption through solvent. Thus, in order to enhance recovery performance while chemical absorption method is applied, the following points should be taken into account:

1. In terms of the used solvent:
   (a) If the utilized solvent for the absorption process is the amine- based ones, blend alkanolamines e.g., MDEA, AMP exhibit superior features in comparison with single alkanolamines e.g., MEA, DEA and TEA. In addition, if one aims to boost absorption capacity, the primary and secondary alkanolamines should be used as the additives for blending, whereas if the objective is to diminish energy requirement for the regeneration, the suitable additive

is the tertiary alkanolamine, suggesting that to enhance the recovery process in both aspects blending of the three types of alkanolamine is advised.

(b) One of solvent that can substitute amine-based solvents is NaOH aqueous solution, which not only improves CO_2 absorption capacity compared to MEA, but also is more plentiful, cheaper, and a less volatile substance with lower vapor pressure. However, the regeneration process is more difficult owing to high solubility of $NaHCO_3$ in aqueous solution.

(c) Another alternative for amine-based solvents is K_2CO_3, which is an attractive candidate owing to low toxicity, solvent cost, and corrosiveness. However, this type of solvent has the drawback of low reaction rate which should be enhanced through adding some promoters.

(d) One other choice for amine-based solvents is NaOH aqueous solution. It is reported that using ammonia can boost CO_2 recovery efficiency to 95–100%, whereas that of MEA is 90%.

2. In term of using membrane technology: This technology can either be used for both absorption and desorption process, simultaneously, or it can be joined to the conventional absorption/desorption process; i.e., membrane processes flash, electrodialysis and electrolysis can be substitute with the desorption section of conventional process.

   (a) If the objective is to use membrane technology alone for both absorption and desorption process, among the different membrane types polymeric, ceramic, metallic and microporous solids, the polymeric type demonstrates superior features for CO_2 recovery process.

   (b) If the goal is to join conventional process with the membrane technology, flash, electro dialysis and electrolysis membrane processes can be applied.

   b.1 Flash process: It is reported that the amount of utilized waste energy in membrane flash process can be higher than that of conventional desorption process. Also, the electrical energy needed for the flash process is much lower than that of conventional one.

   b.2 Electro dialysis process: The utilized solvent in electro dialysis membrane process is NaOH, and both CO_2 and NaOH are recovered through three steps. This process has the benefit of operating in room temperature and ambient pressure. However, it is stated that this process can be more costly than the conventional one unless the costs relating to membrane are reduced.

   b.3 Electrolysis process: The electrolysis process is conducted through utilization of ion-exchange membrane, and K_2CO_3 is used as the solvent. It is reported that this process has lower energy requirement for regeneration process up to 16.6%, and lessen the overall energy needed for CO_2 capture and compression process up to 25.8%.

References

Abu Zahra MRM, Jansens PJ (2009) Carbon dioxide capture from flue gas: development and evaluation of existing and novel process concepts. https://doi.org/10.4233/uuid:6d1689f3-7b1a-4355-81b0-af3d19fae469

Abu-Zahra MRM, Schneiders LHJ, Niederer JPM, Feron PHM, Versteeg GF (2007) CO_2 capture from power plants: part I. A parametric study of the technical performance based on monoethanolamine. Int J Greenhouse Gas Control 1:37–46. https://doi.org/10.1016/S1750-5836(06)00007-7

Adeosun A, Abu-Zahra MRM (2013) Evaluation of amine-blend solvent systems for CO_2 post-combustion combustion capture applications. Energy Procedia 37:211–218. https://doi.org/10.1016/j.egypro.2013.05.104

Alie C, Backham L, Croiset E, Douglas PL (2005) Simulation of CO_2 capture using MEA scrubbing: a flowsheet decomposition method. Energy Convers Manag 46:475–487. https://doi.org/10.1016/j.enconman.2004.03.003

Aroonwilas A, Veawab A (2004) Characterization and comparison of the CO_2 absorption performance into single and blended alkanolamines in a packed column. Ind Eng Chem Res 43:2228–2237. https://doi.org/10.1021/ie0306067

Aroonwilas A, Veawab A (2007) Integration of CO2 capture unit using single-and blended-amines into supercritical coal-fired power plants: implications for emission and energy management. Ind Eng Chem Res 1:143–150. https://doi.org/10.1016/S1750-5836(07)00011-4

Bai H, Yeh AC (1997) Removal of CO2 greenhouse gas by ammonia scrubbing. Ind Eng Chem Res 36:2490–2493. https://doi.org/10.1021/ie960748j

Baltus R, DePaoli D (2002) Separation of CO_2 using room temperature ionic liquids. In: Oak Ridge National Laboratory Environmental Sciences Division Carbon Management Seminar Series

Ban ZH, Keong LK, Mohd Shariff A (2014) Physical absorption of CO_2 capture: a review. Adv Mater Res 917:134–143. Trans Tech Publ. https://doi.org/10.4028/www.scientific.net/AMR.917.134

Bao J, Lu WH, Zhao J, Bi XT (2018) Greenhouses for CO_2 sequestration from atmosphere. Carbon Resour Conv 1:183–190. https://doi.org/10.1016/j.crcon.2018.08.002

Barchas R, Davis R (1992) The Kerr-McGee/ABB Lummus crest technology for the recovery of CO2 from stack gases. Energy Convers Manag 33:333–340. https://doi.org/10.1016/0196-8904(92)90028-U

Bello A, Idem RO (2005) Pathways for the formation of products of the oxidative degradation of CO2-loaded concentrated aqueous monoethanolamine solutions during CO_2 absorption from flue gases. Ind Eng Chem Res 44:945–969. https://doi.org/10.1021/ie049329

Bishnoi S, Rochelle GT (2000) Absorption of carbon dioxide into aqueous piperazine: reaction kinetics, mass transfer and solubility. Chem Eng Sci 55:5531–5543. https://doi.org/10.1016/S0009-2509(00)00182-2

Booth NJ (2005) Secondment to the international test centre for CO_2 capture (ITC), University of Regina, Canada. January–March 2005 Report No. 6:798. https://doi.org/10.1016/j.ijggc.2015.06.026

Borhani TNG, Azarpour A, Akbari V, Alwi SRW, Manan ZA (2015) CO_2 capture with potassium carbonate solutions: a state-of-the-art review. Int J Greenhouse Gas Control 41:142–162. https://doi.org/10.1016/j.ijggc.2015.06.026

Bruder P, Grimstvedt A, Mejdel T, Svendsen H (2011) CO_2 capture into aqueous solutions of piperazine activated 2-amino-2-methyl-1-propanol. Chem Eng Sci 66:6193–6198. https://doi.org/10.1016/j.ces.2011.08.051

Burr B, Lyddon L (2008) A comparison of physical solvents for acid gas removal. Bryan Research & Engineering, Inc. In 87th annual gas processors association convention, Grapevine, TX, March 2–5

Chakma A, Tontiwachwuthikul P (1999) CO_2 separation from combustion gas streams by chemical reactive solvents. Proceedings of Combustion Canada

Chakravarty T, Phukan UK, Weilund RH (1985) Reaction of acid gases with mixtures of amines. Chem Eng Prog 81:4

Chang H, Shih CM (2005) Simulation and optimization for power plant flue gas CO_2 absorption-stripping systems. Sep Sci Technol 40:877–909. https://doi.org/10.1081/SS-200048014

Change O (2015) Climate change 2014: mitigation of climate change. Cambridge University Press, Cambridge

Chapel DG, Mariz CL, Ernest J (1999) Recovery of CO_2 from flue gases: commercial trends. Canadian society of chemical engineers annual meeting vol. 4. Saskatchewan Canada

Choi JH, Oh SG, Jo M, Yoon YI, Jeong SK, Nam SC (2012) Absorption of carbon dioxide by the mixed aqueous absorbents using 2-methylpiperidine as a promoter. Chem Eng Sci 72:87–93. https://doi.org/10.1016/j.ces.2012.01.001

Ciferno JP, Fout TE, Jones AP, Murphy JT (2009) Capturing carbon from existing coal-fired power plants, pp 105–133

Cullinane JT, Rochelle GT (2004) Carbon dioxide absorption with aqueous potassium carbonate promoted by piperazine. Chem Eng Sci 59:3619–3630. https://doi.org/10.1016/j.ces.2004.03.029

Darde V, Thomsen K, Van Well W, Stenby E (2009) Chilled ammonia process for CO_2 capture. Energy Procedia 1:1035–1042. https://doi.org/10.1016/j.egypro.2009.01.137

Davidson RM (2007) Post-combustion carbon capture from coal fired plants: solvent scrubbing. IEA Clean Coal Centre London

Davis J, Rochelle G (2009) Thermal degradation of monoethanolamine at stripper conditions. Energy Procedia 1:327–333. https://doi.org/10.1016/j.egypro.2009.01.045

Davison J (2007) Performance and costs of power plants with capture and storage of CO_2. Energy 32:1163–1176. https://doi.org/10.1016/j.energy.2006.07.039

Dugas RE, Rochelle GT (2011) CO_2 absorption rate into concentrated aqueous monoethanolamine and piperazine. Chem Eng Data 56:2187–2195. https://doi.org/10.1021/je101234t

Edenhofer O, Pichs-Madruga R, Sokona Y, Farahani E, Kadner S, Seyboth K, Adler A, Baum I, Brunner S, Eickemeier P, Kriemann B, Savolainen J, Schlmer S, von Stechow S, Zwickel T, Minx JC (2014) Contribution of working group III to the fifth assessment report of the intergovernmental panel on climate Change (IPCC). Cambridge University Press, Cambridge/New York

Eisaman MD, Alvarad L, Larner D, Wang P, Garg B, Littau KA (2011a) CO_2 separation using bipolar membrane electrodialysis. Energy Environ Sci 4:1319–1328. https://doi.org/10.1039/C0EE00303D

Eisaman MD, Alvarado L, Larner D, Wang P, Littau KA (2011b) CO_2 desorption using high-pressure bipolar membrane electrodialysis. Energy Environ Sci 4:4031–4037. https://doi.org/10.1039/C1EE01336J

Farla JCM, Hendriks CA, Blok K (1995) Carbon dioxide recovery from industrial processes. Clim Chang 29:439–461. https://doi.org/10.1007/BF01092428

Figueroa JD, Fout T, Plasynski S, McIlvried H, Srivastava RD (2008) Advances in CO_2 capture technology—the US department of energy's carbon sequestration program. Int J Greenhouse Gas Control 2:9–20. https://doi.org/10.1016/S1750-5836(07)00094-1

Freeman SA, Dugas R, Van Wagener DH, Nguyen T, Rochelle GT (2010) Carbon dioxide capture with concentrated, aqueous piperazine. Int J Greenhouse Gas Control 4:119–124. https://doi.org/10.1016/j.ijggc.2009.10.008

Frumhoff PC, Heede R, Oreskes N (2015) The climate responsibilities of industrial carbon producers. Clim Chang 132:157–171. https://doi.org/10.1007/s10584-015-1472-5

Gielen D (2003) The energy policy consequences of future CO_2 capture and sequestration technologies. Proceedings of the 2nd annual conference on carbon sequestration. Citeseer, Alexandria

Göttlicher G, Pruschek R (1997) Comparison of CO_2 removal systems for fossil-fuelled power plant processes. Energy Convers Manag 38:S173–S178. https://doi.org/10.1016/S0196-8904(96)00265-8

Heda PK, Dollimore D, Alexander KS, Chen D, Law E, Bicknell P (1995) A method of assessing solid state reactivity illustrated by thermal decomposition experiments on sodium bicarbonate. Thermochim Acta 255:255–272. https://doi.org/10.1016/0040-6031(94)02154-G

Heede R (2014) Tracing anthropogenic carbon dioxide and methane emissions to fossil fuel and cement producers. Clim Change 122:229–241. https://doi.org/10.1007/s10584-013-0986-y

Heede R, Oreske N (2016) Potential emissions of CO_2 and methane from proved reserves of fossil fuels: an alternative analysis. Glob Environ Chang 36:12–20. https://doi.org/10.1016/j.gloenvcha.2015.10.005

Henni A, Tontiwachwuthiku P, Chakma A (2005) Solubilities of carbon dioxide in polyethylene glycol ethers. Can J Chem Eng 83:358–361. https://doi.org/10.1002/cjce.5450830224

Huertas JI, Gomez MD, Giraldo N, Garzón J (2015) CO_2 absorbing capacity of MEA. J Chemother. https://doi.org/10.1155/2015/965015

Iizuka A, Hashimoto K, Nagasawa H, Kumagai K, Yanagisawa Y, Yamasaki A (2012) Carbon dioxide recovery from carbonate solutions using bipolar membrane electrodialysis. Sep Purif Technol 101:49–59. https://doi.org/10.1016/j.seppur.2012.09.016

Isa F, Zabiri H, Ng NKS, Shariff AM (2018) Purification of CO_2 removal via promoted potassium carbonate: a review on modelling & simulation techniques. Int J Greenhouse Gas Control 76:236–265. https://doi.org/10.1016/j.ijggc.2018.07.004

Kentish SE, Scholes CA, Stevens GW (2008) Carbon dioxide separation through polymeric membrane systems for flue gas applications. Recent Patents Chem Eng 1:52–66. https://doi.org/10.2174/2211334710801010052

Khalilpour R, Mumford K, Zhai H, Abbas A, Stevens G, Rubin ESJ (2015) Membrane-based carbon capture from flue gas: a review. J Clean Prod 103:286–300. https://doi.org/10.1016/j.jclepro.2014.10.050

Khanipour M, Mirvakili A, Bakhtyari A, Farniaei M, Rahimpour MR (2017) Enhancement of synthesis gas and methanol production by flare gas recovery utilizing a membrane based separation process. Fuel Process Technol 166:186–201. https://doi.org/10.1016/j.fuproc.2017.06.008

Kim YS, Yang SM (2000) Absorption of carbon dioxide through hollow fiber membranes using various aqueous absorbents. Sep Purif Technol 21:101–109. https://doi.org/10.1016/S1383-5866(00)00195-7

Kim YE, Choi JH, Nam SC, Yoon YIJ (2012) CO_2 absorption capacity using aqueous potassium carbonate with 2-methylpiperazine and piperazine. J Ind Eng Chem 18:105–110. https://doi.org/10.1016/j.jiec.2011.11.078

Kim YE, Lim JA, Jeong SK, Yoon YI, Bae ST, Nam SC (2013) Comparison of carbon dioxide absorption in aqueous MEA, DEA, TEA, and AMP solutions. J Ind Eng Chem 34:783–787. https://doi.org/10.5012/bkcs.2013.34.3.783

Kittel J, Idem R, Gelowitz D, Tontiwachwuthikul P, Parrain G, Bonneau A (2009) Corrosion in MEA units for CO_2 capture: pilot plant studies. Energy Procedia 1:791–797. https://doi.org/10.1016/j.egypro.2009.01.105

Kohl AL, Nielsen R (1997) Gas purification, Elsevier

Koonaphapdeelert S, Tan X, Wu Z, Li KJ (2008) Solvent distillation by ceramic hollow fibre membrane contactors. J Membr Sci 314:58–66. https://doi.org/10.1016/j.memsci.2008.01.022

Kothandaraman A (2010) Carbon dioxide capture by chemical absorption: a solvent comparison study. Phd thesis in department of Chemical Engineering. Massachusetts Institute of Technology

Kovvali AS, Sirkar KK (2002) Carbon dioxide separation with novel solvents as liquid membranes. Ind Eng Chem Res 41:2287–2295. https://doi.org/10.1021/ie010757e

Li MH, Chang BCJ (1994) Solubilities of carbon dioxide in water+ monoethanolamine+ 2-amino-2-methyl-1-propanol. Chem Eng Data 39:448–452. https://doi.org/10.1021/je00015a010

Li T, Keener TC (2016) A review: desorption of CO2 from rich solutions in chemical absorption processes. Int J Greenhouse Gas Control 51:290–304. https://doi.org/10.1016/j.ijggc.2016.05.030

Lin CC, Liu WT, Tan CS (2003) Removal of carbon dioxide by absorption in a rotating packed bed. Ind Eng Chem Res 42:2381–2386. https://doi.org/10.1021/ie020669

Ma'mun S, Svendsen HF, Hoff KA, Juliussen O (2007) Selection of new absorbents for carbon dioxide capture. Energy Convers Manag 48:251–258. https://doi.org/10.1016/B978-008044704-9/50006-9

Mandal B, Guha M, Biswas A, Bandyopadhyay S (2001) Removal of carbon dioxide by absorption in mixed amines: modelling of absorption in aqueous MDEA/MEA and AMP/MEA solutions. Chem Eng Sci 56:6217–6224. https://doi.org/10.1016/S0009-2509(01)00279-2

Mangalapally H, Hasse H (2010) Pilot plant experiments with mea and new solvents for post combustion CO2 capture by reactive absorption. Int J Greenhouse Gas Control 6:84–112. https://doi.org/10.1016/j.ijggc.2011.11.004

Mansourizadeh A, Ismail AF (2009) Hollow fiber gas–liquid membrane contactors for acid gas capture: a review. J Hazard Mater 171:38–53. https://doi.org/10.1016/j.jhazmat.2009.06.026

Mariz CL (1998) Carbon dioxode recovery: large scale design trends. J Can Pet Technol 37:42–47. https://doi.org/10.2118/98-07-04

Martin FJ, Kubic WL (2007) Green freedom: a concept for producing carbon-neutral synthetic fuels and chemicals. Los Alamos Report La-UR-07-7897

Mattisson T, Lyngfelt A, Cho P (2001) The use of iron oxide as an oxygen carrier in chemical-looping combustion of methane with inherent separation of CO_2. Fuel 80:1953–1962. https://doi.org/10.1016/S0016-2361(01)00051-5

McKee B (2002) Solutions for 21st century, zero emissions technologies for fossil fuels, Technology Status report

Meisen A, Shuai X (1997) Research and development issues in CO_2 capture. Energy Convers Manag 38:S37–S42. https://doi.org/10.1016/S0196-8904(96)00242-7

Metz B, Davidson O, De Coninck H (2005a) Carbon dioxide capture and storage: special report of the intergovernmental panel on climate change. Cambridge University Press, Cambridge

Metz B, Davidson O, De Coninck H, Loos M, Meyer L (2005b) IPCC special report on carbon dioxide capture and storage. Prepared by working group III of the intergovernmental panel on climate change 442. Cambridge University Press

Moomaw WR (1996) Industrial emissions of greenhouse gases Industrial emissions of greenhouse gases 24:951–968. https://doi.org/10.4172/2157-7617.1000147

Nagasawa H, Yamasaki A, Iizuka A, Kumagai K, Yanagisawa Y (2009) A new recovery process of carbon dioxide from alkaline carbonate solution via electrodialysis. Environ Energy Eng 55:3286–3293. https://doi.org/10.1002/aic.11907

Nguyen T, Hilliard M, Rochelle GT (2010) Amine volatility in CO_2 capture. Int J Greenhouse Gas Control 4:707–715. https://doi.org/10.1016/j.ijggc.2010.06.003

Ogawa T, Ohashi Y, Yamanaka S, Miyaike K (2009) Development of carbon dioxide removal system from the flue gas of coal fired power plant. Energy Procedia 1:721–724. https://doi.org/10.1016/j.egypro.2011.01.019

Okabe K, Mano H, Fujioka Y (2008) Separation and recovery of carbon dioxide by a membrane flash process. Int J Greenhouse Gas Control 2:485–491. https://doi.org/10.1016/j.ijggc.2008.06.004

Okabe K, Kodama S, Mano H, Fujioka Y (2009) Separation and recovery of carbon dioxide by a membrane flash process utilizing waste thermal energy. Int J Greenhouse Gas Control 1:1281–1288. https://doi.org/10.1016/j.egypro.2009.01.168

Okabe K, Mano H, Fujioka Y (2010) Preliminary estimations of energy and cost for CO_2 recovery by a membrane flash process utilizing waste thermal energy. Int J Greenhouse Gas Control 4:597–602. https://doi.org/10.1016/j.ijggc.2010.01.005

Ozturk B, Hughes R (2012) Evaluation of mass transfer characteristics of non-porous and microporous membrane contactors for the removal of CO_2. Chem Eng J 195:122–131. https://doi.org/10.1016/j.cej.2012.04.085

Pacheco MA, Kaganoi S, Rochelle GT (2000) CO_2 absorption into aqueous mixtures of diglycolamine and methyldiethanolamine. Chem Eng Sci 55:5125–5140. https://doi.org/10.1016/S0009-2509(00)00104-4

Park JY, Yoon SJ, Lee H (2003) Effect of steric hindrance on carbon dioxide absorption into new amine solutions: thermodynamic and spectroscopic verification through solubility and NMR analysis. Environ Sci Technol 37:1670–1675. https://doi.org/10.1021/es0260519

Park SW, Choi BS, Lee JW (2006) Chemical absorption of carbon dioxide with triethanolamine in non-aqueous solutions. Korean J Chem Eng 23:138–143. https://doi.org/10.1007/BF02705705

Raksajati A, Ho MT, Wiley DE (2013) Reducing the cost of CO_2 capture from flue gases using aqueous chemical absorption. Ind Eng Chem Res 52:16887–16901. https://doi.org/10.1021/ie402185h

Raksajati A, Ho MT, Wiley DE (2016) Understanding the impact of process design on the cost of CO2 capture for precipitating solvent absorption. Ind Eng Chem Res 55:1980–1994. https://doi.org/10.1021/acs.iecr.5b03633

Rao AB, Rubin ES (2002) A technical, economic, and environmental assessment of amine-based CO2 capture technology for power plant greenhouse gas control. Environ Sci Technol 36:4467–4475. https://doi.org/10.1021/es0158861

Rao AB, Rubin ES, Berkenpas MB (2004) An integrated modeling framework for carbon management technologies. U.S. Department of Energy Office of Scientific and Technical Information 15213:3890. https://doi.org/10.2172/836715

Reddy S, Scherffius J, Freguia S, Roberts C (2003) Fluor's econamine FG plus SM technology. Proceedings of the second annual conference on carbon sequestration, pp 5–8

Resnik KP, Yeh JT, Pennline HWJ (2004) Aqua ammonia process for simultaneous removal of CO2, SO2 and NOx. Environ Technol Manage 4:89–104

Rochelle GT (2009) Amine scrubbing for CO_2 capture. Science 325:1652–1654. https://doi.org/10.1126/science.1176731

Saha AK, Bandyopadhyay SS, Biswas AK (1995) Kinetics of absorption of CO_2 into aqueous solutions of 2-amino-2-methyl-1-propanol. Chem Eng Sci 50:3587–3598. https://doi.org/10.1016/0009-2509(95)00187-A

Sander MT, Mariz CL (1992) The Fluor Daniel® econamine FG process: past experience and present day focus. Energy Convers Manag 33:341–348. https://doi.org/10.1016/0196-8904(92)90029-V

Seo DJ, Hong WH (1996) Solubilities of carbon dioxide in aqueous mixtures of diethanolamine and 2-amino-2-methyl-1-propanol. Chem Eng Data 41:258–260. https://doi.org/10.1021/je950197o

Singh P, Versteeg GF (2008) Structure and activity relationships for CO_2 regeneration from aqueous amine-based absorbents. Process Saf Environ Prot 86:347–359. https://doi.org/10.1016/j.psep.2008.03.005

Singh D, Croiset E, Douglas PL, Douglas MA (2003) Techno-economic study of CO_2 capture from an existing coal-fired power plant: MEA scrubbing vs. O_2/CO_2 recycle combustion. Energy Convers Manag 44:3073–3091. https://doi.org/10.1016/S0196-8904(03)00040-2

Siriwardane RV, Robinson C, Shen M, Simonyi T (2007) Novel regenerable sodium-based sorbents for CO_2 capture at warm gas temperatures. Energy Fules 21:2088–2097. https://doi.org/10.1021/ef070008v

Smith KH, Anderson CJ, Tao W, Endo K, Mumford KA, Kentish SE, Qader A, Hooper B, Stevens GWJ (2012) Pre-combustion capture of CO_2—results from solvent absorption pilot plant trials using 30 wt% potassium carbonate and boric acid promoted potassium carbonate solvent. Int J Greenhouse Gas Control 10:64–73. https://doi.org/10.1016/j.ijggc.2012.05.018

Suda T, Fujii M, Yoshida K, Iijima M, Seto T, Mitsuoka S (1992) Development of flue gas carbon dioxide recovery technology. Energy Convers Manag 33:317–324. https://doi.org/10.1016/0196-8904(92)90026-S

Takahashi N, Mano H, Okabe K, Nakamura M, Fujioka Y, Mimura T, Yagi Y (2008) Experimental and theoretical analyses of effects of membrane properties on CO_2 absorption performance of a hollow fiber membrane contactor. Kagaku Kogaku Ronbunshu 34:76–84

Tepe JB, Dodge BF (1943) Absorption of carbon dioxide by sodium hydroxide solutions in a packed column. Inst Chem Eng 39:255–276

Tontiwachwuthikul P, Meisen A, Lim CJ (1992) CO_2 absorption by NaOH, monoethanolamine and 2-amino-2-methyl-1-propanol solutions in a packed column. Chem Eng Sci 47:381–390. https://doi.org/10.1016/0009-2509(92)80028-B

Vaidya PD, Kenig EY (2007) CO2-alkanolamine reaction kinetics: a review of recent studies. Chem Eng Technol 30:1467–1474. https://doi.org/10.1002/ceat.200700268

Veawab A, Aroonwilas A, Chakma A, Tontiwachwuthikul P (2001) Solvent formulation for CO_2 separation from flue gas streams. First National Conference on Carbon Sequestration, Washington, DC

Vega F, Cano M, Camino S, Fernández LMG, Portillo E, Navarrete B (2018) Solvents for carbon dioxide capture. Carbon Dioxide Chemistry, Capture and Oil Recovery. IntechOpen. https://doi.org/10.5772/intechopen.71443

Wall TF (2007) Combustion processes for carbon capture. Proc Combust Inst 31:31–47. https://doi.org/10.1016/j.proci.2006.08.123

Wang M, Lawal A, Stephenson P, Sidders J, Ramshaw C (2011) Post-combustion CO_2 capture with chemical absorption: a state-of-the-art review. Chem Eng Res Des 89:1609–1624. https://doi.org/10.1016/j.cherd.2010.11.005

Wang Z, Fang M, Yan S, Yu H, Wei CC, Luo Z (2013) Optimization of blended amines for CO_2 absorption in a hollow-fiber membrane contactor. Ind Eng Chem Res 52:12170–12182. https://doi.org/10.1021/ie401676t

Xu GW, Zhang CF, Qin SJ, Gao WH, Liu HB (1998) Gas– liquid equilibrium in a CO_2– MDEA–H2O system and the effect of piperazine on it. Ind Eng Chem Res 37:1473–1477. https://doi.org/10.1021/ie9506328

Yamasaki A (2003) An overview of CO_2 mitigation options for global warming—emphasizing CO2 sequestration options. J Chem Eng Japan 36:361–375. https://doi.org/10.1252/jcej.36.361

Yeh JT, Resnik KP, Rygle K, Pennline HW (2005) Semi-batch absorption and regeneration studies for CO2 capture by aqueous ammonia. Fuel Process Technol 86:1533–1546. https://doi.org/10.1016/j.fuproc.2005.01.015

Yeon SH, Sea B, Park YI, Lee KH (2003) Determination of mass transfer rates in PVDF and PTFE hollow fiber membranes for co2 absorption. Sep Sci Technol 38:271–293. https://doi.org/10.1081/SS-120016575

Yu Y, Li Y, Li Q, Jiang J, Zhang Z (2009) An innovative process for simultaneous removal of CO_2 and SO2 from flue gas of a power plant by energy integration. Energy Convers Manag 50:2885–2892. https://doi.org/10.1016/j.enconman.2009.07.003

Zhang Z (1996) Research on physical chemistry characters and production increasing mechanism of long effective ammonium bicarbonate. China Sci 26:453

Zhang HY, Wang R, Liang DT, Tay JH (2008) Theoretical and experimental studies of membrane wetting in the membrane gas–liquid contacting process for CO_2 absorption. J Membr Sci 308:162–170. https://doi.org/10.1016/j.memsci.2007.09.050

Zhao Z, Cui X, Ma J, Li RJ (2007) Adsorption of carbon dioxide on alkali-modified zeolite 13X adsorbents. Int J Greenhouse Gas Control 1:355–359. https://doi.org/10.1016/S1750-5836(07)00072-2

Zhao X, Liu N, Wang Y, Fei W, Stevens GW (2011) Study on the mechanism and energy consumption of CO2 regeneration process by membrane electrolysis. Ind Eng Chem Res 50:8620–8631. https://doi.org/10.1021/ie1024529

Zoghi AT, Feyzi F, Zarrinpashneh SJ (2012) Experimental investigation on the effect of addition of amine activators to aqueous solutions of N-methyldiethanolamine on the rate of carbon dioxide absorption. Int J Greenhouse Gas Control 7:12–19. https://doi.org/10.1016/j.ijggc.2011.12.001

Chapter 10
Cryogenic CO_2 Capture

Amineh Keshavarz, Maryam Ebrahimzadeh Sarvestani, and Mohammad Reza Rahimpour

Abstract To alleviate the global warming issue arising from emission of greenhouse gases involving CO_2 through burning of fossil fuels CO_2 recovery and its utilization and storage has been known as an appropriate choice. With this purpose in mind, various CO_2 recovery methods have been applied recently including adsorption, membrane, hydrating, chemical looping, biofixation, and absorption. Among these strategies, cryogenic CO_2 capture technique has been become more attractive. However, there are some challenges including capture cost, impurities and cold energy sources which encounter utilization of this technique.

In this investigation, the major technologies and strategies of capturing CO_2 which is exhausted from the combustion of fossil fuels are summarized, and also the features of cryogenic routes for CO_2 capture have been reviewed. Also, the future prospect of the improved cryogenic processes is discussed. According to the consequences of this study, cryogenic CO_2 capture methods can be utilized for the industrial scale and eliminate the problems associated with physical sorbents and chemical solvents. Also the integration of two or more conventional CO_2 capture technologies can minimize the disadvantages of standalone process. These processes, named as hybrid system, compared to the standalone technology indicated the superiority in terms of energy penalty, the installation investment, and CO_2 recovery.

Keywords CO_2 capture · Usage and storage · Cryogenic CO_2 capture technique · The combustion of fossil fuels · Cryogenic routes · Cold energy sources · Hybrid system · Energy penalty · The installation investment · CO_2 recovery

A. Keshavarz · M. Ebrahimzadeh Sarvestani · M. R. Rahimpour (✉)
Department of Chemical Engineering, Shiraz University, Shiraz, Iran
e-mail: rahimpor@shirazu.ac.ir

Inamuddin et al. (eds.), *Sustainable Agriculture Reviews 38*, Sustainable Agriculture Reviews 38, https://doi.org/10.1007/978-3-030-29337-6_10

Abbreviations

CFZ	Controlled Frosting Zone
CO	Carbon monoxide
CO_2	Carbon dioxide
$DeNO_X$	The process of NOx reduction
EOR	Enhanced Oil Recovery
H_2	Hydrogen
H_2O	Water
LNG	Liquefied Natural Gas
MEA	Monoethanolamine
O_2	Oxygen
VPSA	Vacuum Pressure Swing Adsorption

10.1 Introduction

The climate change arising from greenhouse effect has attracted global concern, recently. Therefore, it is necessary to utilize fossil fuels in a way that environmental challenges including global warming can be restricted (Bozzano and Manenti 2016). Generated CO_2, as one of the greenhouse gases, from combustion of fossil fuel contributing significantly to the climate change (Zhang 2018). Figure 10.1 demonstrates that among various emission sources, CO_2 is mainly produced and exhausted from Coal-fired power station throughout the world (Song et al. 2019).

A promising approach to alleviate CO_2 emissions is to capture it from large sources such as iron-steel, cement industry and coal-fired power station (Li et al. 2013). CO_2 emissions could be decreased about 80–90% if a traditional coal-fired power station is modified by providing CO_2 recovery units (Metz et al. 2005; Stewart and Hessami 2005). Researchers have investigated different technologies of CO_2 capture and separation, and the major issue is the improvement of novel structures

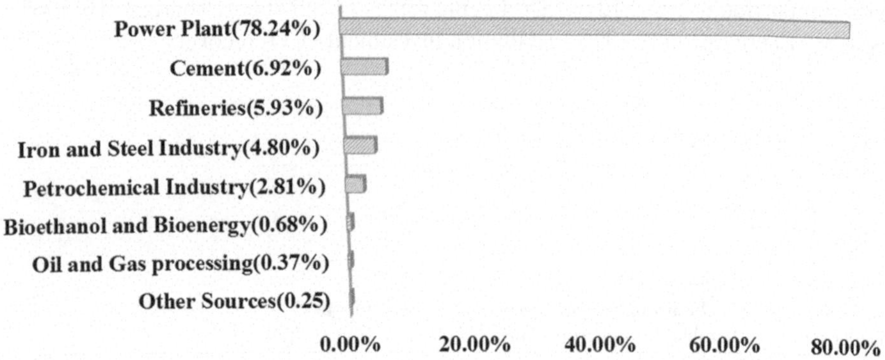

Fig. 10.1 Breakdown of the dominant CO_2 emission sources in 2014

or materials that can permit an optimal capture efficiency and the minimum energy penalty. CO_2 capture strategies generally includes adsorption, membrane, absorption, hydrate and cryogenic (Amrollahi et al. 2012; Liu et al. 2017; Miller and Stöcker 1989; Wang et al. 2015).

The main target of CO_2 capture process is its removal and storage. The flue gas is mostly comprised of nitrogen and 5–20 vol% carbon dioxide. Moreover, it includes water and impurities such as nitrogen oxide and sulphur, depending on the process and feedstock. It can be extremely costly to compress and store the flue gas completely. Thus, before storage in geological formation, it is essential to obtain purified form of CO_2. Around 50% of the entire cost can be attributed to the capture stage of the CO_2 recovery process. Therefore, the development of capture technologies has come into sharp focus of many researchers. The selection of suitable capture method highly depends on particular discharge condition (Olajire 2010). The flue gas condition, as the primary criterion, contains its temperature, composition, CO_2 concentration, and flow rate (Porter et al. 2015). Additionally, the characteristics of the targeted production including transport pressure, and CO_2 purity or discharged standards including H_2S, NO_X, and SO_X also have an important effect on the capture technology selectivity (Song et al. 2017a, b). For instance, the captured CO_2 have to be compressed to 110 bar to ease transportation and storage (Ali et al. 2017; Skaugen et al. 2016).

One conventional route for CO_2 separation is the application of cryogenic techniques, which has dominant benefits such as preventing the chemical solvents usage and no pollution. In order to develop an efficient cryogenic methods, a comprehensive evaluation of current processes in terms of advantages and disadvantages is required (Berstad et al. 2013; Maqsood et al. 2017; Pan et al. 2013; Song et al. 2019; Thomas and Denton 1988). In this study, the applications of the CO_2 capture process and its potential are discussed. In general, CO_2 which is mainly produced through burning of carbon-based fossil fuels can be captured through three main technologies. These strategies are divided into pre-combustion, oxy-fuel combustion and post-combustion.

10.2 Pre-combustion Carbon Dioxide Capture

The carbon existed in the fuel gas reacts with H_2O and O_2 to produce CO_2, CO and H_2 via partial oxidation or autothermal reforming. Besides, CO_2 and H_2 are produced through water-gas shift reaction which contains approximately 20–40% CO_2 and 60–80% H_2. Subsequently, CO_2 is captured and H_2 is passed through the fuel cell or combustion chamber (Deng et al. 2010). Coal gasification is mostly conducted through pre-combustion via a cyclic gasification system. One of the main benefits of this capture technique is the H_2 production which can be used as a green source of energy because combustion of hydrogen only creates water (Ma et al. 2017). Furthermore, it can be applied in many fields including in energy vehicles, fuel batteries, chemical industry (Deng et al. 2010).

10.3 Oxy-Fuel Combustion Carbon Dioxide Capture

The combustion of fuel in pure O_2 occurs in the process of oxy-fuel combustion, leading to a high concentration of CO_2 in the flue gas. Therefore, compared to the post-combustion technology, the CO_2 purification in this technology is easier (Mathekga et al. 2016; Stanger et al. 2015). Furthermore, lesser proportion of NOx exists in the flue gas when this process is applied. Oxy-fuel combustion integrated with fluidized bed burners and powdered coal combustion could be a satisfactory alternative strategy. For a given input of coal, a fluidized bed burner has the ability to decrease the flue gas flow while keeping a high temperature of furnace which is the important advantage of this technique in comparison to the powdered coal combustion (Mathekga et al. 2016). If oxy-fuel carbon capture facilities were retrofitted due to the air separation, the compression and purification stages, the coal-fired power stations' efficiency would be decrease about 10–12% (Escudero et al. 2016). In order to carry out the CO_2 capture process efficiently through oxy-fuel combustion, minimization of energy requirement should be conducted through optimization of the process.

10.4 Post Combustion Carbon Dioxide Capture

Fuel burns in the air in the process of post-combustion, and produce CO_2. Nevertheless, the concentration of produced CO_2 from the typical post combustion process is approximately low usually about 4–14%, and the process is carried out in atmospheric pressure. Thus, a disadvantage of this method for separation is the reduction of driving force because of the diluted CO_2 and low pressures. As a result, in order to recover CO_2 appropriately to obtain CO_2 concentration of more than 95.5% which made it suitable for storage and transport, high energy is required (Leung et al. 2014).. However, Post-combustion units of CO_2 capture can be put with available coal-fired power station directly by a bit retrofitting, and this process has been applied at pilot-scale plants compared to the oxy-fuel combustion and pre-combustion (Abu-Zahra et al. 2007; Oh et al. 2016).

10.5 Low Temperature Carbon Dioxide Capture Strategies

Cryogenic processes separate CO_2 from feed gases via utilizing different desubliamtion and condensation instruments. This technology can gain higher CO_2 purity and recovery than other separation methods. Several types of CO_2 recovery processes through cryogenic method have developed until now. By way of illustration, the advanced cryogenic can be exemplified with the advantage of inhibiting blockage of undesired compounds involving H_2O as well as enhancing energy efficiency through reducing energy requirements (Berstad et al. 2013; Brunetti et al. 2010).

10.5.1 Cryogenic Distillation CO_2 Capture

Among the most prevalent cryogenic processes for CO_2 recovery is cryogenic distillation can be exemplified. In this regard, Holmes and Ryan proposed a conventional process of cryogenic distillation (Holmes and Ryan 1982). Figure 10.2 demonstrates the summarized fractionation process in which the feed stream enters the distillation section after two stages of cooling. The distillation column consists of packing materials or trays which are contacted devices of vapor-liquid. Methane as the top product is removed by a partial condenser, and desublimated CO_2 as the bottom product is gathered. The stream of rich CO_2 is passed through reboiler to be recycled and provide heat of vaporization. The recovered CO_2 is then further purified.

Operational conditions for the conventional cryogenic distillation are high pressures and super low temperatures which separate components based on the differences in their boiling temperatures. Vapor CO_2 or liquefied CO_2 are manufactured during the process at high pressure. By this technique, storage cost decreases. However, in the range of higher and lower pressure at the top of columns, column choking and solid formation cause operating problems.

An extractive distillation technology was proposed by Holmes et al. (1983) and Holes and Rayan (1982). They added heavier hydrocarbons in the distillation condenser to prevent CO_2 solidification. Valencia and Denton added helium to improve the CO_2 capture from methane and prevent CO_2 solidification, but helium removal from methane was quite difficult (Valencia and Denton 1985). For the CO_2 removal from natural gas, Atkinson et al. presented a dual-pressure distillation system (Atkinson et al. 1988). This process was comprised of two distillation columns that operated at different pressures to inhibit CO_2 solidification. The product of the sec-

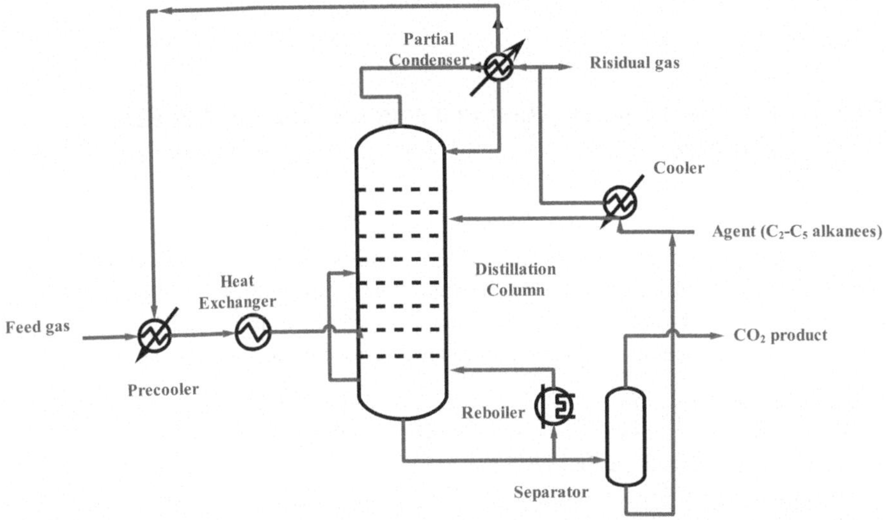

Fig. 10.2 Process flow diagram of cryogenic CO_2 capture via Rayan Holmes technology

ond column was methane, however, to prevent CO_2 solidification, the pressure of the columns must be accurately selected. Berstad et al. simulated a three distillation columns network via adding pentane, using Aspen HYSYS to separate CO_2 from natural gas (Berstad et al. 2012).

Masqsood et al. presented a detailed classification of the cryogenic processes based on condensation and sublimation (Maqsood et al. 2014a, b, c). The cryogenic technologies are divided to the non-conventional, conventional, and hybrid systems. They forecasted the solid formation in the distillation process inside the loop. According to the investigation results, the solid formation can be kept off by increasing the pressure of distillation column or utilizing extractive distillation. Also, a hybrid cryogenic network conformation for CO_2 capture from natural gas was proposed by Maqsood et al. (Maqsood et al. 2014a, b, c). The obtained results demonstrated a significant decrease in the size of equipment, energy requirement and methane loss by utilizing the networks of intensified integration cryogenic distillation. The optimized cryogenic network enhances the overall profit to 69.24%.

Although cryogenic distillation has major benefits and extensive industrial application, energy requirement of this process for the plant is nearly 50% of operating costs (Ebrahimzadeh et al. 2016; Li and Bai 2012). Process intensification and hybrid technologies such as thermally coupled columns, reactive distillation, cyclic distillation, and a hybrid distillation column have been developed to improve energy efficiency (Jana 2016; Kazemi et al. 2016).

Yousef et al. proposed a process of cryogenic distillation via biogas upgrading. This method included two distillation columns that operated at different pressures (Yousef et al. 2018). The optimum distillation pressures for the first and second columns were 4983 kpa and 4763 kpa. For the first column, 13 trays at 2.8 reflux ratio, and for the second one, 11 trays at 2.8 reflux ratio were the optimum numbers. The advantages of this process were high purity of biomethane nearly about 97.12%, mol and liquid CO_2 about 99.92% as by-products.

10.5.2 External Cooling Loop Cryogenic Carbon Dioxide Capture

Energy source is one of the main challenges for capturing of CO_2 based on cryogenic. In cryogenic technologies, the effective method for reduction of energy consumption is to utilize a process of thermal swing, an exterior cooling loop, or a system of carbon extraction and reuse cold energy from sources. Baxter et al. developed a hybrid cryogenic system via an exterior cooling loop which involved:

1. flue gas was dried and cooled.
2. flue gas was compressed and cooled until CO_2 changed to the solid form.
3. gas was expanded to further chill it.
4. an amount of CO_2 precipitated.
5. CO_2 was pressurized.

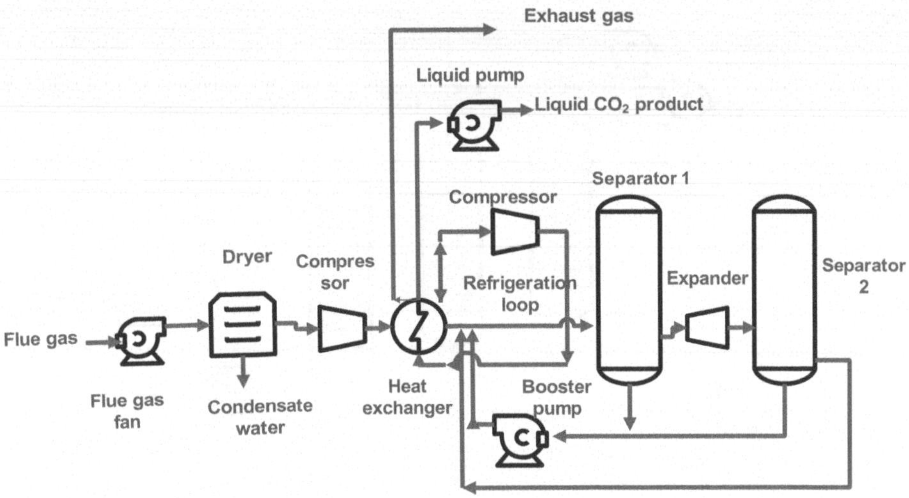

Fig. 10.3 The schematic flow diagram of CO_2 capture via low temperature process

6. CO_2 and residual flue gas were heated again by the incoming gases (Baxter et al. 2009; Jensen 2015).

According to Fig. 10.3 at the end of the process, CO_2 was captured as the liquid phase, and a stream of N_2-rich was discharged. This technology has some conformations that can save energy of liquefied natural gas (Jensen 2015). This ability handles the energy requirement for CO_2 removal via applying a stored refrigerant. Furthermore, this method becomes profitable when integrates with the traditional power plants by renewable power source. The average total energy consumption of External cooling loop cryogenic carbon dioxide capture is around 0.98 $MJ_{elecrtrical}$/kg CO_2, which is less energy-intensive than other traditional processes (Safdarnejad et al. 2015).

10.5.3 Cryogenic Packed Bed

Tuinier et al. developed cryogenic CO_2 capture process using dynamically cooled packed bed(Tuinier et al. 2010). The basis of separation is the differences in the dew points and the composition of components in the natural gas. There are no plugging problems or pressure drops in this technology. Cryogenic packed bed contains moving interface to separate CO_2 and water, simultaneously. A process cycle includes three main steps: cooling down the bed, capturing CO_2 and recovering steps, as shown in Fig. 10.4. During the first step, refrigerated N_2 gas flow is fed to the cooling cycles. Feed gas is passing through the packed bed to capture water and CO_2. A packed bed, containing borosilicate tubes, is encompassed by a vacuum jacket. At the surface of packing bed, CO_2 and water desublimate and condense by cooling

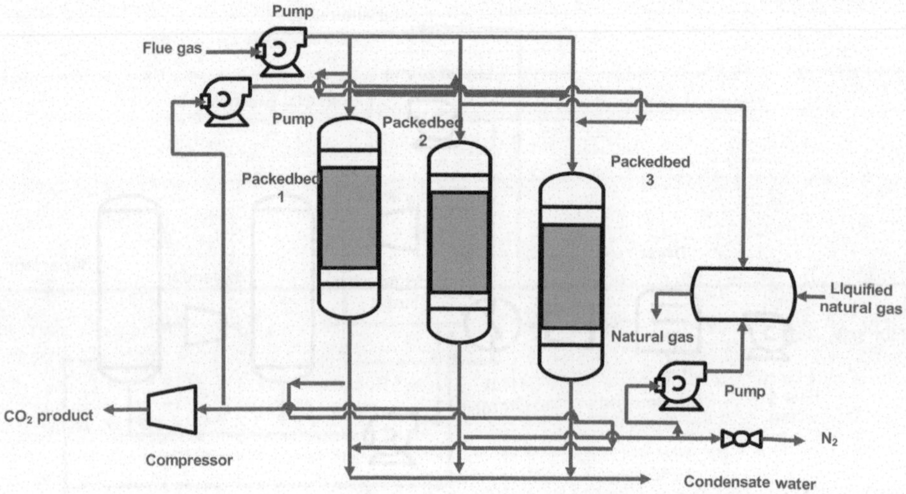

Fig. 10.4 Process flow diagram of cryogenic CO_2 capture via packed bed

down of the flue gas. In the last step, CO_2 by recycling CO_2 and water via recycling air are recovered. All these steps should operate parallel in the beds, to reach continuous performance of the process. The separation process is performed in the atmospheric pressure. Experimental validation is described by 1D pseudo-homogeneous model, and axial temperature, mass deposition and concentration are compared with numerical model.

Dynamically operated packed beds based on cryogenic CO_2 removal were proposed by Tuinier et al. Liquefied natural gas was utilized as the cold energy and the packed beds are made of steel monolith(Tuinier et al. 2011). Simultaneous separation of both carbon dioxide and water based on the sublimation points is the benefit of this method. Also, pressure drop and clogging could be avoided in the dynamic packed bed based on cryogenic separation and by this process, there is no need for chemical absorbent. Tuinier and his team utilized the packed bed to upgrade biogas at cryogenic condition(Tuinier and van Sint Annaland 2012). Volume percentage of the captured water and CO_2 from the flue gas is 1% and 10%, respectively, and compared to the conventional method, the productivity and CH_4 recovery in the cryogenic technology can be increased to 94.3%. Installation investments and energy consumption is lower than the process of vacuum pressure swing adsorption. (Tuinier et al. 2011).

Abulhasan et al. developed the cryogenic packed bed system for high concentration and pressure of CO_2 (Ali et al. 2014). According to the experimental validation, energy requirements with a low CO_2 concentration for both cryogenic systems including packed-bed and distillation are similar and with a high CO_2 concentration, the cryogenic distillation separation consumes more energy than the cryogenic packed-bed technology. In addition, for the cryogenic packed bed technology CO_2 capture concept, a simulation and nucleation investigation with validated results was presented utilizing 1-D pseudo homogenous model. Abulhasan et al. used series packed-bed pipelines to capture and dehydrate CO_2 from natural gas, simultane-

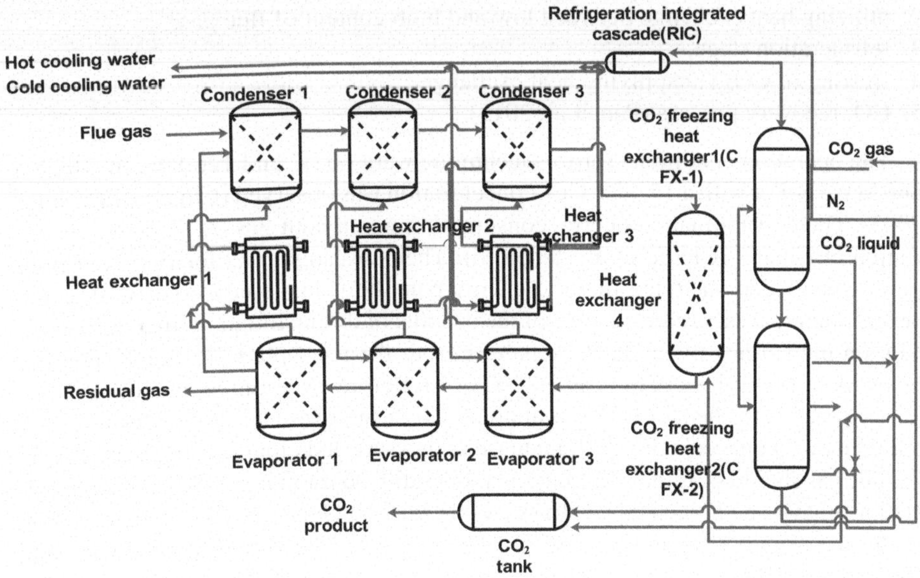

Fig. 10.5 The schematic flow diagram of CO_2 capture by de-sublimation technology

ously (Ali et al. 2016). The advantages of this strategy are clean and efficient pipelines in the packed beds with low cost. Also, in the next study, optimization of cryogenic packed bed system was carried out to investigate the temperature and pressure effect of each bed (Ali et al. 2018). Product purity can reach up to 94% and hydrocarbon losses decrease from 39% to 16%.

Although the process of cryogenic packed-bed for CO_2 capture compared to vacuum pressure swing adsorption (VPSA) processes and monoethanolamine (MEA) absorption has higher potential, several drawbacks involving thermal insulation and the amount of H_2S in raw biogas, require to be dealt with before the commercial utilization. In order to obtain high removal of H_2S, and thereby enhancing the operating costs, it is necessary to keep the packed backed bed in a low temperature about −150 °C. The cryogenic processes require the cold energy, and liquefied natural gas as the cold source is economical (Abatzoglou and Boivin 2009).

10.5.4 CO_2 Cryogenic De-sublimation

An efficient CO_2 capture based on cryogenic process was developed by Clodic et al. and Clodic and Younes, in which CO_2 was condense on the surface of heat exchangers (Clodic et al. 2005a, b). According to the Fig. 10.5, the overall process contains five sections:

1. cleaning up the flue gas with condensation removal and chilling to $-40^{\circ}C$,

2. utilizing heat exchange between low and high content of flue gas,
3. refrigeration stage,
4. frosting of CO_2 via applying heat exchange,
5. CO_2 recovery process (Pan et al. 2013).

The operation conditions for this technique were 60 °C and 120 kpa, in which a 660 MWe boiler with a 15.47% CO_2 concentration was observed (Clodic and Younes 2003). The de-sublimation energy consumption was a bout 3.8–7.2% of the power station efficiency (Clodic et al. 2005a, b). Their investigations for two coal-fired power plants demonstrated that the energy penalty of this study compared to the absorption processes reduced. One of the benefits of this technique is that defrosting CO_2 on the surface of the heat exchanger from the solid phase to liquid, permits recovery of fusion heat (Clodic et al. 2005a, b). Before evaporation, to chill mixture of refrigerants, the heat of fusion is applied (Clodic and Younes 2003). During operation, to avoid unacceptable high increase in pressure and clogging, the humidity of the flue gas should be removed. Materials were utilized for making the heat exchangers must have good mechanical stress and thermal conductivity. The freezing point of flue gas usually containing 15% CO_2, was −99.3 °C from a coal-fired power plant (Song et al. 2014).

Cryogenic de-sublimation method for capturing CO_2 is a compact process with no contamination. To study the dynamic de-sublimation process for cryogenic CO_2 capture in the gas mixture, Wang et al. proposed a transient model which includes a tube-in-tube heat exchanger with the nitrogen coolant, three control volumes, and the wall containing the layer of solid CO_2 and the mixture (Wang et al. 2018). The model has been validated by experimental results. The results show that the solid layer generated via de-sublimation should be considered during simulation, and before capture stage, the wall is not required to be precooled. During the process, low energy consumption and high mass diffusion can be gained at low velocity flow and high concentration of CO_2 supply.

Also Yu et al. developed the numerical and analytical model of cryogenic CO_2 capture technology in a de-sublimating heat exchanger to calculate the rate of desubliamted CO_2 through the heat exchanger, outlet temperature of gas mixture and liquid temperature, CO_2 mole fraction and temperature distribution (Yu et al. 2017).

10.5.5 Stirling Cooler Strategy

A stirling cooler strategy for capturing CO_2 at cryogenic condition with heat recovery was proposed by Song et al. (2012a, b). Figure 10.6 demonstrated this system which contains three sections: a pre-frosting section, main frosting section and storage, and each section is cooled via a free-piston of Stirling cooler. For the separation of H_2O

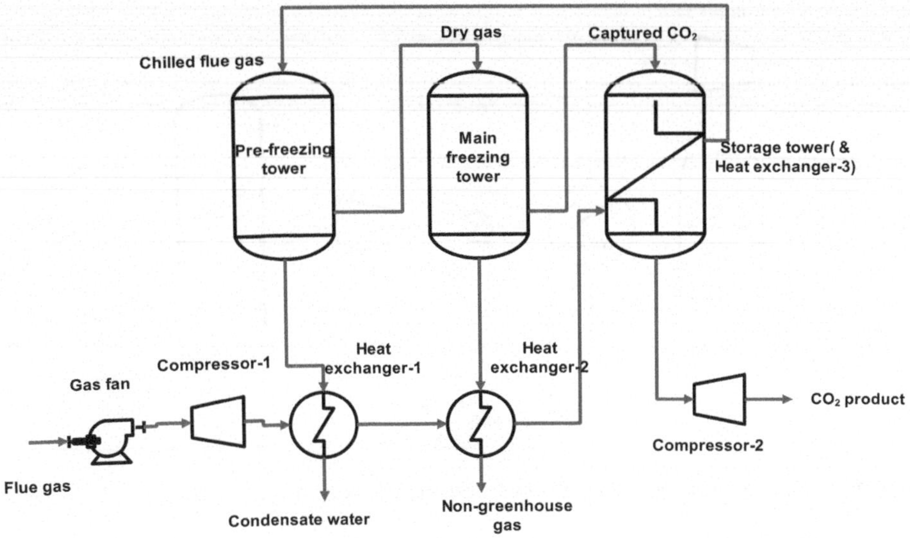

Fig. 10.6 Process flow diagram of cryogenic CO_2 capture by stirling cooler strategy with heat recovery

in the feed gas by condensation, the pre-frosting section is cooled via the Stirling cooler-1. The extra heat of the main-frosting tower is taken away via Stirling cooler-2 at the temperature of de-sublimation. Moreover, the freeze CO_2 is collected in the storage tower which is cooled via Stirling cooler-3, and The emission gas is sent to the environment. The details of this route can be observed in reports of Song and his team (Song et al. 2012a, b, 2018a, b). This process hasn't been applied for a large-scale or pilot plant, because simulated gas condition greatly differs from the flue gas emitted from the industry for instance, CO_2 concentration, flow rate impurities, and temperature. Thus, in the future works, investigators should focus more on this challenge of Stirling coolers process of cryogenic CO_2 capture (Song et al. 2017a, b).

10.5.6 CryoCell System

Cool Energy Ltd. developed the Cryocell system and examined with other industrial companies in Western Australia (Hart and Gnanendran 2009). Figure 10.7 described the CryoCell process configuration. According to this method, the first step is dehydration of feed gas for lowing water characterizations nearly about 5 ppm, in order to manage the downstream of cryogenic systems. Then, the dehydrated gas passes through heat exchanger to chill the CO_2 until achieving the freezing point. Before

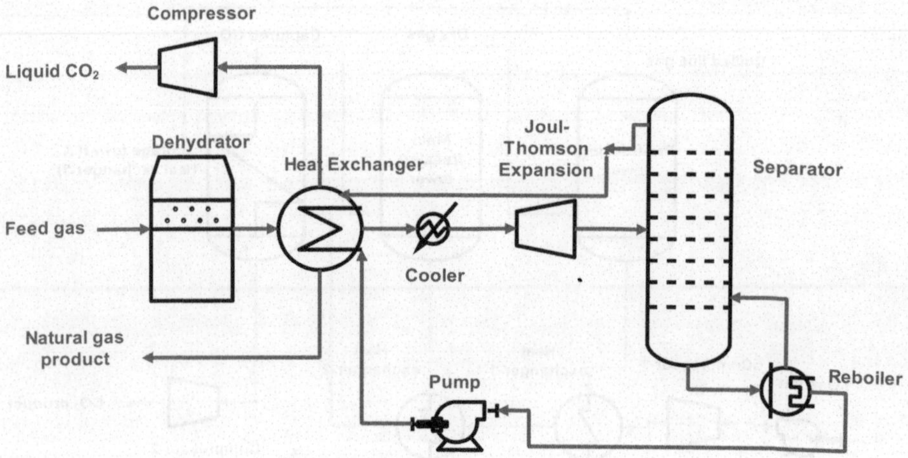

Fig. 10.7 The schematic flow diagram of CO_2 capture via the Cryocell system

liquid is passed through the CryoCell separator, it is expanded through a Joule-Thomson valve.

10.5.7 Controlled Frosting Zone

Some developing countries such as China, recommended replacing natural gas, as a green source of energy, with coal to reduce the air pollution made by coal combustion (Rufford et al. 2012). However, sour impurities such as H_2S and CO_2 in natural gas must be removed since these compounds influence its combustion and transport. CO_2 to H_2S ratio differs enormously through geographic regions (Kelley et al. 2011). One of the most efficient methods of removing CO_2 and H_2S from the sour gas is a controlled frosting zone technology. Figure 10.8 shows the whole process in which the utilized tower can be divided into three zones including upper distillation tower, down distillation tower and controlled frosting zone (CFZ). First, the feed stream is passed through a precooler. Then, the low temperature feed stream is delivered to the controlled frosting zone via a spray nozzle. The company of ExxonMobil Research applied this technology successfully through a commercial manifestation plant of Wyoming. ExxonMobil company utilized majority of the captured CO_2 for Enhanced Oil Recovery (EOR) (Parker et al. 2011).

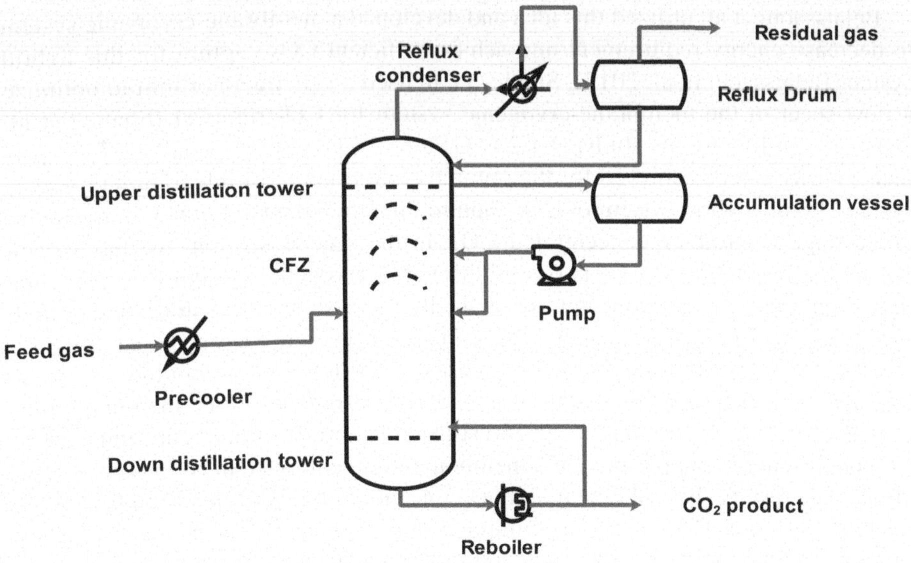

Fig. 10.8 Process flow diagram of cryogenic CO_2 capture by controlled frosting zone technology

10.6 Hybrid Methods for CO_2 Capture

CO_2 capture methods such as cryogenics, adsorption and membrane have some limitations. For example, when high purity and high recovery CO_2 ratio is needed, utilization of membrane technology is very expensive (Wu et al. 2017). Also, when concentration of CO_2 is low, the feed stream of cryogenic process becomes very expensive in terms of energy requirement. It is reported that cryogenic process is suitable when the CO_2 concentration in the feed gas is more than 40 v/v% (Tuinier et al. 2011). Likewise, adsorption process for capturing CO_2 is viable in the high concentration of CO_2 stream.

From all the statements above, it can generally be concluded that the mentioned processes can be regarded economical if the concentration of CO_2 in the feed stream is high enough. Therefore, the major purpose of a hybrid system is integration of various methods and materials to decrease energy consumption, and thereby achieving higher efficiency for CO_2 capture process.

Hybrid processes are divided into 5 technologies including membrane-absorber, membrane-cryogenic, membrane-adsorption, chemical absorption-electrolysis and the solar thermal electrochemical technique (Asif et al. 2018). Cryogenic process in comparison to the membrane process based on the phase change properties. Percentage of CO_2 is one of the factor that influences the temperature of CO_2 desublimation in the cryogenic separation. Integration of membrane process with cryogenic increases CO_2 capture rate, and energy consumption of the cryogenic process decreases by enhancement of liquefaction temperature (Song et al. 2017a, b).

Belaissaoui et al. utilized this idea and developed a membrane-cryogenic system to decrease energy requirement and achieve efficient CO_2 capture for this hybrid system(Belaissaoui et al. 2012). Shafiee et al. used a genetic algorithm to optimize a flow sheet in the membrane-cryogenic system for CO_2 removal (Shafiee et al. 2017). According to this study, at 1.249 GJ/ton of CO_2 energy penalty, 97.9% permeate purity can be achieved for the optimal conformation of integration or hybrid system. In this hybrid system of CO_2 capture, first the inlet stream is fed to a membrane stage to produce concentrated CO_2. In the second section, the gas stream containing concentrated CO_2 is passed through a cryogenic system to achieve the desired product. Energy consumption in the hybrid system can be attributed to both membrane and cryogenic stages. The consequences demonstrated that for 85% recovery of CO_2 ratio and 89% CO_2 purity, 3 GJ/ton CO_2 minimum energy is required, and therefore this system is economical compared to the membrane processes. Moreover, Belaissaaui et al. investigated the effect of inlet concentration of CO_2 on the energy requirement of a membrane-cryogenic system (Belaissaoui et al. 2012). The results showed that at an inlet concentration of CO_2 between 0.1 to 0.3, for 85% purity of CO_2, energy consumption of membrane-cryogenic system is less than other single-stage processes including cryogenic separation, membrane process and conventional absorption process through MEA solvent.

Jean and his team studied a hybrid system process to determine the optimum range of operating conditions for the process (Fong et al. 2016). This integration technique has three sections: vacuum swing adsorption, membrane section and cryogenic. They installed the cryogenic separation instead of the multi stage compression of CO_2 to obtain high purity of products and manufacture liquid carbon dioxide. For optimization of the hybrid system, a new methodology combining with exergy analysis was used in their applied hybrid system, flow rate of refrigerant, multi-stage compression and decreasing temperature of cryogenic separation can influence the rate of recovery. The results demonstrated that for a recovery percentage of 94.4%, a minimum loss of energy of 1.6 GJ/ton of CO_2 was obtained.

A novel strategy using a low-temperature of membrane-cryogenic system instead of a conventional hybrid system or capturing of CO_2 was proposed by Song et al. (2018a, b). The consequences showed that CO_2 recovery and energy requirements have been improved by this novel process. Moreover, the energy requirement of a conventional membrane-cryogenic integration system for CO_2 capture could be decreased to 1.7 MJ/kg CO_2 by utilizing this strategy for a coal-combustion power plant.

Song et al. presented an improved process of cryogenic CO_2 capture utilizing heat generation to further reduction of total energy (Song et al. 2013). This technology consists of three parts: pre-freezing tower, main freezing tower and storage tank. These three parts are cooled by stirling coolers. Also, Hasse et al. implemented the same idea using air liquid membrane in which membrane sections were based on a hollow fiber integrated with cryogenic separation (Hasse et al. 2013). Their studied indicated that a good integration of CO_2 selectivity and permeability happened utilizing air liquid at low temperature.

A hybrid process including production of liquefied natural gas(LNG), oxy-fuel power generation system and cryogenic CO_2 capture was proposed by Mehrpooya et al. (2017a, b). In their study, the effect of operating parameters in the system and its components were investigated through thermodynamic simulation. The energy efficiency for the oxy-fuel power system and liquefaction process of natural gas obtained are 69% and 51%, respectively. According to the obtained results, this process could be used for the actual plant designation.

10.6.1 *Cryogenic-Hydrate Technologies*

High pressure and low temperature are the main conditions for both hydrate and cryogenic CO_2 capture processes (Surovtseva et al. 2011). In the first step of this process, feed gas is condensed via cryogenic at −55 °C in which concentration of CO_2 is expected to be decreased to 75 mol%. In comparison with the traditional standalone cryogenic distillation methods, the cryogenic-hydrate CO_2 capture process operates at lower energy requirement and cost. Also captured CO_2 as the form of solid hydrate at nearly 1 °C decreases concentration of CO_2 down to 7 mol% (Sreenivasulu et al. 2015). Promoters such as tetrabutylammonium chloride, tetrabutylammonium bromide, tetrabutylammonium nitrate, tetrabutylammonium fluoride, tetrabutylphosphonium chloride, and tetrabutylphosphonium bromide are usually applied in hydrate unit not only to intensify formation rate of hydrate, but also to lower equilibrium condition such as high pressure (Ma et al. 2016).

10.6.2 *Cryogenic –Membrane Technologies*

The performance of gas separation in the membrane process depends strongly on the ambient temperature. Mostly, with the reduction of operating temperature, the permeability of membrane reduces while the selectivity enhances (Hasse et al. 2013). When operating temperature was below −20 °C, it demonstrated two or four times enhancement in the selectivity of CO_2/N_2 with minimum CO_2 permeation loss. Long term of bench-scale test containing the mixture of CO_2/N_2 at sub temperature conditions has affirmed the increased separation performance observed at lab-scale translated to the commercial modules of membrane (Hasse et al. 2013; Kulkarni et al. 2012).

10.6.3 Low Temperature Absorption Technologies

Monoethanolamine (MEA) can be possibly utilized in the early step of commercial scale plants of post-combustion CO_2 capture (Wang et al. 2011). Using MEA is accompanied by high energy requirement due to thermal degradation, oxidation made by O_2, heat-stable salts' formation, and rich solvent regeneration (Pfaff et al. 2010; Qi et al. 2013). Thus, to tackle the mentioned drawbacks, the chilled ammonia process can be a good alternative for MEA absorption process (Hanak et al. 2015). The CO_2 absorption process using chilled ammonia is usually conducted in the temperature of 2 to 10 °C. In general NH_3 exhibits superior features than CH_4 in terms of low energy requirement for regeneration, non-corrosive, commercial availability, comparatively low price, and relatively high capacity of CO_2 absorption (Hanak et al. 2015; Zhao et al. 2012). Additionally, the integration of low temperature stations can effectively prevent NH_3 volatility in the absorber (Strube and Manfrida 2011).

A summary of different cryogenic processes based on the hybrid CO_2 capture technologies and their application for various gas sources is shown in Tables 10.1, and 10.2 demonstrates the benefits and limitations of hybrid processes for cryogenic CO_2 capture. A combination of cryogenic and hydrate methods as Cryogenic-hydrate hybrid can obtain high CO_2 recovery in the pre-combustion CO_2 capture. Nevertheless, to apply this method for an industrial scale, more comprehensive studies are required. Cryogenic CO_2 capture technology, integrating with membrane CO_2 capture process, requires less energy (Sun and Kang 2016). When chilled ammonia absorption applied into post-combustion CO_2 capture, a high recovery of

Table 10.1 Summary of hybrid processes for cryogenic CO_2 capture

Hybrid technologies	Product purity	Phase state	Gas source	Energy consumption	References
Cryogenic-hydrate	95 mol% to 97 mol% CO_2	Liquid	Integrated gasification combined cycle power plant	–	Surovtseva et al. (2011)
Cryogenic-absorption	93.8 wt%	Gas	Coal fired power plant	1.163GJ/t CO_2	Hanak et al. (2015)
Cryogenic-membrane	98.3 mol%	Liquid	Coal fired power plants	1.215GJ/t CO_2	Anantharaman et al. (2014), Scholz et al. (2013), Shafiee et al. (2017), and Zhao et al. (2012))
			Biogas	–	
			Coal fired power plants	1.249 GJ/t CO_2	
			Coal fired power plants	48 £$_{2008}$/t CO_2	
Cryogenic-adsorption	99.9992% CO_2 and 99.99955% CH_4	Gas	Raw natural gas	–	Grande and Blom (2014)

Table 10.2 Benefits and limitations of hybrid processes for cryogenic CO_2 capture technologies

Hybrid technologies	Benefits	Limitations	References
Cryogenic-membrane	Large enhances in CO_2/N_2 selectivity	Humidity sensitivity	Clodic and Younes (2003) and Hasse et al. (2014)
		Capital cost enhancement	
	Long term stability(about 8 months)	Reduction of the membrane module productivity	
Cryogenic-hydrate	Low energy demand and capture cost	Low CO_2 recovery	Bak et al. (2015), Ramezan et al. (2007), and Surovtseva et al. (2011)
	High CO_2 recovery	Lab scale	
Chilled-adsorption	Low theoretical regeneration heat (0.67 GJ/t CO_2 vs. 1.92 GJ/t CO_2 for MEA)	Flammability and toxicity	Hanak et al. (2015)
	Avoid ammonia loss because of evaporation	High volatility	

CO_2 is obtained. But, the energy consumption for regeneration have to be more decreased.

CO_2 can be obtained in various phases, such as vapor-solid, vapor-liquid or a combination such as CO_2 slurry in the hybrid processes of cryogenic CO_2 capture. CO_2 product as liquid form requires both high pressure condition and the cryogenic temperature. CO_2 product as solid form is mostly captured under lower temperature in comparison to the liquid form (Song et al. 2018a, b).

Transportation and utilization stage for CO_2 capture technologies have to be considered. For transportation, the easiest form of CO_2 product is liquid phase with high pressure (Clodic et al. 2005a, b). Although, in the de-sublimation method, capture cost is low, the transport cost[5] goes up significantly in solid form (Tuinier et al. 2010). Thus, before transportation and heat recovery, there should be a suitable scheme to transform the produced CO_2 into the liquid or gas phase. Some researchers assessed the benefits of hybrid CO_2 capture technologies in comparison to the standalone technologies. (Fong et al. 2014, 2016).

10.7 Advantages and Limitations of CO_2 Capture Methods Based on Cryogenic Process

The usual routes for high content of CO_2 are capture and usage or storage. To deliver the CO_2 product to specified sites, pipeline transport and compression are necessary. Because of the long distance between the CO_2 recovery unit and the unit where it needs to be delivered for utilization and storage, it is vitally important to purify the recovered CO_2 as much as possible. In fact, if the purity of recovered CO_2 is high enough, it is feasible to compact it to the desirable pressure before it is transported.

CO_2 is captured as different phases, namely, solid, liquid or a combination of both, during the cryogenic technologies. Cryogenic CO_2 capture methods can reach higher purity of CO_2 over 99.9% in comparison with other separation routes, owing to desublimation inherent properties. Consequently, with such purity of CO_2, production of various chemical substances that can be used in other industries including fertilizer, industrial food, is more possible through biological or catalytic reactions for instance, artificial photosynthesis and steam methane reforming. Besides, when CO_2 is recovered through the cryogenic technique, not only the need for extra compression processes is eliminated and thereby lower energy is consumed, but also it is a cold source of energy, meaning that it is useful to be combined with the processes at which low temperature is required, such as membrane, liquefaction of natural gas. The energy storage for the process of CO_2 capture based on cryogenic by integrating a cryogenic technique with a refrigeration loop of an open natural gas, was designed by Baxter's group (Hasse et al. 2013; Kulkarni et al. 2012). The total process consists of three sections:

1. energy saving, in which production of liquefied natural gas(LNG) enhances by 40% to utilize excess energy.
2. equilibrium, in which production of LNG equalizes the requirement from the plant to operate the process of carbon capture based on cryogenic.
3. energy recovery, in which production of LNG reduces by 70% to decrease the parasitic loss.

Table 10.3 indicates the summary of benefits and challenges of the CO_2 removal technologies based on low temperature via various cold sources. Installing cryogenic system is extremely restricted by the accessible cryogenic sources, because the energy requirements are considerable for cryogenic methods. Refrigeration and compression are usual routes utilized to supply cryogenic condition. Although liquefied natural gas could be a good option for the energy source, it is greatly subjected to the scale, location, and other characteristics of the natural gas plant. Also, one of the most important cryogenic sources are coolers. The total energy demand increases, when the chilling in the cryogenic systems is supplied by refrigeration. The performance coefficient of a common refrigerator which chills to −140 °C is usually 0.5, indicating that the used electrical energy is approximately double of the thermal energy (Tuinier et al. 2011). One of the solution to decrease the consumed energy of refrigeration and compression is to recover the heat of the cold products of CO_2 and residual gas (Hasse et al. 2013; Sreenivasulu et al. 2015).

Capture cost is the principal challenge related to the large-scale organization of CO_2 cryogenic capture. For instance, around 80% of the overall cost is attributed to compression and capture, 10% for storage, and 10% for transportation. (Hanak et al. 2015; Qi et al. 2013). Siriwardhana et al. and Raksajati et al. denoted that for a conventional process of MEA absorption, the CO_2 capture expense is nearly 55 USD$/ton CO_2 (Siriwardana et al. 2011). In comparison, in Europe and Australia are 3.5 €/ton CO_2 and 23 AUD$/ton CO_2, respectively, that limits severely the commercial application of cryogenic process (Raksajati et al. 2013).

Table 10.3 Advantages and challenges of cryogenic CO_2 capture technologies

Catcgory	Advantages	Challenges	Energy source	References
Distillation	Prevent compression cost		Compressor and cooler	Baxter et al. (2009)
	Be pumped to storage units easily			
	Water saving possible			
	No corrosion	High instillation cost		
	No foaming possible	Capital cost for pressure difference		
	Energy saving potential			
	Remove other pollutants(SOX, HCL, Hg, NO2) simultaneously			
Packed-bed	Surface area to volume ratio of the column		Liquid nitrogen gas	Lively et al. (2012) and Tuinier and van Sint Annaland (2011))
	Remove H_2O and CO_2 simultaneously	Lab scale		
		Directly be effected by the availability of LNG		
	Atmosphere			
	Preventing high pressure drop			
De-subliamtion	Pilot demonstration	Undesired mechanical stresses	Liquefied natural gas	Clodic et al. (2005a, b)
	Lower energy consumption than MEA absorption	Freeze CO_2 adversely influences heat conduction		
	Atmosphere	H_2O can't be tolerated		
		Directly be effected by natural gas location		
Stirling cooler	Remove H_2O and CO_2 simultaneously	Exergy loss because of temperature difference	Stirling cooler	Song et al. (2012a, b)
	Energy storage possible	Lab scale		
	Lower energy consumption than MEA absorption	Difficulty of freeze layer scrapping		
	Atmosphere			

(continued)

Table 10.3 (continued)

Category	Advantages	Challenges	Energy source	References
Cryocell	No process heating system required	High compression power need	Chiller	Hart and Gnanendran (2009)
		More suitable for high concentration of CO_2(more than 20%)		

Table 10.4 Comparison of capturing performance for cryogenic CO2 capture technologies

Cryogenic technologies	Phase state	CO_2 recovery	Gas sources	Energy consumption	References
External cooling loop	Liquid	95.6 mol%	Raw natural gas	1.401 kWh/kg CO_2	Ebrahimzadeh et al. (2016)
Packed-bed	Liquid	99.%	Coal fired power plants(10 vol% CO_2, 89 vol% N_2, 1 vol% H_2O)	1.8 $MJ_{electrical}$/ kg CO_2	Tuinier and van Sint Annaland (2011)
De-Subliamtion	Liquid	90%	Coal fired power plant(12% CO_2)	1.18 $GJ_{electrical}$/ ton CO_2	Pan et al. (2013)
Stirling cooler	Solid	85%	Coal fired power plants (13% CO_2 and 87% N_2)	3.4 $MJ_{thermal}$/kg CO_2	Song et al. (2018a, b)
Cryocell	Liquid	–	Raw natural gas (20–35 mol% CO_2)	–	Hart and Gnanendran (2009)
Controlled frosting zone	Liquid	98–99%CH_4	Raw natural gas(8–71% CO_2)	–	Carpenter (2015)

Table 10.4 demonstrates the capture operating of existing CO_2 capture based on cryogenic. For a common CO_2 capture process, there are two kinds of costs involving capital cost and operational cost. The major energy-intensive instruments in cryogenic CO_2 methods, are pumps and compressors, and also heat exchangers for CO_2 liquefaction and condensation (Clodic et al. 2005a, b; Tuinier et al. 2010). On one hand, the capital cost of conventional adsorption and absorption processes is lower than that of cryogenic processes because their equipment sizes (in the absence of using a large adsorption/absorption tower) are smaller. On the other hand, the operational cost of cryogenic routes is lower than conventional adsorption and absorption due to eliminating sorbent/solvent regeneration costs. However, different technologies of CO_2 capture performance by various types of energy characters. (Song et al. 2012a, b; Zhao et al. 2012).

Operating temperature is one of the main factor that highly influences the efficiency of cryogenic CO_2 capture techniques. As expected, the efficiency enhances with reducing operating temperature. The efficiency of CO_2 capture based on cryogenic condition is extremely affected by the networks of heat exchanger. The multi heat exchangers with optimal thermal conductivity and viscosity can help a reduction in the exergy destruction.

The difference of temperature between the ambient condition and the cryogenic process would cause heat loss, and consequently reduce the total efficiency. To decrease energy loss, a vacuum interlayer could be an efficient route. The heat exchangers' materials must be carefully chosen to ensure its proper performance at low temperatures, as well as sustaining mechanical stresses and heat conductivity. The frozen CO_2 on the heat exchanger surface might adversely influence heat transfer between the solid and gas phase. While many researchers have studied the process of frost formation, most of these research is referred to humidity. There is a difference between the behavior of frost CO_2 and humidity, and the overall process is comprised of three step:

1. crystal emergence,
2. frost emergence,
3. formation of frost.

The existence of other compounds is another key factor that influences the CO_2 freezing process. Thus, frost CO_2 of flue gas in comparison with that of CO_2, consists of a complicated deposition process, and additional efforts would be required.

When water converts to the ice under low temperature condition, it can cause the clogging of equipment and pipes. Also, any humidity should be eliminated via the pretreatment units to reach a high efficiency of capturing in cryogenic methods. Tuinier et al. proposed the process of cryogenic packed bed based on the differences in the sublimation and dew points of various components. Respectively, CO_2 and H_2O can be desublimated and condensed on the packed bed at different situations (Tuinier et al. 2011).

Additionally, corrosion of the entire facilities and the heat exchanger can occur by acidic impurities. $DeNO_X$ treatment and an efficient desulfurization must be carried out prior to cryogenic CO_2 capture that would cause an increase in the cost of process. Some researchers have designed a novel process of cryogenic CO_2 capture that could eliminate other pollutants such as NO_X, PM, SO_X, and mercury, simultaneously to alleviate the negative impact of the impurities. Fazlollahi et al. provided an advanced method of cryogenic process to remove all species which are less volatile compared to CO_2 and desublimate CO_2 and other pollutants (Fazlollahi et al. 2016).

10.8 Conclusions

In this review, the development of CO_2 capture based on cryogenic is summarized. This process has some advantages in comparison with other technologies including adsorption, chemical looping, membrane, hydrating, and absorption. There is no need of chemical adsorbents or absorbents and chemical solvent for recovery plants in the cryogenic technology. CO_2 can be reached at high pressure and can be utilized for enhanced oil recovery. By this technology, operating cost decreases because heating systems are not required. One of the beneficial features of cryogenic CO_2 capture technologies is that CO_2 with high purity can be produced in either solid form or liquid form, which is profitable for utilization, transport and storage. Additionally, exergy destruction because of high temperature differences is one obstacle that encounters industrial scale utilization of cryogenic technique. Thus, this method should be studied more comprehensively to provide higher efficiency.

References

Abatzoglou N, Boivin S (2009) A review of biogas purification processes. Biofuels Bioprod Biorefin 3(1):42–71. https://doi.org/10.1002/bbb.117

Abu-Zahra MR, Schneiders LH, Niederer JP, Feron PH, Versteeg G (2007) CO_2 capture from power plants: part I. A parametric study of the technical performance based on monoethanolamine. Int J Greenhouse Gas Control 1(1):37–46. https://doi.org/10.1016/S1750-5836(07)00032-1

Ali A, Maqsood K, Syahera N, Shariff AB, Ganguly S (2014) Energy minimization in cryogenic packed beds during purification of natural gas with high CO_2 content. Chem Eng Technol 37:1675–1685. https://doi.org/10.1002/ceat.201400215

Ali A, Maqsood K, Redza A, Hii K, Shariff AB (2016) Performance enhancement using multiple cryogenic desublimation based pipeline network during dehydration and carbon capture from natural gas. Chem Eng Res Des 109:519–531. https://doi.org/10.1016/j.cherd.2016.01.020

Ali U, Font-Palma C, Akram M, Agbonghae EO, Ingham DB, Pourkashanian M (2017) Comparative potential of natural gas, coal and biomass fired power plant with post-combustion CO_2 capture and compression. Int J Greenhouse Gas Control 63:184–193. https://doi.org/10.1016/j.ijggc.2017.05.022

Ali A, Maqsood K, Shin LP, Sellappah V, Garg S, Shariff AB, Ganguly S (2018) Synthesis and mixed integer programming based optimization of cryogenic packed bed pipeline network for purification of natural gas. J Clean Prod 171:795–810. https://doi.org/10.1002/ceat.201400215

Amrollahi Z, Ystad PAM, Ertesvåg IS, Bolland O (2012) Optimized process configurations of post-combustion CO_2 capture for natural-gas-fired power plant–Power plant efficiency analysis. Int J Greenhouse Gas Control 8:1–11. https://doi.org/10.1016/j.ijggc.2012.01.005

Anantharaman R, Berstad D, Roussanaly S (2014) Techno-economic performance of a hybrid membrane–liquefaction process for post-combustion CO_2 capture. Energy Procedia 61:1244–1247. https://doi.org/10.1016/j.egypro.2014.11.1068

Asif M, Suleman M, Haq I, Jamal SA (2018) Post-combustion CO_2 capture with chemical absorption and hybrid system: current status and challenges. Greenhouse Gases 8:998–1031. https://doi.org/10.1002/ghg.1823

Atkinson TD, Lavin JT, Linnett D (1988) Separation of gaseous mixtures. In: Google Patents

Bak C, Asif M, Kim W (2015) Experimental study on CO_2 capture by chilled ammonia process. Chem Eng J 265:1–8. https://doi.org/10.1016/j.cej.2014.11.145

Baxter L, Baxter A, Burt S (2009) Cryogenic CO_2 capture as a cost-effective CO_2 capture process. Paper presented at the International Pittsburgh Coal Conference

Belaissaoui B, Le Moullec Y, Willson D, Favre E (2012) Hybrid membrane cryogenic process for post-combustion CO_2 capture. J Membr Sci 415–416:424–434. https://doi.org/10.1016/j.memsci.2012.05.029

Berstad D, Nekså P, Anantharaman R (2012) Low-temperature CO_2 removal from natural gas. Energy Procedia 26:41–48. https://doi.org/10.1016/j.egypro.2012.06.008

Berstad D, Anantharaman R, Nekså P (2013) Low-temperature CO_2 capture technologies–applications and potential. Int J Refrig 36:1403–1416. https://doi.org/10.1016/j.ijrefrig.2013.03.017

Bozzano G, Manenti F (2016) Efficient methanol synthesis: perspectives, technologies and optimization strategies. Prog Energy Combust Sci 56:71–105. https://doi.org/10.1016/j.pecs.2016.06.001

Brunetti A, Scura F, Barbieri G, Drioli E (2010) Membrane technologies for CO_2 separation. J Membr Sci 359:115–125. https://doi.org/10.1016/j.memsci.2009.11.040

Carpenter C (2015) Controlled-freeze-zone technology for the distillation of high-CO_2 natural gas. J Pet Technol 67:100–101. https://doi.org/10.2118/1115-0100-JPT

Clodic D, Younes M (2003) A new method for CO_2 capture: frosting CO_2 at atmospheric pressure. Paper presented at the greenhouse gas control technologies-6th international conference. https://doi.org/10.1016/B978-008044276-1/50025-8

Clodic D, El Hitti R, Younes M, Bill A, Casier F (2005a) CO_2 capture by anti-sublimation Thermo-economic process evaluation. Paper presented at the 4th annual conference on carbon capture and sequestration

Clodic D, Younes M, Bill A (2005b) Test results of CO_2 capture by anti-sublimation capture efficiency and energy consumption for boiler plants. In Greenhouse Gas Control Technologies 1775–1780. https://doi.org/10.1016/B978-008044704-9/50210-X

Deng X, Wang H, Huang H, Ouyang M (2010) Hydrogen flow chart in China. Int J Hydrog Energy 35:6475–6481. https://doi.org/10.1016/j.ijhydene.2010.03.051

Ebrahimzadeh E, Matagi J, Fazlollahi F, Baxter L (2016) Alternative extractive distillation system for CO_2–ethane azeotrope separation in enhanced oil recovery processes. Appl Therm Eng 96:39–47. https://doi.org/10.1016/j.applthermaleng.2015.11.082

Escudero AI, Espatolero S, Romeo LM, Lara Y, Paufique C, Lesort AL, Liszka M (2016) Minimization of CO_2 capture energy penalty in second generation oxy-fuel power plants. Appl Therm Eng 103:274–281. https://doi.org/10.1016/j.applthermaleng.2016.04.116

Fazlollahi F, Bown A, Ebrahimzadeh E, Baxter L (2016) Transient natural gas liquefaction and its application to CCC-ES (energy storage with cryogenic carbon capture™). Energy 103:369–384. https://doi.org/10.1016/j.energy.2016.02.109

Fong J, Anderson C, Hooper B, Xiao G, Webley P, Hoadley A (2014) Multi-objective optimisation of hybrid CO_2 capture processes using exergy analysis. Chem Eng Trans 39:1501–1506. https://doi.org/10.3303/CET1439251

Fong J, Anderson CJ, Xiao G, Webley PA, Hoadley A (2016) Multi-objective optimisation of a hybrid vacuum swing adsorption and low-temperature post-combustion CO_2 capture. J Clean Prod 111:193–203. https://doi.org/10.1016/j.jclepro.2015.08.033

Grande CA, Blom R (2014) Cryogenic adsorption of methane and carbon dioxide on zeolites 4A and 13X. Energy Fuel 28:6688–6693. https://doi.org/10.1021/ef501814x

Hanak DP, Biliyok C, Manovic V (2015) Rate-based model development, validation and analysis of chilled ammonia process as an alternative CO_2 capture technology for coal-fired power plants. Int J Greenhouse Gas Control 34:52–62. https://doi.org/10.1016/j.ijggc.2014.12.013

Hart A, Gnanendran N (2009) Cryogenic CO_2 capture in natural gas. Energy Procedia 1:697–706. https://doi.org/10.1016/j.egypro.2009.01.092

Hasse D, Kulkarni S, Sanders E, Corson E, Tranier J (2013) CO_2 capture by sub-ambient membrane operation. Energy Procedia 37:993–1003. https://doi.org/10.1016/j.egypro.2013.05.195

Hasse D, Ma J, Kulkarni S, Terrien P, Tranier JP, Sanders E, Brumback J (2014) CO_2 capture by cold membrane operation. Energy Procedia 63:186–193. https://doi.org/10.1016/j.egypro.2014.11.019

Holmes AS, Ryan JM (1982) Cryogenic distillative separation of acid gases from methane. In: Google Patents

Holmes A, Price B, Ryan J, Styring R (1983) Pilot tests prove out cryogenic acid-gas/hydrocarbon separation processes. Oil Gas J 81(26):85–86

Jana A (2016) A new divided-wall heat integrated distillation column (HIDiC) for batch processing: feasibility and analysis. Appl Energy 172:199–206. https://doi.org/10.1016/j.apenergy.2016.03.117

Jensen M (2015) Energy process enabled by cryogenic carbon capture

Kazemi A, Hosseini M, Mehrabani-Zeinabad A, Faizi V (2016) Evaluation of different vapor recompression distillation configurations based on energy requirements and associated costs. Appl Therm Eng 94:305–313. https://doi.org/10.1016/j.applthermaleng.2015.10.042

Kelley B, Valencia J, Northrop P, Mart C (2011) Controlled freeze zone™ for developing sour gas reserves. Energy Procedia 4:824–829. https://doi.org/10.1016/j.egypro.2011.01.125

Kulkarni S, Hasse D, Sanders E, Chaubey T (2012) CO_2 capture by sub-ambient membrane operation. American Air Liquide Incorporated

Leung DY, Caramanna G, Maroto-Valer M (2014) An overview of current status of carbon dioxide capture and storage technologies. Renew Sust Energ Rev 39:426–443. https://doi.org/10.1016/j.rser.2014.07.093

Li G, Bai P (2012) New operation strategy for separation of ethanol–water by extractive distillation. Ind Eng Chem Res 51(6):2723–2729. https://doi.org/10.1021/ie2026579

Li L, Zhao N, Wei W, Sun Y (2013) A review of research progress on CO_2 capture, storage, and utilization in Chinese Academy of Sciences. J Fuel 108:112–130. https://doi.org/10.1016/j.fuel.2011.08.022

Liu Y, Deng S, Zhao R, He J, Zhao L (2017) Energy-saving pathway exploration of CCS integrated with solar energy: a review of innovative concepts. Renew Sust Energ Rev 77:652–669. https://doi.org/10.1016/j.rser.2017.04.031

Lively RP, Koros WJ, Johnson J (2012) Enhanced cryogenic CO_2 capture using dynamically operated low-cost fiber beds. Chem Eng Sci 71:97–103. https://doi.org/10.1016/j.ces.2011.11.042

Ma Z, Zhang P, Bao H, Deng S (2016) Review of fundamental properties of CO_2 hydrates and CO_2 capture and separation using hydration method. Renew Sust Energ Rev 53:1273–1302. https://doi.org/10.1016/j.rser.2015.09.076

Ma Y, Guan G, Hao X, Cao J, Abudula A (2017) Molybdenum carbide as alternative catalyst for hydrogen production–a review. Renew Sust Energ Rev 75:1101–1129. https://doi.org/10.1016/j.rser.2016.11.092

Maqsood K, Ali A, Shariff AB, Ganguly S (2014a) Synthesis of conventional and hybrid cryogenic distillation sequence for purification of natural gas. J Appl Sci 14(21):2722–2729. https://doi.org/10.3923/jas.2014.2722.2729

Maqsood K, Mullick A, Ali A, Kargupta K, Ganguly S (2014b) Cryogenic carbon dioxide separation from natural gas: a review based on conventional and novel emerging technologies. Rev Chem Eng 30(5):453–477. https://doi.org/10.1515/revce-2014-0009

Maqsood K, Pal J, Turunawarasu D, Pal AJ, Ganguly S (2014c) Performance enhancement and energy reduction using hybrid cryogenic distillation networks for purification of natural gas with high CO_2 content. Korean J Chem Eng 31(7):1120–1135. https://doi.org/10.1007/s11814-014-0038-y

Maqsood K, Ali A, Shariff AB, Ganguly S (2017) Process intensification using mixed sequential and integrated hybrid cryogenic distillation network for purification of high CO_2 natural gas. Chem Eng Res Des 117:414–438. https://doi.org/10.1016/j.cherd.2016.10.011

Mathekga H, Oboirien B, North B (2016) A review of oxy-fuel combustion in fluidized bed reactors. Int J Energy Res 40(7):878–902. https://doi.org/10.1002/er.3486

Mehrpooya M, Ansarinasab H, Sharifzadeh M, Rosen M (2017a) Process development and exergy cost sensitivity analysis of a hybrid molten carbonate fuel cell power plant and carbon dioxide capturing process. J Power Sources 364:299–315. https://doi.org/10.1016/j.jpowsour.2017.08.024

Mehrpooya M, Esfilar R, Moosavian SA (2017b) Introducing a novel air separation process based on cold energy recovery of LNG integrated with coal gasification, transcritical carbon dioxide power cycle and cryogenic CO_2 capture. J Clean Prod 142:1749–1764. https://doi.org/10.1016/j.jclepro.2016.11.112

Metz B, Davidson O, De Coninck H (2005) Carbon dioxide capture and storage: special report of the intergovernmental panel on climate change. Cambridge University Press, New York

Miller GQ, Stöcker J (1989) Selection of a hydrogen separation process. National Petroleum Refiners Association, Washington, DC

Oh SY, Binns M, Cho H, Kim J (2016) Energy minimization of MEA-based CO_2 capture process. Appl Energy 169:353–362. https://doi.org/10.1016/j.apenergy.2016.02.046

Olajire A (2010) CO_2 capture and separation technologies for end-of-pipe applications–a review. J Energy 35(6):2610–2628. https://doi.org/10.1016/j.energy.2010.02.030

Pan X, Clodic D, Toubassy J (2013) CO_2 capture by antisublimation process and its technical economic analysis. J Greenhouse Gases 3(1):8–20. https://doi.org/10.1002/ghg.1313

Parker PME, Northrop S, Valencia JA, Foglesong RE, Duncan W (2011) CO_2 management at ExxonMobil's LaBarge field, Wyoming, USA. Energy Procedia 4:5455–5470. https://doi.org/10.1016/j.egypro.2011.02.531

Pfaff I, Oexmann J, Kather A (2010) Optimised integration of post-combustion CO_2 capture process in greenfield power plants. J Energy 35(10):4030–4041. https://doi.org/10.1016/j.energy.2010.06.004

Porter RT, Fairweather M, Pourkashanian M, Woolley R (2015) The range and level of impurities in CO_2 streams from different carbon capture sources. Int J Greenhouse Gas Control 36:161–174. https://doi.org/10.1016/j.ijggc.2015.02.016

Qi G, Wang S, Yu H, Wardhaugh L, Feron P, Chen C (2013) Development of a rate-based model for CO_2 absorption using aqueous NH3 in a packed column. Int J Greenhouse Gas Control 17:450–461. https://doi.org/10.1016/j.ijggc.2013.05.027

Raksajati A, Ho MT, Wiley D (2013) Reducing the cost of CO_2 capture from flue gases using aqueous chemical absorption. Ind Eng Chem Res 52(47):16887–16901. https://doi.org/10.1021/ie402185h

Ramezan M, Skone TJ, Nsakala N, Liljedahl G, Gearhart L, Hestermann R, Rederstorff B (2007) Carbon dioxide capture from existing coal-fired power plants. National Energy Technology Laboratory, DOE/NETL Report (401/110907)

Rufford TE, Smart S, Watson GC, Graham B, Boxall J, Da Costa JD (2012) The removal of CO_2 and N_2 from natural gas: a review of conventional and emerging process technologies. J Pet Sci Eng 94:123–154. https://doi.org/10.1016/j.petrol.2012.06.016

Safdarnejad SM, Hedengren JD, Baxter LL (2015) Plant-level dynamic optimization of cryogenic carbon capture with conventional and renewable power sources. Appl Energy 149:354–366. https://doi.org/10.1016/j.apenergy.2015.03.100

Scholz M, Frank B, Stockmeier F, Falß S, Wessling M (2013) Techno-economic analysis of hybrid processes for biogas upgrading. Ind Eng Chem Res 52(47):16929–16938. https://doi.org/10.1021/ie402660s

Shafiee A, Nomvar M, Liu Z, Abbas A (2017) Automated process synthesis for optimal flowsheet design of a hybrid membrane cryogenic carbon capture process. J Clean Prod 150:309–323. https://doi.org/10.1016/j.jclepro.2017.02.151

Siriwardana M, Meng S, McNeill J (2011) The impact of a carbon tax on the Australian economy: results from a CGE model. Business, Economics Public Policy Working Papers 2

Skaugen G, Roussanaly S, Jakobsen J, Brunsvold A (2016) Techno-economic evaluation of the effects of impurities on conditioning and transport of CO_2 by pipeline. Int J Greenhouse Gas Control 54:627–639. https://doi.org/10.1016/j.ijggc.2016.07.025

Song C, Kitamura Y, Li SH, Ogasawara K (2012a) Design of a cryogenic CO_2 capture system based on Stirling coolers. Int J Greenhouse Gas Control 7:107–114. https://doi.org/10.1016/j.ijggc.2012.01.004

Song CF, Kitamura Y, Li S (2012b) Evaluation of Stirling cooler system for cryogenic CO_2 capture. Appl Energy 98:491–501. https://doi.org/10.1016/j.apenergy.2012.04.013

Song C, Kitamura Y, Li S, Lu J (2013) Deposition CO_2 capture process using a free piston Stirling cooler. Ind Eng Chem Res 52(42):14936–14943. https://doi.org/10.1021/ie401026f

Song C, Kitamura Y, Li S (2014) Energy analysis of the cryogenic CO_2 capture process based on Stirling coolers. J Energy 65:580–589. https://doi.org/10.1016/j.energy.2013.10.087

Song C, Liu Q, Ji N, Deng S, Zhao J, Kitamura Y (2017a) Advanced cryogenic CO_2 capture process based on Stirling coolers by heat integration. Appl Therm Eng 114:887–895. https://doi.org/10.1016/j.applthermaleng.2016.12.049

Song C, Liu Q, Ji N, Deng S, Zhao J, Li Y, Kitamura Y (2017b) Reducing the energy consumption of membrane-cryogenic hybrid CO_2 capture by process optimization. J Energy 124:29–39. https://doi.org/10.1016/j.energy.2017.02.054

Song C, Liu Q, Ji N, Deng S, Zhao J, Li Y (2018a) Alternative pathways for efficient CO_2 capture by hybrid processes—a review. Renew Sust Energ Rev 82:215–231. https://doi.org/10.1016/j.rser.2017.09.040

Song C, Sun Y, Fan Z, Liu Q, Ji N, Kitamura Y (2018b) Parametric study of a novel cryogenic-membrane hybrid system for efficient CO_2 separation. Int J Greenhouse Gas Control 72:74–81. https://doi.org/10.1016/j.ijggc.2018.03.009

Song C, Liu Q, Deng S, Li H, Kitamura Y (2019) Cryogenic-based CO_2 capture technologies: state-of-the-art developments and current challenges. Renew Sust Energ Rev 101:265–278. https://doi.org/10.1016/j.rser.2018.11.018

Sreenivasulu B, Gayatri D, Sreedhar I, Raghavan K (2015) A journey into the process and engineering aspects of carbon capture technologies. Renew Sust Energ Rev 41:1324–1350. https://doi.org/10.1016/j.rser.2014.09.029

Stanger R, Wall T, Spörl R, Paneru M, Grathwohl S, Weidmann M, Ritvanen J (2015) Oxyfuel combustion for CO_2 capture in power plants. Int J Greenhouse Gas Control 40:55–125. https://doi.org/10.1016/j.ijggc.2015.06.010

Stewart C, Hessami M (2005) A study of methods of carbon dioxide capture and sequestration—the sustainability of a photosynthetic bioreactor approach. Energy Convers Manag 46(3):403–420. https://doi.org/10.1016/j.enconman.2004.03.009

Strube R, Manfrida G (2011) CO_2 capture in coal-fired power plants—impact on plant performance. Int J Greenhouse Gas Control 5(4):710–726. https://doi.org/10.1016/j.ijggc.2011.01.008

Sun Q, Kang Y (2016) Review on CO_2 hydrate formation/dissociation and its cold energy application. Renew Sust Energ Rev 62:478–494. https://doi.org/10.1016/j.rser.2016.04.062

Surovtseva D, Amin R, Barifcani A (2011) Design and operation of pilot plant for CO_2 capture from IGCC flue gases by combined cryogenic and hydrate method. Chem Eng Res Des 89(9):1752–1757. https://doi.org/10.1016/j.cherd.2010.08.016

Thomas E, Denton R (1988) Conceptual studies for CO_2/natural gas separation using the controlled freeze zone (CFZ) process. Gas Sep Purif 2(2):84–89. https://doi.org/10.1016/0950-4214(88)80017-3

Tuinier M, van Sint Annaland M (2011) A novel process for cryogenic CO_2 capture using dynamically operated packed beds—an experimental and numerical study. Int J Greenhouse Gas Control 5(4):694–701. https://doi.org/10.1016/j.ijggc.2010.11.011

Tuinier MJ, van Sint Annaland M (2012) Biogas purification using cryogenic packed-bed technology. Ind Eng Chem Res 51(15):5552–5558. https://doi.org/10.1021/ie202606g

Tuinier M, van Sint Annaland M, Kramer GJ, Kuipers (2010) Cryogenic CO_2 capture using dynamically operated packed beds. Chem Eng Sci 65(1):114–119. https://doi.org/10.1016/j.ces.2009.01.055

Tuinier M, Hamers H, van Sint Annaland M (2011) Techno-economic evaluation of cryogenic CO_2 capture—a comparison with absorption and membrane technology. Int J Greenhouse Gas Control 5(6):1559–1565. https://doi.org/10.1016/j.ijggc.2011.08.013

Valencia JA, Denton R (1985) Method and apparatus for separating carbon dioxide and other acid gases from methane by the use of distillation and a controlled freezing zone. In: Google Patents

Wang M, Lawal A, Stephenson P, Sidders J, Ramshaw C (2011) Post-combustion CO_2 capture with chemical absorption: a state-of-the-art review. Chem Eng Res Des 89(9):1609–1624. https://doi.org/10.1016/j.cherd.2010.11.005

Wang M, Joel AS, Ramshaw C, Eimer D, Musa N (2015) Process intensification for post-combustion CO_2 capture with chemical absorption: a critical review. Appl Energy 158:275–291. https://doi.org/10.1016/j.apenergy.2015.08.083

Wang Y, Pfotenhauer J, Zhi X, Qiu L, Li J (2018) Transient model of carbon dioxide desublimation from nitrogen-carbon dioxide gas mixture. Int J Heat Mass Transf 127:339–347. https://doi.org/10.1016/j.ijheatmasstransfer.2018.07.068

Wu H, Gao L, Jin H, Li S (2017) Low-energy-penalty principles of CO_2 capture in polygeneration systems. Appl Energy 203:571–581. https://doi.org/10.1016/j.apenergy.2017.06.012

Yousef AM, El-Maghlany WM, Eldrainy YA, Attia A (2018) New approach for biogas purification using cryogenic separation and distillation process for CO_2 capture. J Energy 156:328–351. https://doi.org/10.1016/j.energy.2018.05.106

Yu Z, Miller F, Pfotenhauer JM (2017) Numerical modeling and analytical modeling of cryogenic carbon capture in a de-sublimating heat exchanger. Paper presented at the IOP Conference Series: Materials Science and Engineering. https://doi.org/10.1088/1757-899X/278/1/012032

Zhang X (2018) Current status of stationary fuel cells for coal power generation. Clean Energy 2(2):126–139. https://doi.org/10.1093/ce/zky012

Zhao B, Su Y, Tao W, Li L, Peng Y (2012) Post-combustion CO_2 capture by aqueous ammonia: a state-of-the-art review. Int J Greenhouse Gas Control 9:355–371. https://doi.org/10.1016/j.ijggc.2012.05.006

Index

Inamuddin et al. (eds.), *Sustainable Agriculture Reviews 38*, Sustainable Agriculture Reviews 38, https://doi.org/10.1007/978-3-030-29337-6

The manufacturer's authorised representative in the EU is Springer Nature Customer Service Centre GmbH, Europaplatz 3, 69115 Heidelberg, Germany. If you have any concerns regarding our products, please contact ProductSafety@springernature.com

Printed and bound by CPI Group (UK) Ltd, Croydon, CR0 4YY
30/09/2026
02225645-0017